Mastering SAS Programming for Data Warehousing

An advanced programming guide to designing and managing Data Warehouses using SAS

Monika Wahi

BIRMINGHAM—MUMBAI

Mastering SAS Programming for Data Warehousing

Commissioning Editor: Richa Tripathi
Acquisition Editor: Karan Gupta
Senior Editor: Nitee Shetty
Content Development Editor: Ruvika Rao
Technical Editor: Gaurav Gala
Copy Editor: Safis Editing
Project Coordinator: Deeksha Thakkar
Proofreader: Safis Editing
Indexer: Tejal Daruwale Soni
Production Designer: Aparna Bhagat

First published: October 2020

Production reference: 1141020

Published by Packt Publishing Ltd.
Livery Place
35 Livery Street
Birmingham
B3 2PB, UK.

ISBN 978-1-78953-237-1

www.packt.com

This book is dedicated to my mother, Carol Wahi. Although happily retired now, she spent her career programming in COBOL with assembler, and teaching data management skills. Here's to the next generation of big data women!

`Packt.com`

Subscribe to our online digital library for full access to over 7,000 books and videos, as well as industry leading tools to help you plan your personal development and advance your career. For more information, please visit our website.

Why subscribe?

- Spend less time learning and more time coding with practical eBooks and videos from over 4,000 industry professionals

- Improve your learning with Skill Plans built especially for you

- Get a free eBook or video every month

- Fully searchable for easy access to vital information

- Copy and paste, print, and bookmark content

Did you know that Packt offers eBook versions of every book published, with PDF and ePub files available? You can upgrade to the eBook version at `packt.com` and, as a print book customer, you are entitled to a discount on the eBook copy. Get in touch with us at `customercare@packtpub.com` for more details.

At `www.packt.com`, you can also read a collection of free technical articles, sign up for a range of free newsletters, and receive exclusive discounts and offers on Packt books and eBooks.

Contributors

About the author

Monika Wahi, MPH, CPH, is a well-published epidemiologist, biostatistician, informaticist, and data scientist. For over 20 years, Monika has worked at various governmental organizations and non-profits, and led consulting projects in academia and for governments both in the United States and internationally. She is President of DethWench Professional Services (DPS), which offers consulting and training in data science, specializing in public health and healthcare. Monika is proficient in SAS, R, Excel, and SQL, and is the author of many articles and online courses in data science and health data analytics.

I'd like to thank the following people at Packt who made this book possible: Karan Gupta, who was the first to express faith in me, and Afshaan Khan, Ruvika Rao, and Prajakta Naik, who worked tirelessly with me to improve my drafts. I'd also like to thank the reviewer, Sunil Gupta, an admirable SAS author himself, for his helpful advice and encouragement.

About the reviewer

Sunil Gupta, MS, is an international speaker, best-selling author of five SAS books, and a global SAS and CDISC corporate trainer. Sunil has over 25 years' experience in the pharmaceutical industry. Most recently, Sunil has been involved in several CDISC and PhUSE working groups and has taught his CDISC online class at the University of California at San Diego.

In 2019, Sunil published his fifth book, *Clinical Data Quality Checks for CDISC Compliance Using SAS*, and, in 2011, he launched his unique SAS mentoring blog for smarter SAS searches. Sunil has an MS in bioengineering from Clemson University and a BS in applied mathematics from the College of Charleston.

Packt is searching for authors like you

If you're interested in becoming an author for Packt, please visit `authors.packtpub.com` and apply today. We have worked with thousands of developers and tech professionals, just like you, to help them share their insight with the global tech community. You can make a general application, apply for a specific hot topic that we are recruiting an author for, or submit your own idea.

Table of Contents

4

Managing ETL in SAS

5

Managing Data Reporting in SAS

Section 2:
Using SAS for Extract-Transform-Load (ETL) Protocols in a Data Warehouse

6
Standardizing Coding Using SAS Arrays

7
Designing and Developing ETL Code in SAS

8
Using Macros to Automate ETL in SAS

9
Debugging and Troubleshooting in SAS

Section 3:
Using SAS When Serving Warehouse Data to Users

10
Considering the User Needs of SAS Data Warehouses

11
Connecting the SAS Data Warehouse to Other Systems

12
Using the ODS for Visualization in SAS

Assessments

Other Books You May Enjoy

Index

Preface

SAS is used for various functions in the development and maintenance of data warehouses because of its reputation of being able to handle so-called big data. SAS software has been in existence a long time, and has been implemented in many large, data-intensive environments, including data warehouses.

This book provides end-to-end coverage of the practical programming considerations to make when involving SAS in a data warehouse environment. Complete with step-by-step explanations of essential concepts, practical examples, and self-assessment questions, the book helps you begin to make decisions about the roles SAS should play in your data warehouse. It will teach you how to design arrays and macros to standardize extract-transform-load protocols, as well as how to develop strategies to optimally serve data warehouse customers.

You will learn the pros and cons of storing data in SAS, how to document and design ETL protocols for SAS processes, and how the use of SAS arrays and macros can help improve input/output (I/O) efficiency. You will also examine approaches to serving up data using SAS, and how to connect SAS to other systems to enhance the data warehouse user's experience. By the end of this book, you will have a foundational understanding of the roles SAS can play in a warehouse environment, and be able to choose wisely when designing your data warehousing processes involving SAS.

Who this book is for

This book is aimed at programmers using SAS products who are working on a data lake, data mart, or data warehouse. It is also aimed at managers heading up projects involving using SAS to maintain a data lake, data mart, or data warehouse. To benefit from this book, it is helpful to have a background in working on data projects that require serving data or reports to customers. Also, some experience of working with big datasets will be helpful in understanding this book.

What this book covers

Chapter 1, Using SAS in a Data Mart, Data Lake, or Data Warehouse, explains the origins of SAS, and how data **input/output (I/O)** are managed in SAS. It also provides context for how SAS products are used today, in modern data warehouses.

Chapter 2, Reading Big Data into SAS, covers how to read data in different formats into SAS. It also talks about **SAS data formats**, and packaging data for **import** and **export** in SAS.

Chapter 3, Helpful PROCs for Managing Data, provides an introduction to PROC CONTENTS, PROC SQL, and PROC PRINT, and describes how to deal with SAS **formats** and **labels**. It also provides different strategies for **viewing data** in SAS.

Chapter 4, Managing ETL in SAS, explains how to prepare an analytic environment, including developing **naming conventions**, and SAS format and label policies. It also describes the designation of data storage and **user groups**.

Chapter 5, Managing Data Reporting in SAS, introduces you to the **output delivery system (ODS)**, and explains how the ODS is used for outputting graphics files from SAS. This chapter also covers how to use PROCs that were developed specifically for the ODS, including PROC TABULATE and PROC SGPLOT.

Chapter 6, Standardizing Coding Using SAS Arrays, explains how to do array processing in a SAS data warehouse, how to add **conditions** to arrays, and how to deal with naming conventions in arrays. In SAS, because of I/O limitations, the use of **arrays** is usually necessary in ETL code.

Chapter 7, Designing and Developing ETL Code in SAS, goes over how to plan ETL code, using PROC UNIVARIATE and PROC FREQ to study our data and help us plan how to serve up variables. The second part of the chapter focuses on how to develop optimal ETL code based on our plans.

Chapter 8, Using Macros to Automate ETL in SAS, describes how to convert data step code used in ETL to **SAS macro language** in order to automate the process. It also covers how to store and call macros, and how to use them to load transformed data.

Chapter 9, Debugging and Troubleshooting in SAS, covers **debugging** approaches in SAS. Advice for forming and formatting code is given, and special attention is given to debugging **do loop code** and macros.

Chapter 10, Considering the User Needs of SAS Data Warehouses, describes a method by which to classify users, and then apply **data stewardship policies** that help ensure their needs are met. For analyst users, providing data access, **foreign keys**, and **crosswalk variables** is described. For developer users, providing **data curation** and other support is delineated.

Chapter 11, Connecting the SAS Data Warehouse to Other Systems, talks about serving SAS to other data systems, which is typically done asynchronously. Next, it describes connecting SAS to other data systems, which is typically done synchronously through an **open database connectivity (ODBC)** protocol using **SAS/ACCESS**.

Chapter 12, Using the ODS for Visualization in SAS, describes the differences with using the ODS and visualization in SAS when done in print compared to on the web. Next, ways to serve SAS data to the web using the **SAS Enterprise Guide** aided by **SAS Viya** are explained, and how to visualize SAS data in other programs, such as **R** and **Tableau**, is described.

To get the most out of this book

You will need access to a version of SAS. If you do not have access to a SAS server environment or PC SAS, you can use the free version of SAS, called SAS University Edition (available here: https://www.sas.com/en_us/software/university-edition/download-software.html). SAS University Edition is available for Windows, macOS, and Linux. All code examples have been tested using SAS University Edition in Windows, but they should work on any version of SAS.

Example data curation files in this book were developed using Microsoft Word, Excel, and PowerPoint. These files can be developed in the same or comparable software.

If you are using the digital version of this book, we advise you to type the code yourself or access the code via the GitHub repository (link available in the next section). Doing so will help you avoid any potential errors related to the copying/pasting of code.

You may benefit from following the author on YouTube (https://www.youtube.com/channel/UCCHcm7rOjf7Ruf2GA2Qnxow) and LinkedIn (https://www.linkedin.com/in/dethwench/), where she posts video tutorials and information about data science.

Download the example code files

You can download the example code files for this book from GitHub at `https://github.com/PacktPublishing/Mastering-SAS-Programming-for-Data-Warehousing`. In case there's an update to the code, it will be updated on the existing GitHub repository.

We also have other code bundles from our rich catalog of books and videos available at `https://github.com/PacktPublishing/`. Check them out!

Download the color images

We also provide a PDF file that has color images of the screenshots/diagrams used in this book. You can download it here: `https://static.packt-cdn.com/downloads/9781789532371_ColorImages.pdf`.

Conventions used

There are a number of text conventions used throughout this book.

`Code in text`: Indicates code words in text, database table names, folder names, filenames, file extensions, pathnames, dummy URLs, user input, and Twitter handles. Here is an example: "We will map `LIBNAME` to `X`, with `X` being the folder where we put the dataset."

A block of code is set as follows:

```
LIBNAME X "/folders/myfolders/X";
PROC CONTENTS data=X.Chap5_1;
run;
```

When we wish to draw your attention to a particular part of a code block, the relevant lines or items are set in bold:

```
ODS TRACE ON / label;
PROC UNIVARIATE data=X.chap5_1;
    var _AGE80;
run;
ODS TRACE OFF;
```

Bold: Indicates a new term, an important word, or words that you see on screen. For example, words in menus or dialog boxes appear in the text like this. Here is an example: "If you are using SAS University Edition, the **RESULTS** tab will display the graphic."

> **Tips or important notes**
> Appear like this.

Get in touch

Feedback from our readers is always welcome.

General feedback: If you have questions about any aspect of this book, mention the book title in the subject of your message and email us at customercare@packtpub.com.

Errata: Although we have taken every care to ensure the accuracy of our content, mistakes do happen. If you have found a mistake in this book, we would be grateful if you would report this to us. Please visit www.packtpub.com/support/errata, selecting your book, clicking on the Errata Submission Form link, and entering the details.

Piracy: If you come across any illegal copies of our works in any form on the internet, we would be grateful if you would provide us with the location address or website name. Please contact us at copyright@packt.com with a link to the material.

If you are interested in becoming an author: If there is a topic that you have expertise in, and you are interested in either writing or contributing to a book, please visit authors.packtpub.com.

Reviews

Please leave a review. Once you have read and used this book, why not leave a review on the site that you purchased it from? Potential readers can then see and use your unbiased opinion to make purchase decisions, we at Packt can understand what you think about our products, and our authors can see your feedback on their book. Thank you!

For more information about Packt, please visit packt.com.

Section 1: Managing Data in a SAS Data Warehouse

This first section focuses on the basics of managing data in a data warehouse in SAS. First, we focus heavily on the process of data **input/output (I/O)** in SAS, which has not changed since SAS was originally created. Then, we see how to use **data steps** and **PROCs**, or SAS procedures, to read big data into SAS in various formats, given SAS's distinct data I/O processes.

After that, we are introduced to PROCs in SAS that can help manage data, especially with respect to I/O. These include PROCs that allow you to **view** and **profile** the dataset, including PROC CONTENTS and PROC PRINT.

Then, we see how to prepare for **extract, transform, and load (ETL)** processes by setting **naming conventions**, designating **user groups**, and setting other policies. Lastly, we are introduced to SAS's **output delivery system (ODS)** and see how **reporting** is done in SAS.

This section comprises the following chapters:

- *Chapter 1, Using SAS in a Data Mart, Data Lake, or Data Warehouse*
- *Chapter 2, Reading Big Data into SAS*
- *Chapter 3, Helpful PROCs for Managing Data*
- *Chapter 4, Managing ETL in SAS*
- *Chapter 5, Managing Data Reporting in SAS*

1
Using SAS in a Data Mart, Data Lake, or Data Warehouse

The purpose of this chapter is to showcase how SAS has been used in data warehousing over its lifetime, and how that history impacts SAS data warehousing today. It provides an opportunity to see how slight changes in coding in SAS **data steps** can greatly impact data **input/output (I/O)**. It also covers how SAS data is managed, and how **Base SAS**, the analytic component, interacts with stored data.

As SAS developed, there became a need to set **indexes** on variables, and to use **SQL** coding in SAS. How PROC SQL in SAS compares with data steps and other SQL programming will be reviewed in this chapter. I will also explain strategies to deal with memory issues in SAS, and how it has evolved to now be used with data in the cloud.

In this chapter, we are going to cover the following main topics:

- How early versions of SAS handled data
- Different ways to access data in SAS
- Considerations in improving I/O in SAS
- Dealing with storage and memory issues in SAS
- Using SAS in modern data warehousing

> **Note:**
> The links to all the white papers and other sources mentioned in the chapter are provided in the *Further reading* section toward the end of the chapter.

Technical requirements

The dataset used as a demonstration in this chapter, in `*.csv` format, can be found online on GitHub: `https://github.com/PacktPublishing/Mastering-SAS-Programming-for-Data-Warehousing/blob/master/Chapter%201/Datasets/Chap%201_1_Infile.csv`.

The code bundle for this chapter is available on GitHub here: `https://github.com/PacktPublishing/Mastering-SAS-Programming-for-Data-Warehousing/tree/master/Chapter%201`.

Using original versions of SAS

Initially, SAS data had to be input through code into memory whenever analysis code was to be run on the data. This section covers the following:

- How to enter data into **SAS datasets** using SAS
- The early **PROCs** developed, such as PROC PRINT and PROC MEANS
- Improvements to data handling made in **Base SAS**

In this section, you will learn how SAS's data management processes were initially developed. The processes impact how SAS runs today.

Initial SAS data handling

As described on SAS's website (`https://www.sas.com/en_us/company-information/profile.html`), SAS was invented in 1966 as the **Statistical Analysis System**, developed under a grant from the **United States (US) National Institutes of Health (NIH)** to eight universities. The immediate need was to develop a computerized system that could analyze the large amount of agriculture data being collected through the **US Department of Agriculture (USDA)**.

According to the SAS history listed on the Barr Systems website (`http://www.barrsystems.com/about_us/the_company/professional_history.asp`), Anthony J. Barr was in the Statistics Department of North Carolina State University and was recruited to help program SAS. He was responsible for developing the first **analysis of variance (ANOVA)** and regression programs in SAS and created the software for inputting and transforming data.

Even today, it is relevant here to reflect on Barr's early development of what would later be called **data step** language in SAS. This is because current data import processes in SAS continue to use roughly the same approach, which presents both opportunities and limitations in data warehouse management.

In the early data step code, data was entered as part of the code, which still can be done today. Let's consider a modern example of using data step code to enter data in SAS by referring to the **2018 BRFSS Codebook** listed in the technical requirements for this chapter. Each year, the **United States Centers for Disease Control and Prevention (CDC)** organizes an annual anonymous phone survey of approximately 450,000 residents asking about health conditions and risk factors. This survey is called the **Behavioral Risk Factor Surveillance System (BRFSS)**. The 2018 BRFSS Codebook describes the 2018 version of a SAS dataset from a survey in the US that is conducted by phone every year.

The *codebook* describes specifications about each variable in the dataset, including the following:

- Variable name
- Allowable values
- Frequencies in the dataset for each value

The BRFSS Codebook is quite extensive and can be confusing for an analyst without a background in the dataset to understand it. In *Chapter 3, Helpful PROCs for Managing Data,* we will look closely at an example from the BRFSS Codebook. For now, let's review a codebook that is easier for the beginner to interpret. Here is an example of a codebook entry from the online codebook for the **US National Health and Nutrition Examination Survey (NHANES)**:

BMXWT - Weight (kg)

Variable name:	BMXWT
SAS Label:	Weight (kg)
English Text:	Weight (kg)
Target:	Both males and females 0 YEARS - 160 YEARS

Code or Value	Value Description	Count	Cumulative
3.6 to 198.9	Range of Values	9445	9445
.	Missing	99	9544

Figure 1.1 – Example of a codebook entry from the US NHANES

The following table represents how three of the variables in the BRFSS Codebook – _STATE, SEX1, and _AGE80 – could be represented in three lines of data:

_STATE	SEX1	_AGE80
12	1	72
25	2	25
27	2	54

Table 1.1 – Example of three variable values for three respondents in the 2018 BRFSS dataset

Here, the state of residence of the respondent is recorded under X_STATE according to its corresponding numerical **Federal Information Processing System (FIPS)** number, and SEX1 is coded as 1 for **male** and 2 for **female**, 7 for **don't know/not sure**, and 9 for **refused** (to decode state FIPS numbers, please see the link for the FIPS state codes list in the *Further reading* section). The _AGE80 variable refers to the age of the respondent imputed from other data (with ages over 80 collapsed). Using the codebook to decode the preceding data, we see the three rows represent a 72-year-old man from Florida (FL), a 25-year-old woman from Massachusetts (MA), and a 54-year-old woman from Minnesota (MN).

Let's look at an example of using data step code to create this table in SAS:

```
DATA THREEROWS;
    INFILE CARDS;
    INPUT _STATE SEX1 _AGE80;
    CARDS;
12 1 72
25 2 25
27 2 54
;
RUN;
```

Let's go through the code:

1. The THREEROWS dataset is created in the **WORK directory** of SAS. The WORK directory, simply called WORK, is the working directory for the SAS session, which means when the session is over, the data in WORK will be erased.

2. As is typical in SAS programming, each of the programming lines ends with a semi-colon, except each of the data lines.

3. The next line, INFILE CARDS;, indicates to SAS that data will now be entered from cards (although it is possible to replace this with the more modern version of the command, datalines).

4. When Barr designed this process, the next step would be for SAS to input punch cards that held the data. The next line, INPUT _STATE SEX1 _AGE80;, designates that the data that will be input from the cards has these headers: _STATE, SEX1, and _AGE80.

5. The next line, CARDS;, indicates that it is time for the cards to start to be read. What follows in the code is our modern representation of entering the data represented in the table into SAS using CARDS.

6. By the time SAS processes the CARDS statement, it already knows from the INPUT statement to expect three columns – _STATE, SEX1, and _AGE80 – so even without formatting the lines in three rows, SAS would read the values sequentially and assemble the dataset with three columns, ending when it hits the semi-colon at the end.

7. These three variables are numeric by default unless '$' is included in the INPUT statement.

> **Note:**
>
> It is not a good idea to store actual data values in SAS code today. They can easily be lost, and if the data is private, it can create privacy issues around the code. Further more, many of the datasets used today are extremely large, and it would not be practical to store them as actual data values in SAS; instead, they might be stored in a database system such as Oracle, or in an Excel file.

Early SAS data handling

In September 1966, the conceptual ideas behind SAS were presented by Barr and others to the Committee on Statistical Software of the **University Statisticians of Southeast Experiment Station (USSERS)** at their meeting held in Athens, GA. Barr began working with others, including the current SAS CEO, James Goodnight, on developing the first worldwide release of SAS.

> **Note:**
>
> The first worldwide release of SAS in 1972 consisted of 37,000 lines of code, 65% written by Barr, and 32% written by Goodnight.

Improvements implemented in the 1972 worldwide release of SAS focused on procedures known as **PROCs**. Procedures are applied to SAS datasets. Some PROCs are for data editing and handling, but most are focused on data visualization and analysis, with the data handling typically done using data steps. Barr developed some basic PROCs still used today to assist in data handling, including PROC SORT and PROC PRINT.

Let's look at an example of PROC SORT and PROC PRINT. Coming back to the data we entered earlier, the rows were sorted according to the value in the _STATE variable:

- If we wanted to sort the dataset in order of the respondent's age, or _AGE80, we could use PROC SORT with the _AGE80 command.

- Following that with a PROC PRINT would then print the resulting dataset to the screen.

This is shown in the following code:

```
PROC SORT data=THREEROWS;
by _AGE80;
PROC PRINT;
RUN;
```

While Barr also developed analysis PROCs such as `PROC ANOVA`, which conducts an ANOVA, Goodnight developed PROCs aimed principally at analysis, such as `PROC CORR` for correlations and `PROC MEANS` for calculating means. The most ideal situation at the time for data handling was to have the data already stored on the cards, essentially in the format it needed to be in for analysis. However, data step code was available for editing the data.

At this time, NIH discontinued funding the project, but the consortium of universities that had worked on the project agreed to provide funding support, allowing the programmers to continue building SAS. Barr, Goodnight, and others continued to develop the software, adding mainly statistical functions rather than data management functions, and released a 1976 version. For the 1976 version, Barr rewrote the *internals* of SAS, including the following:

- Data management functions
- Report-writing functions
- The compiler

This was the first big rewrite of SAS's processing functions.

In 1976, SAS Institute, Inc. was incorporated, with ownership split between Barr, Goodnight, and two others:

Over 100 organizations including pharmaceutical companies, insurance companies, banks, governmental entities, and members of the academic community were using SAS.

More than 300 people attended the first SAS users conference in 1976.

Reflecting on this short history, it is understandable that even today, SAS maintains the reputation of being the only statistical software that can comprehensively handle **big data**. While in some regards this statement remains true, it is also necessary to revisit a more subtle point, which is that SAS was initially developed for data analysis – not for data storage. Even with improvements, SAS data handling is still limited by some of the features originally developed in this early era.

SAS data handling improvements

Both the SAS programs and the data SAS analyzed were initially stored on punch cards. These were physical cards with hole punches in them to indicate instructions to the computer. The following photograph shows a real punch card that was used for an IBM 1130 program:

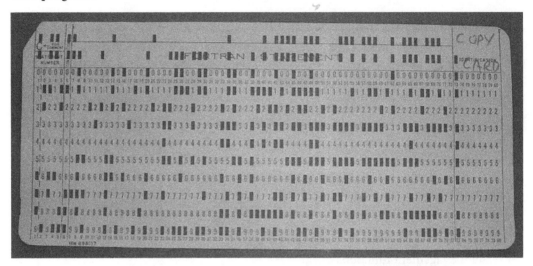

Figure 1.2 – This card contains a self-loading w:IBM 1130 program that copies the deck of cards placed after it in the input hopper. Photograph by Arnold Reinhold, CC BY-SA 2.5 (https://creativecommons.org/licenses/by-sa/2.5/deed.en)

In his 2005 report titled *Programming with Punch Cards*, Dale Fisk explains how creating a set of punch cards to run a computer program was a multi-step, labor- and time-intensive process:

1. First, cards had to be punched by hand.

2. Next, the program punched into the cards had to be compiled through a computer, which would produce a printed list of errors if there were any.

3. If the program compiled, the computer would print a set of cards with the compiled program.

4. This new set of cards would be the ones used to launch the program.

SAS was the first application that could feasibly handle running programs to analyze large datasets using punch cards. The positive result of the development of early SAS programs was the ability to use these punch cards to run complex regressions that could never have been attempted before.

But punch cards also created challenges. The foundation SAS component, called **Base SAS**, was about 300,000 lines of code. This program was stored in 150 boxes of cards that would stand 40 feet high. These boxes were separate from boxes of cards of data that SAS would be used to analyze.

This meant that card storage was an issue. A lot of space was required already for the computers themselves, which could take up entire rooms. In addition, the cards were unwieldy and required their own set of error handling procedures, including the LOST CARD error that can still be displayed in SAS today under particular circumstances when there is an error reading in data.

Nevertheless, SAS continued to reach out and recruit new customers. From the beginning, SAS has always prided itself on its customer service. As of 1978, there were 600 SAS customer sites, but only 21 employees. The climate was that of everyone pitching in to help fill customer needs, and Goodnight was known for recognizing the value of employees to the company.

Accessing data in SAS

This section covers how accessing data in SAS changed over the years:

- First, SAS data storage moved from punch cards to **mainframes**.
- Next, the invention of **personal computers (PCs)** led to reconfiguring how SAS data was accessed.
- Consequently, reading data into SAS from external data files became more common.

In this section, we will discuss how to read data in SAS from an external file, as well as the opportunities and limitations of how SAS processes data.

Upgrading to mainframes

In 1979, Databank of New Zealand adapted SAS to run under IBM's VM/CMS system using IBM's **disk operating system (DOS)**, thus solving the punch card problem and establishing SAS as mainframe software that was remotely hosted. This represented essentially the second rewrite of SAS since its 1976 rewrite. This upgrade made SAS more easily accessible to more customers. It also facilitated the ability for SAS to include more sophisticated components to add onto Base SAS. At the same time, it created new challenges for efficient data **input and output (I/O)**.

In 1979, Barr resigned from SAS, and SAS moved into its current headquarters in Cary, North Carolina (NC). In 1980, SAS added the components SAS/GRAPH for the presentation of graphics, and SAS/ETS for econometric and time series analysis. Prior to SAS/GRAPH, plots were developed using text characters. SAS/GRAPH allowed the output to be displayed as graphics rather than text.

New PROCs added as part of the new SAS/GRAPH component included `PROC GCHART` and `PROC GPLOT`. In the current SAS University Edition, these PROCs are no longer available and have been replaced with updated versions. At the time, however, SAS/GRAPH was considered a great improvement in graphical display over what had been available previously.

> **Note:**
>
> For examples of `PROC GCHART` and `PROC GPLOT` output, read Mike Kalt and Cynthia Zender's white paper on SAS graphics for the non-statistician (included in the *Further reading* section).

During the 1980s, SAS as a company grew dramatically; its campus expanded to 18 buildings that included a training center, publications warehouse, and video studio. By the end of the 1980s, SAS had nearly 1,500 worldwide employees and had established new offices on four continents. But as the 1980s wore on, PCs were becoming popular, and customers were demanding a way of running SAS on PCs.

Therefore, SAS had to iterate again in order to keep up with the pace of technological innovation in the background. Even though SAS was now running on mainframes, SAS's effort in innovation had been concentrated on the PROCs, and less attention was paid to optimizing the data management functions provided by the data steps. The response by both the company and SAS was to find ways to optimize the functioning of the SAS system, rather than rebuilding PROCs or data steps.

Transitioning to personal computers

To accommodate PC users, in the 1980s, SAS was rewritten in the C language, which was popular at the time for PC applications. At the same time, SAS developed a new software architecture to run across multiple platforms, which SAS is still known for today. At the time, this PC functionality was introduced as a **micro-to-mainframe** link, allowing customers to store a dataset on a mainframe while running programs from their PCs.

Having an application running on PCs afforded SAS the opportunity to improve the user experience. SAS developed a **graphical user interface (GUI)** that resembled the Macintosh and Windows environments that were popular at the time and continued to move away from a numbers-centric format for data display and toward enhanced graphics and visualizations.

> **Note:**
>
> PC SAS does not run on Macintosh computers. But during the 1980s, SAS developed **JMP*** (pronounced *jump*), which is a statistical program that can perform many of the same tasks as SAS on macOS.

SAS still had the limitations that its data steps were sequential; so it still read data line by line, just as it had done with punch cards. Data steps were the main functions used to export data out of SAS, and therefore, SAS exported data line by line. This created a lack of flexibility in the format of output files. So, to get around this, the **Output Delivery System (ODS)** was created. This system allows the user to format output in a variety of formats and is still used currently in SAS today, such as Excel, `*.pdf`, or `*.rtf` files, with a specific component for delivering graphics called **ODS Graphics**.

While this period of SAS's evolution brought many innovations, they were mostly in the area of improving the user experience, rather than focusing on data handling. In terms of the micro-to-mainframe link, the development was mostly focused on the *micro* rather than the *mainframe* component. This focus on user experience seemed consistent with SAS's values of putting customers and employees first. During the 1980s, the company was recognized by its customers as helping them make sense out of their vast amount of data and helping them have the results of their data analysis guide their decisions. It also innovated in the area of employee wellness, opening an on-site childcare center in 1981, followed by establishing an on-site fitness center, health care center, and café.

Reading external files

With the movement to PCs, customers wanted to import external data files into local copies of SAS, rather than using `CARDS` or `DATALINES` to input data, or connecting to a mainframe data source. This was accomplished through revisions to the `INFILE` statement, the use of external file references, and the setting of options. Today, SAS has created an automated way to read files using a **graphical user interface (GUI)** which launches a wizard that creates `PROC IMPORT` code.

If you are using the University Edition of SAS, you can place the data file for this chapter named `Chap_1_1_Infile.csv` into your `myfolders` folder, and if you run the following `PROC IMPORT` code, the file should be imported into `WORK`:

```
%web_drop_table(WORK.IMPORT);
FILENAME REFFILE '/folders/myfolders/Chap 1 1 Infile.csv';
PROC IMPORT DATAFILE=REFFILE
    DBMS=CSV
    OUT=WORK.IMPORT;
    GETNAMES=YES;
RUN;
PROC CONTENTS DATA=WORK.IMPORT;
RUN;
%web_open_table(WORK.IMPORT);
```

When SAS code is run, it produces a **log file**. In current SAS applications, the log file opens in a separate window. The log file repeats the code that has been run and includes messages providing feedback, including error messages. It is important to always review the log file to make sure errors and key warnings do not exist, as well as to confirm any assumptions by SAS. Using the point-and-click GUI in SAS University Edition, SAS will create the preceding code and then run it to import the `Chap 1_1_Infile.csv` file. This is evident because this is the code that is displayed in the log file.

Notice that the code refers to the following items:

- An external reference file using the `REFFILE` command
- A connection between the data file being created and the reference file through `DATAFILE = REFFILE`
- The specification that the input file is a **comma-separated values (CSV)** file through `DBMS = CSV`
- The `OUT` specification to make SAS output the resulting dataset named `IMPORT` into `WORK` through `OUT = WORK.IMPORT`
- The automatic placement of a `PROC CONTENTS` command to display the contents of the dataset

`PROC IMPORT` code like the preceding code does not provide the opportunity for a programmer to specify details about how they want the resulting dataset formatted, column by column. This type of specification can be achieved using a series of commands.

Let's consider the source file, which is Chap 1_1_Infile.csv. This source file has the rows from the BRFSS 2018 dataset for FL, MA, and MN (_STATE equals 12, 25, or 27). It also has these columns: _STATE, SEX1, _AGE80, and _BMI5, which is the respondent's body mass index (BMI) stored as a four-position integer that should have a decimal placed between the second and third integer. The following table displays three records from the source data:

_STATE	SEX1	_AGE80	_BMI5
12	2	76	3175
25	1	67	2525
27	1	43	2999

Table 1.2 – Example of four variables from three records of BRFSS source data

Using the following code, we can read in the *.csv file and specify details about the column formats:

```
data Chap_1_1_Infile;
%let _EFIERR_ = 0;
infile '/folders/myfolders/Chap 1_1_Infile.csv' delimiter = ','
firstobs=2
    MISSOVER DSD lrecl=32767;
    informat    _STATE 2.;
    informat    SEX1 2.;
    informat    _AGE80 2.;
    informat    _BMI5 2.2;
    format      _STATE 2.;
    format      SEX1 2.;
    format      _AGE80 2.;
    format      _BMI5 4.1;
input
    _STATE
    SEX1
    _AGE80
    _BMI5
;
if _ERROR_ then call symputx('_EFIERR_',1);
RUN;
```

The preceding code provides an opportunity to look at various aspects of the way SAS reads in data using the INFILE statement:

- The code opens with a data step specifying the output file to be named Chap_1_1_Infile and placed in WORK.

- The %let _EFIERR_ = 0; and if _ERROR_ then call symputx('_ EFIERR_',1); commands are used for error handling.

- The INFILE command has many options that can be used. The preceding code uses delimiter = ',' to indicate that the source file is comma-delimited, and firstobs=2 to indicate that the values in the first observation are on row 2 (as the column names are in row 1).

- The INFORMAT command provides the opportunity to specify the format of the source data being read in. This is important to make sure data is read correctly without losing any information. Notice how _BMI5 is specified at 2.2, meaning two numbers before the decimal, and two numbers after it.

- The FORMAT command allows the ability to specify the format of the data output into the Chap_1_1_Infile file. Note that formats do not change how the data is stored, but only control how they are displayed. This can be confusing to new SAS programmers. Notice how _BMI5 is specified at 4.1, so it should result in a variable with four numbers before the decimal and one number after it.

- As with when we used CARDS and DATALINES, the INPUT statement signals the point where SAS should start reading in the data, and names each column in order.

The resulting dataset in WORK, named Chap_1_1_Infile, looks like this:

_STATE	SEX1	_AGE80	_BMI5
12	2	76	31.8
25	1	67	25.2
27	1	43	30.0

Table 1.3 – Example of the same source data formatted using the FORMAT command

The ability to specify details about importing data was necessary for SAS users to be able to read **flat files** that were exported out of another system. The INFILE approach with FORMAT and INFORMAT allowed the necessary flexibility in programming to allow conditionals to be placed in code to facilitate SAS reading only parts of the files, and the ability to direct SAS to specific coordinates on raw datafiles and direct it to read those values a certain way. But while using INFILE and related commands increased the flexibility behind the use of big data in SAS (because the data step functioning was still based on the sequential read approach used with the punch cards), there were limited opportunities for the programmer to improve I/O.

Improving I/O

Although SAS has created many features to improve data warehousing, it is still necessary to improve I/O through the strategic use of SAS code. This section will cover the following:

- Features for warehousing that have been developed by SAS

- The importance of using the WHERE rather than the IF clause in data processing

- How sorting and indexing can be done to improve I/O in SAS

Developing warehouse environments

The 1990s saw people working with SAS and big data to find creative solutions to improve data I/O. In his 1997 SAS white paper (available under *Further reading*), Ian Robertson describes the benefits of his case study migrating the **Wisconsin Department of Transportation Traffic Safety and Record-keeping System (TSRS)** from a mainframe SAS setup to one where data was served up to analysts through a **local area network (LAN)**.

By this time, SAS had been reconfigured to run on their LAN's operating system, OS/2, so his team was able to save processing costs by moving the SAS analysis and reporting functions away from the mainframe and onto the PCs of the analysts. One of the innovations that enabled this was **SAS/Connect**, a component that allowed local PCs to connect to a mainframe storing data elsewhere. Using SAS/Connect, Robertson's team downloaded the data onto their LAN, making a local copy for analysis. Over a 2-day period, they were able to transfer 1.6 GB of TSRS data from the years 1988 through 1995 and 30 MB of source code from the mainframe storage system to the LAN.

In SAS's history of data warehousing mentioned on their website, the differences between a **data warehouse, data mart**, and **data lake** are explained:

- A **data warehouse** stores a large amount of enterprise data covering many topics.

- A **data mart** stores a smaller amount of data focused on one topic, usually sourced in a data warehouse.

- **Data warehouses** and **data marts** consist of raw datasets that have been restructured for the purposes of analysis and reporting.

- By contrast, a **data lake** stores a large amount of raw data that has not been processed.

Robertson's local version of the TSRS could be seen as a data warehouse, in that all the data from the source system had been moved to the LAN. By moving the data from the mainframe to the LAN, the data was now in the same physical place as the application accessing it, and this reduced not only I/O time but CPU cost. However, the group encountered a few difficulties after moving the data that required revising their source code.

Using the WHERE clause

Robertson's team used a few different strategies to improve the efficiency of their source code as an approach to **performance tuning**. Robertson's team reviewed their SAS processing code and found that they were using a lot of WHERE clauses to subset data into regions. They began looking into ways to improve the efficiency of their coding with respect to the use of WHERE clauses.

The purpose of the WHERE clause is to subset datasets. The WHERE clause becomes important in data warehousing in two major areas: data management and data reporting. For this reason, WHERE is used in both data steps and PROCs.

Consider our example dataset, Chap_1_1_Infile, and imagine we wanted to create a dataset of just the Massachusetts records (_STATE = 25). Even though the source dataset, Chap_1_1_Infile, has 38,901 observations, by declaring a WHERE clause in our data step, we can avoid reading all those records and only process the ones where _STATE = 25 comes into the data step:

```
DATA Massachusetts;
    set Chap_1_1_Infile;
    WHERE _STATE = 25;
RUN;
```

The first line of the log file says this:

```
NOTE: There were 6669 observations read from the data set WORK.
CHAP_1_1_INFILE.
      WHERE _STATE=25;
```

This indicates that as SAS was processing the file, it skipped over reading the rows where it saw anything other than 25 in _STATE. On my computer, the log file said this operation only took 0.01 seconds of CPU time. However, at the time Robertson was writing, the difference between using and not using a WHERE clause could really impact CPU time, which drove up data warehousing costs.

WHERE can also speed up the reporting of large datasets. Consider the use of PROC FREQ on our example dataset, Chap_1_1_Infile, to get the frequency of the gender variable, SEX1, in Massachusetts (_STATE=25):

```
PROC FREQ data=Chap_1_1_Infile;
    Where _STATE = 25;
    Tables(SEX1);
RUN;
```

Looking at the log file, again, we see that this frequency calculation only considered the 6,669 records from Massachusetts in its processing, thus saving processing time by not considering the whole file. On my computer, the log file says that the CPU time used was 0.10 seconds – still a very small number, but 10 times the number seen in the data step processing. This demonstrates the extra processing power needed for the frequency calculations produced by PROC FREQ.

Using IF compared to WHERE

In 2003, Nancy Croonen and Henri Theuwissen published a white paper providing tips on reducing CPU time through more efficient data step and PROC programming (available under *Further reading*), as many SAS users were still operating in a mainframe environment. In addition to advocating the use of the WHERE clause, the authors did studies to compare the CPU time for different ways of accomplishing the same tasks in SAS. They discussed trade-offs between using the WHERE and IF clauses.

Earlier, we used WHERE in a data step to subset the Chap_1_1_Infile dataset to just the records from Massachusetts and named the dataset Massachusetts. Let's do the same thing again, but this time, we'll use IF instead of WHERE:

```
DATA Massachusetts;
    set Chap_1_1_Infile;
    IF _STATE = 25;
RUN;
```

On my computer, I notice no difference in processing time – it is still 0.01 seconds. However, the first line of the log file is different:

```
NOTE: There were 38901 observations read from the data set
WORK.CHAP_1_1_INFILE.
```

This indicates that SAS read in all 38,901 records before processing the rest of the code, which was instructions to only keep records in the Massachusetts dataset IF _STATE = 25. From this demonstration, it seems like WHERE is superior to IF when trying to make efficient code.

However, this is not always true. There is only a meaningful reduction in processing time by using WHERE instead of IF in case the variable in the statement has many different values, otherwise known as **high cardinality**. Croonen and Theuwissen show an example comparing the use of WHERE and IF to subset a large dataset by a nominal variable with seven levels, which would have **low cardinality**. In their example, there was only a small reduction in processing time using WHERE compared to IF.

> **Note:**
> When the WHERE clause is used, SAS creates a simple index on that variable and searches the index. When the IF clause is used, SAS does a line-by-line sequential search of the dataset. Note that IF must be used when applied to temporary variables not in the dataset. For a deeper discussion on the use of the WHERE and the IF clauses, please see the SAS white paper by Sunil Gupta (under *Further reading*).

Robertson's team managing the TSRS data warehouse had already optimized their code using WHERE for subsetting in their data processing and reporting. However, they found that after they moved their data to a local LAN, they needed to further improve the efficiency of their code, so they started looking into approaches to **indexing**.

Sorting in SAS

An **index** is a separate file from a dataset that is like an address book for SAS to use when looking up records in a large dataset. Unlike **structured query language (SQL)**, SAS does not create indexes automatically in processing. There are ways the programmer can index variables in a dataset as well. Given certain data processing code, placing indexes on particular variables can speed up processing.

As discussed earlier, it is helpful to identify what variables are used in WHERE clauses that could benefit from indexing, and also, whether they are high or low cardinality. If a certain high-cardinality variable is used repeatedly in a WHERE clause, it is a good candidate for an index.

A **simple index** is an index made of one variable (such as _STATE). A **composite index** is one made of two or more variables (such as _STATE plus SEX1). Because SAS processes records sequentially, the easiest way a programmer can simulate an index on a SAS dataset is by sorting the dataset by that variable. It is not unusual for SAS datasets in a data lake to be stored sorted by a particular high-cardinality variable (such as a unique identification number of the row).

According to the BRFSS 2018 Codebook, in the source dataset, 53 regions are represented under the _STATE variable. If this variable was used in warehouse processing, then it would be logical to store the dataset sorted by _STATE, as shown in the following code:

```
PROC SORT data=Chap_1_1_Infile;
    by _STATE;
RUN;
```

Sorting itself takes some time; on my computer, it took 0.05 seconds of CPU time. Compared to the 0.01 seconds it took to read in the dataset of about 36,000 rows and the 0.10 seconds it took to do a PROC FREQ, this shows that if there is to be a policy in the data warehouse that datasets are to be stored sorted by a particular variable, sorting them will take some time to execute.

While sorting is a simple way of placing an index on a SAS variable, it may not be adequate. Using our example dataset, imagine we wanted to know the mean age (_AGE80) of respondents by gender (SEX1). We could use PROC MEANS to do this using the following code:

```
PROC MEANS data=Chap_1_1_Infile;
    by SEX1;
RUN;
```

With our dataset, Chap_1_1_Infile, in the state it is in at the beginning of the PROC, meaning sorted by the _STATE variable and not the SEX1 variable, the preceding PROC MEANS code will not run. The code produces the following error:

```
ERROR: Data set WORK.CHAP_1_1_INFILE is not sorted in ascending
sequence. The current BY group has SEX1 = 9 and the next BY
group has SEX1 = 1.
```

As described in the error wording, SAS is expecting the dataset to already be sorted by the BY variable, which is SEX1 in the PROC MEANS code. Solving this problem can easily be accomplished by resorting the dataset on the SEX1 variable prior to running the preceding code. But this will cause the dataset to no longer be sorted by _STATE, and we will lose the benefit of being able to efficiently use _STATE in a WHERE clause. It is in these more complex situations that we cannot rely on using sorting for a simple index, and should consider placing an index on certain variables.

Setting indexes on variables

Robertson's team ran into a similar problem, where different variables were used in WHERE clauses throughout their programming. Therefore, they could not simply sort their datasets by one variable and rely on that indexed variable to speed up their processing.

One of the ways they dealt with this was to deliberately place **indexes** on certain SAS variables using a method other than sorting. As described earlier with sorting, indexes can help improve SAS's performance when extracting a small subset of data from a larger dataset. According to Michael Raithel, who wrote a SAS white paper about indexing, if the subset comprises up to 20% of the dataset, then an index should improve program performance (white paper available under *Further reading*). But if it is larger, it may not impact or even worsen performance. We saw this situation earlier when comparing the efficiency of processing between using a WHERE versus an IF clause.

SAS continued to release new enterprise versions with upgraded PROCs and data steps and new functionality. Starting in the 1980s, main upgrade versions were released and stated as an integer (for example, version 6), but in reality, these versions were upgraded regularly, with each upgrade designated by two digits after the decimal, (for example, version 6.01). SAS had tried to build indexing features into its version 6 but found that there were performance problems, according to Diane Olson's white paper on the topic (available under *Further reading*). The version 7 releases, otherwise known as the Nashville releases and first available in 1998, fixed these problems.

Let's create an index on the _STATE variable using our example dataset, Chap_1_1_Infile. One way we could have created an index on _STATE was in the original data step we used to read in the data. Notice the same code we used before follows, but with the addition of the index command in the first line of the data step:

```
data Chap_1_1_Infile (index=(_STATE));
```

This is often the most efficient way to place an index, but datasets that have already been read into WORK can have indexes set on a variable using various approaches. One way is to use PROC DATASETS, which is demonstrated here:

```
PROC DATASETS nolist;
    modify Chap_1_1_Infile;
          index create _STATE;
RUN;
```

The nolist option suppresses the printout of the dataset, and the modify statement is used to tell SAS to modify the Chap_1_1_Infile dataset to create an index on the _STATE variable. In both of these examples, a simple index was created. Imagine we wanted to create a composite index including both _STATE and SEX1. We could do that using a data step, or we could do it using PROC DATASETS.

Using a data step, we could set the composite index by replacing the first line of our data step code shown earlier with this code:

```
data Chap_1_1_Infile (index=(STATE_SEX = (_STATE SEX1)));
```

Notice the differences between when we set a simple index on _STATE in the data step and the preceding code:

- Because we are setting a composite index, we have to actually name the index a name that is different than the variables in the dataset. We are using the name STATE_SEX for the index.

- Then, in parentheses, we specify – in order – the two variables in the composite index, which are _STATE and SEX1.

> **Note:**
> Some analysts prefer to use the _IDX suffix when naming indexes to indicate they are indexes.

To create the same index using PROC DATASETS, we would use the same code as we did for our simple index on _STATE, only replacing the index create line with this code:

```
index create STATE_SEX = (_STATE SEX1);
```

In the development of the TSRS data warehouse, Robertson's team leveraged indexes in their performance tuning. First, since indexes are used in WHERE clauses and not IF clauses, they rewrote their code to strategically switch IF clauses with WHERE clauses to improve performance. Then, they set indexes on the variables that were used in WHERE clauses, and only saw a 6% storage overhead.

Dealing with storage and memory issues

This section will cover issues with storage and memory when using SAS for big data. It will cover the following:

- How SAS dealt with competition from **structured query language (SQL)** for data storage

- How PROC SQL works and can be used in data warehouse processing

- Considerations about memory and storage that need to be made when using SAS in a data warehouse in modern times

- How SAS can work in the cloud

Avoiding memory issues

Even as SAS got more powerful, datasets kept getting bigger, and there were always challenges with running out of memory during processing. For example, using WHERE instead of IF when reading in data would not only reduce CPU usage and the time it took for code to run, it would also prevent unnecessary usage of memory. Even today, tuning SAS code may be necessary to avoid memory issues.

In a data warehouse, mart, or lake, datasets that were transformed in SAS may be stored outside of SAS in SAS format. This makes them easy to read into SAS. However, this format can be very large, so the option to COMPRESS the dataset was created. Curtis Smith reported on his test compressing SAS files in his white paper (available under *Further reading*), and found that depending upon the dataset, compressing datasets could make them take up half the space.

Smith recommended not only compressing datasets but also deleting unneeded variables to make datasets smaller. In a data warehouse, mart, or lake, source datasets contain **native variables**. In a data lake, these datasets may remain relatively unprocessed. However, in a data warehouse or data mart, decisions need to be made about what variables to keep available for analysis in the warehouse. Further more, **transformed variables** may be added during processing to serve the needs of the users of the warehouse.

The team running the data warehouse should ask the following for each raw dataset:

- If native variables should be available for analysis, which ones should be kept?

- If transformed variables should be available for analysis, which ones should be provided?

By carefully answering these questions, only the columns needed from each dataset can be retained in analysis files, thus reducing processing time for warehouse developers and users.

Accommodating Structured Query Language

SQL was developed and deployed by various companies in the 1990s and early 2000s. SQL was aimed at data maintenance and storage using **relational** tables rather than flat files. SQL approaches only became possible in the 1990s due to upgrades in technology that allowed faster processing of data.

SQL languages accomplish the same data editing tasks that data steps do in SAS, but they use a different approach. Unlike SAS, which is a **procedural language**, SQL is a **declarative language**:

- **In SAS**, the programmer must program a data step to do the *procedures* in the most efficient way to optimize data handling.

- **In SQL**, the programmer *declares* what query output they desire using easy-to-understand, simple English statements, and an optimization program (or **optimizer**) running in the background figures out the most efficient way to execute the query.

While using efficient code in SQL can still improve performance, the efficiency of SQL is less dependent upon the programmer's code and more dependent on the function of its optimizer. Hence, maintaining data in a database became easier using SQL rather than SAS data steps. In SQL, programmers had to learn a few basic commands that could perform a variety of tasks when used together. But with SAS data steps, programmers needed to study a broad set of commands, and they also had to learn the most efficient way to assemble those commands together in order to achieve optimal SAS performance.

What SQL cannot do is analyze data the way SAS can. Therefore, over the latter half of the 1990s and early 2000s, while many databases began to be stored and maintained in SQL, SAS could still be used on them for analysis through the **SAS/Access** feature.

A 1995 edition of the periodical Computerworld described current options for SAS users in an article titled *SAS Institute's customers keep the faith* by Rosemary Cafasso (available under *Further reading*). There were two ways to conceive of data storage in SAS at that time:

- Using SAS only for analysis, and connecting to a non-SAS data storage system to do this

- Using SAS for both data storage and analysis

For the first option, SAS/Access features could be used. For the second option, **SAS/ Application Facility (SAS/AF)** was used to create a client/server environment to support both data storage and analysis in SAS. Another term for this setup is **server SAS** (as opposed to **PC SAS**, which is an application that runs entirely on a PC without a client/ server relationship). The advantage of using SAS/AF is that a comprehensive SAS solution could be used that optimized the client/server relationship (through, for example, partitioning the application so it ran on different processors).

Also, in 1995, SQL optimizers had not been improved to the point where they outperformed SAS data steps, so at that time SAS/AF was a better approach than SAS/Access to connect to a SQL database. As noted in the Computerworld article, this led programmers to gravitate toward working either entirely in a SAS environment, or entirely outside of one.

With the visualization tools included in SAS/AF, SAS was now competing with visualization applications as well as data management applications. SAS's users continued to rate it highly, and were very loyal, as moving away from the SAS/AF platform would be very difficult given its dissimilarity to other applications.

Using PROC SQL

SAS's trajectory in general through its release of version 8 in 1999 and later version 9 (the current one) in 2002 has been to build extra functions into its core analysis products, and to also design supporting products to support its functionality. Unlike in the early years, SAS has not revisited data step functioning, nor considered redeveloping its data step language as declarative rather than procedural.

Through the late 1990s and early 2000s, SQL became more predominant, and therefore more programmers were trained in SQL. These SQL programmers had a lot of trouble transferring their skills to use in SAS data steps, so SAS developed a SQL language within SAS called PROC SQL.

PROC SQL has the following features:

- It is a language within SAS, in that PROC SQL code starts with a PROC SQL statement and ends with a quit statement.

- It uses SQL commands, such as CREATE TABLE and SELECT with GROUP BY and WHERE.

- It allows the user to control its use of processors during execution through the THREADED option.

- It includes a WHERE clause and other clauses that use indexes if they are available.

- Unlike other SQLs, it does not have an independent optimizer program, so creating optimized code is important.

Like SQL, PROC SQL is much easier to use than data step language for a few common tasks. One particularly useful task that is much easier in PROC SQL is creating a VIEW of the data, which allows the user to look at a particular section of the dataset.

Imagine we wanted to view the data in our example dataset, Chap_1_1_Infile, but we only wanted to look at the data for women (SEX1 = 2) who live in Massachusetts (_STATE = 25). We could use this PROC SQL code:

```
PROC SQL;
 Select * from Chap_1_1_Infile
     where SEX1 = 2 and _STATE = 25;
 quit;
```

This code produces output in the following structure (with just the first three rows provided):

_STATE	SEX1	_AGE80	_BMI5
25	2	43	36.9
25	2	53	38.1
25	2	66	41.6

Table 1.4 – Output from PROC SQL

To get similar output using SAS commands, the following PROC PRINT code could be used. Note that all variables in the order stored in the dataset are displayed since the VAR statement is excluded:

```
PROC PRINT DATA=Chap_1_1_Infile;
where SEX1 = 2 and _STATE = 25;
RUN;
```

But imagine we did not want to return all the variables – assume we only wanted to return age (_AGE80) and BMI (_BMI5). We could easily replace the asterisk in our PROC SQL code to specify only those two columns:

```
PROC SQL;
Select _AGE80, _BMI5 from Chap_1_1_Infile
    where SEX1 = 2 and _STATE = 25;
quit;
```

In PROC PRINT, to achieve the same output, we would add a VAR statement to our previous code:

```
PROC PRINT DATA=Chap_1_1_Infile;
where SEX1 = 2 and _STATE = 25;
var _AGE80 _BMI5;
RUN;
```

Even in this short example, it is easy to see how SAS PROCs and data steps are more complicated than SQL commands because SQL has fewer, more modular commands. By contrast, SAS has an extensive toolset of commands and options that, when understood and used wisely, can achieve just about any result with big data.

Using SAS today in a warehouse environment

While PROC SQL appears to be a workaround from learning complicated data step language, this is not the case in data warehousing. Because of the lack of optimization of PROC SQL, in many environments, it is very slow and can only be feasibly used with smaller datasets. Even today, when transforming big data in SAS, in most environments, it is necessary to use data step language, and this affords the programmer an opportunity to develop optimized code, as efficiency is always necessary when dealing with data in SAS.

However, when interfacing with another **database management system (DBMS)** where native data are stored in SQL, SAS PROC SQL might be more useful. In his recent white paper on working with big data in SAS, Mark Jordan describes various modern approaches to improving the processing efficiency of both PROC SQL and SAS data steps in both server SAS environments, as well as environments where SAS is used as the analysis engine and connects to a non-SAS DBMS through SAS/Access.

Jordan describes two scenarios for big data storage and SAS:

- **Using a modern server SAS set up**: Server SAS comes with its own OS, and Base SAS version 9.4 includes its own DS2 programming language. These can be used together to create threaded processing that can optimize data retrieval.

- **Using SAS for analysis connected to non-SAS data storage**: In this setup, SAS/Access is used to connect to a non-SAS DBMS and pull data for analysis into the SAS application. This can create a lag, but if SAS and the DBMS are co-located together and the DBMS can use parallel processing, speed can be achieved.

Ultimately, the main bottleneck in SAS processing has to do with I/O, so the easier it is for the SAS analytic engine to interact with the stored data, the faster processing will go. But even in this modern era, limitations surrounding data I/O continue to force SAS users to develop efficient code.

Jordan provides the following tips for thinking about coding for a SAS data warehouse:

- Use WHERE instead of IF wherever possible (due to its increased processing efficiency).

- As stated earlier, reduce columns retained to just the native and transformed variables needed in the warehouse.

- Using the options SASTRACE and SASTRACELOC will echo all the SQL generated to the SAS log file, which can be useful for performance tuning.

- Use PROC SQL and data steps to do the same tasks, and then compare their processing time using information from the SAS log to choose the most efficient code.

- It is especially helpful to compare PROC SQL code performance on summary tasks, such as developing a report of order summaries, because PROC SQL may perform better than PROCs or data steps.

- If using a server SAS setup with DS2 and data steps and if the log from your data steps shows a CPU time close to the program runtime, then your data steps are **CPU-bound**. In those cases, rewriting the data step process in DS2 could be helpful because it could take advantage of threaded processing.

- DS2 has another advantage as it is able to develop results at a higher precision level than data steps.

- DS2 code uses different commands than data step code but can achieve the same results.

- On **massively parallel processing (MPP)** DBMS platforms such as Teradata and Hadoop, DS2 can run as an in-database process using the **SAS In-Database Code Accelerator**. Using this code accelerator can significantly improve the efficiency of data throughput in these environments.

> **Note:**
>
> In his white paper, Mark Jordan compared PROC SQL processing using the SCAN command compared to the LIKE command for retrieving a record with criteria set on a high-cardinality variable and found the LIKE command to be more efficient.

Using SAS in the cloud

In his white paper, Jordan also describes how SAS now has a new **Viya architecture** that offers **cloud analytic services (CAS)**. A CAS library allows the following capabilities:

- Fast-loading data into memory
- Conducting distributed processing across multiple nodes
- Retaining data in memory for use by other processes until deliberately saved to disk

A CAS library has **application programming interfaces (APIs)** that allow actions to be executed from a variety of languages, including **Java**, **Python**, and **R**, and of course, the SAS Version 9.4 client application.

Today, not all warehouse data is stored in the cloud, and many datasets are still stored on traditional servers. Jordan recommended that if the user has an installation of the SAS 9.4M5 application and has access to SAS Viya CAS, and they want to decide whether or not to move to CAS from a traditional server, they should compare the processing time on a subset of data in both environments. Jordan was able to demonstrate cutting the processing time from over 1 minute to 2.35 seconds by moving his data from a traditional server to SAS Viya CAS.

Using SAS in modern warehousing

Today, SAS data warehousing is more complicated than it was in the past because there are so many options. Learning about these options can help the user envision the possibilities, and design a SAS data warehousing system that is appropriate for their organization's needs. This section will cover the following:

- A modern case study that used SAS components for analyzing unstructured text in helpdesk tickets

- A case study of a data SAS warehouse that upgraded an old system to include a new API allowing users more visualization functionality through **SAS Visual Analytics**

- A case study of a legacy SAS shop that began to incorporate R into their system

- A review of how SAS connects with a new cloud storage system, Snowflake

Warehousing unstructured text

In his white paper on warehousing unstructured text in SAS, Nick Evangelopoulos describes how the **IT Shared Services (ITSS)** division at the **University of North Texas (UNT)** used SAS to study their service tickets to try to improve services (link available under *Further reading*). Here are the steps they took:

- They chose to study a set of 9,691 tickets (representing approximately 18 months' worth of tickets) comprised mainly of unstructured text from the native application platform ServiceNow.

- Using the open source statistical application R, they conducted text cleaning. Mostly, this consisted of removing back-and-forth conversations by email that were recorded in the unstructured ticket text.

- Using the text mining component of SAS called **SAS Text Miner** (used within the SAS platform **SAS Enterprise Miner (SAS EM)**), they were able to use text extraction to help classify the tickets by topic.

- Next, the team used Base SAS and the analytics component **SAS STAT** to add indicator variables and other quantitative variables to the topics, thus creating a quantitative dataset that could be analyzed and visualized.

After doing this, the team wondered if SAS EM would classify the tickets under the same topic as the user entering the ticket would. To answer this question, the team analyzed 1,481 new tickets that were classified using SAS EM as well as being classified by the user. They found dramatic differences between how users and SAS EM classified the tickets, suggesting that this classification may need additional development in order to be useful.

Using SAS components for warehousing

A white paper by Li-Hui Chen and Manuel Figallo describes a modern SAS data warehouse using SAS applications (available under *Further reading*). The **US Department of Health and Human Services (DHHS)** has a data warehouse of health indicators called the **Health Indicators Warehouse (HIW)**. They described how they upgraded their SAS data warehouse system to improve performance and customer service using **SAS Visual Analytics (VA)** accessed through an API.

The HIW serves many users over the internet. Prior to the upgrade, SAS datasets were accessed from storage using SAS, and **extract-transform-load (ETL) processes** needed to take place manually on the data before it could be visualized. This made the data in the warehouse difficult to visualize.

With the upgrade, this is the new process:

1. Users obtain permission to access the API, which controls access to the underlying data as well as the VA capabilities.

2. Using the API, which contains a GUI, users indicate which health indicator they want to extract from the HIW, and how they want to visualize it.

3. The API extracts the necessary data from the HIW data store using automated processing.

4. The API performs necessary ETL processes to support visualization.

5. The API then visualizes the results using VA.

Here is a conceptual diagram of the old and new systems:

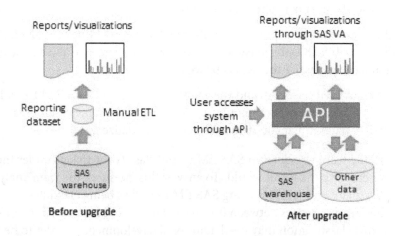

Figure 1.3 – SAS warehousing system before and after adding an API layer

Focusing on I/O, the authors pointed out that ETL in the API is achieved by running SAS **macros**, or code routines developed in the **SAS macro language** that can take user or system inputs and can be run automatically. They pointed out that they can run these macros either through a **stored process** (where the macro can be run on one dataset at a time) or a **batched process** (where the macro is run on several datasets at once). The authors found that they needed to use a batch process when transferring large amounts of HIW data through an API call.

Using other applications with SAS

SAS has been around a long time and has typically been the first choice for warehousing big data. However, since the invention and rise of SQL, there has been competition between SAS and SQL for data storage functions. With the rise of R, open source statistical software known for visualization and an easy web interface, SAS has seen competition with respect to statistical analysis functions.

Over time, SAS responded to competition by building in extra functionality. SAS/Access, SAS VA, and SAS Viya are all examples of this. However, the reality is that SAS is best at analytics, so other applications tend to be superior at these other functions. This has created challenges for legacy SAS warehouses that are now rethinking how they use SAS in their system. Teams are approaching this challenge with a variety of responses.

Dr. Elizabeth Atkinson shared her team's story of moving from a 100% SAS shop to incorporating R for some functions. She leads a biostatistics service at the Mayo Clinic, a famous specialty clinic in the US, which has been a SAS shop since 1974, when punch cards were still being used, and now has a staff of 300 in 3 locations. The service supports data storage and analysis for studies, both large and small.

In 2014, Mayo went to negotiate their SAS license and found that the price had increased significantly. SAS has always been a distinctive product with a high price. According to the Computerworld article, in 1995, a full SAS application development package, when bundled for 10 users, cost $1,575 per seat; this is expensive even by today's standards. However, in 2014, the increase in cost was felt to be unsustainable, and the Mayo team started looking for other options.

They wanted to decrease their dependence on SAS by moving some of their functions to R, and also improving their customer service and satisfaction. They faced the following challenges:

- **SAS infrastructure was entrenched**: All of the training was based on SAS, SAS was integrated into every workflow, and automation used SAS macros. Many users only trusted SAS and did not trust numbers coming out of R. SAS users relied on their personal code repositories.

- **R infrastructure was dynamic**: Unlike SAS, R releases new versions often. R innovates quickly, so it is hard to keep up a stable R environment. **R packages**, which are external components of Base R that can be added, were also upgraded regularly, leading to code that would break without warning, and cause user confusion.

- **Time constraints**: Reworking some SAS functions to be done by R required a lot of effort of deconstructing SAS and constructing R. Both leaders and users had time constraints.

- **Different learning styles and levels of knowledge**: SAS users had spent years learning data steps. R data management is completely different. It was hard for SAS users to learn R, and R users to learn SAS.

- **R support needed**: SAS provides customer support, but that is not available with open source software like R. The organization needed to build its own R support desk. Compared to SAS, R's documentation is less standardized and comprehensive.

To integrate R into their shop, they took the following steps:

- **Committed funding**: Divisional funding was committed to the project.

- **Identified R champions**: This was a group of R users with expertise in R and SAS.

- **Set up an R server**: Having an R server available increased enthusiasm and interest in R.

- **Rebuilt popular local SAS macros in R**: These are the ones that were deconstructed and rebuilt in R. Many of these were for reporting. They took the opportunity to improve reporting when rebuilding these macros.

- **Developed integrated SAS and R training**: Because they are now a combined shop, their training shows how to do the same tasks in SAS and R. They also hold events demonstrating R and providing online examples.

- **Set up an R helpdesk**: This provides on-call, in-house R support. They maintain a distribution list and send out R tips.

Even after offering R as an alternative, many users chose to stay with SAS. The reasons the shop could not completely convert from R to SAS include the following:

- **Time and cost constraints**: It was not possible to move all the small projects already in SAS over to R.

- **Data retrieval and ETL**: R cannot handle big data like SAS. The SAS data steps provide the ability to control procedural data processing in SAS, and this is not possible in R.

- **Analysis limitations**: Certain tasks are much clumsier in R than in SAS. At the Mayo Clinic, they found that mixed effect models were much more challenging in R than in SAS.

One of the overall benefits of this effort was that it opened the larger conversation behind what skills will be needed among analysts in the division in the future. These considerations run parallel to the consideration as to what SAS and non-SAS components will be used in the data system in the near future, what roles they will play, how they will be supported, and how they will work together to improve the user experience.

Connecting to Snowflake

As data gets bigger and bigger, new solutions have been developed to store data in the cloud. Microsoft Azure and **Amazon Web Services (AWS)** are cloud services to help move business operations to the cloud. **Snowflake** (https://www.snowflake.com/) is a relatively new cloud data platform that runs on Microsoft Azure and AWS and may run on other cloud services in the future.

Snowflake enables a programmer to make a virtual data warehouse with little cost, thus solving a data storage problem. However, data still needs to be accessed to be analyzed. Therefore, SAS upgraded its SAS/Access component to now be able to connect directly to Snowflake.

SAS documentation about connecting to Snowflake indicates that Snowflake uses SQL as its query language. Both PROC SQL and regular SAS functions can be passed to Snowflake, but there are cases where SAS and Snowflake function names conflict. Further more, careful settings of options and code tuning are needed to improve I/O from SAS to Snowflake.

Although products like Snowflake can solve the big data storage problem, the issue with SAS will always be I/O. Using the newest and most appropriate technology along with the most efficient coding approaches will always be the best strategy for dealing with the data warehousing of big data in SAS.

Summary

This chapter provided a short history of SAS, focusing on how it has been used for data storage and analysis over the years. Initially, SAS data was stored on punch cards. Once data became electronic, the main challenge to SAS users working with big data was I/O. As SAS environments evolved from being on mainframes to being accessible by PCs, SAS developed new products and services to complement its core analytics and data management functions.

SAS data steps are procedural, and allow the programmer opportunities to greatly improve I/O through the use of certain commands, features, and approaches to programming. When SQL became popular, PROC SQL was invented. This allowed SAS users to choose between using data steps or SQL commands when managing data in SAS.

Today, SAS is still used in data warehousing, but there are new challenges with accessing data in the cloud. SAS data warehouses today can include predominantly SAS components, such as SAS VA and CAS. Or, SAS can be part of a warehouse system that includes other components and applications, such as cloud storage in Snowflake, and supplemental analytic functions provided by R.

Modern SAS data warehousing still seeks to improve I/O and to better serve warehouse users through the development of an efficient system that meets customer needs. Creativity is required in the design of modern SAS data warehouses so that the system can leverage the best SAS has to offer while avoiding its pitfalls.

Although this chapter covers the entire history of SAS for data storage, it is important for the new data scientist to understand this information because the way SAS runs today can often be explained by certain events in its history. Particular terminology and features that are unique to SAS arise from how it has evolved over time, and it is helpful to know this background when communicating with today's SAS data warehouse developers and data scientists.

The next chapter takes a sharp focus on the act of reading data into SAS and will close with strategies that can be used when importing difficult data into SAS.

Questions

1. What is the difference between SAS and SQL with respect to data handling?

2. What is the difference between subsetting datasets using WHERE compared to the IF clause?

3. What is the component of SAS that allows it to connect to non-SAS databases?

4. Under what circumstance should you place an index on a variable in a large dataset?

5. Should you use SAS to enter a small dataset through data steps? State the reason for your answer.

6. What is the main advantage of using all SAS components in your warehouse?

7. What is a good way to decide whether to use a data step or PROC SQL for a particular data editing task?

Further reading

- BRFSS reference: Centers for Disease Control and Prevention (CDC). Behavioral Risk Factor Surveillance System Survey Data. Atlanta, Georgia: U.S. Department of Health and Human Services, Centers for Disease Control and Prevention, 2020.

- *Introduction to ODS Graphics for the Non-statistician*, SAS white paper by Mike Kalt and Cynthia Zender – available at `https://support.sas.com/resources/papers/proceedings11/294-2011.pdf`

- *Programming with Punch Cards* by Dale Fisk – report available at `http://www.columbia.edu/cu/computinghistory/fisk.pdf`

- *How to save $30,000 in 4 Hours": Migrating SAS® systems from the mainframe to the PC* SAS white paper by Ian Robertson – available at `https://support.sas.com/resources/papers/proceedings/proceedings/sugi22/SYSARCH/PAPER304.PDF`

- US National Health and Nutrition Examination Survey (NHANES) codebook – available at `https://wwwn.cdc.gov/Nchs/Nhanes/2015-2016/BMX_I.htm#BMXWT`

- FIPS state codes list – available at `https://transition.fcc.gov/oet/info/maps/census/fips/fips.txt`

- *Reducing the CPU Time of Your SAS® Jobs by More than 80%: Dream or Reality?* SAS white paper by Nancy Croonen and Henri Theuwissen – available at `https://support.sas.com/resources/papers/proceedings/proceedings/sugi28/002-28.pdf`

- *Creating and Exploiting SAS® Indexes* SAS white paper by Michael Raithel – available at `https://support.sas.com/resources/papers/proceedings/proceedings/sugi29/123-29.pdf`

- *Power Indexing: A Guide to Using Indexes Effectively in Nashville Releases* SAS white paper by Diane Olson – available at `https://support.sas.com/resources/papers/proceedings/proceedings/sugi25/25/dw/25p124.pdf`

- *Programming Tricks For Reducing Storage And Work Space* SAS white paper by Curtis Smith – available at `https://support.sas.com/resources/papers/proceedings/proceedings/sugi27/p023-27.pdf`

- *WHERE vs. IF Statements: Knowing the Difference in How and When to Apply* SAS white paper by Sunil Gupta – available at `https://support.sas.com/resources/papers/proceedings/proceedings/forum2007/213-2007.pdf`

- 1995 edition of ComputerWorld – available at `https://books.google.com/books?id=OcZZwenlL_0C&q=SAS+Institute#v=snippet&q=SAS%20Institute&f=false`

- *Working with Big Data in SAS®* SAS white paper by Mark Jordan – available at `https://www.sas.com/content/dam/SAS/support/en/sas-global-forum-proceedings/2018/2160-2018.pdf`

- *From Unstructured Text to the Data Warehouse: Customer Support at the University of North Texas* SAS white paper by Nick Evangelopoulos – available at `https://www.sas.com/content/dam/SAS/support/en/sas-global-forum-proceedings/2018/1900-2018.pdf`

- *Bridging the Gap: Importing Health Indicators Warehouse Data into SAS® Visual Analytics Using SAS® Stored Processes and APIs* SAS white paper by Li-Hui Chen and Manuel Figallo – available at `https://support.sas.com/resources/papers/proceedings16/10540-2016.pdf`

- Video of Dr. Elizabeth Atkinson – available at `https://resources.rstudio.com/rstudio-conf-2018/a-sas-to-r-success-story-elizabeth-j-atkinson`

- SAS documentation about connecting to Snowflake – available at `https://documentation.sas.com/?docsetId=acreldb&docsetTarget=n1d5j8d7wegfeznlirjj3hcrneln.htm&docsetVersion=9.4&locale=en`

2
Reading Big Data into SAS

This chapter will introduce SAS data warehouse developers to the issues and strategies surrounding reading big data into SAS. SAS has native data formats `*.SAS7bdat` and **XPT**, but also reads in non-native formats such as `*.csv` and `*.txt`. There are advantages and disadvantages to storing data in any of these formats, and special considerations need to be made when preparing **transfers of big data** in these formats. SAS warehouse developers are tasked with reading data from multiple different source systems into SAS, and this can be done using `infile` statements, `PROC IMPORT`, or a strategy that combines both techniques. Because SAS has proficiency in handling big data, SAS data warehouses often need to read in large extracts from **legacy systems**, many of which provide **fixed-width** extracts. These can be particularly challenging to read into SAS, and so this chapter also describes approaches to tackling these challenges.

This chapter takes a deep dive into the following topics:

- How to use `LIBNAME` statements in SAS to refer to physical locations that SAS can access to read and write data
- How to read and write data in SAS native formats to and from physical locations
- How to create and unpack XPT files
- How to use `infile` and `PROC IMPORT` for importing non-SAS data formats into SAS

- How to use a file viewer, file editor, and `infile` statements to research difficult datasets and build SAS code to read them in properly
- How to verify whether data files were properly transferred

Technical requirements

The datasets in various formats used for demonstrations in this chapter are available online from GitHub: `https://github.com/PacktPublishing/Mastering-SAS-Programming-for-Data-Warehousing/tree/master/Chapter%202/Data`.

The code bundle for this chapter is available on GitHub here: `https://github.com/PacktPublishing/Mastering-SAS-Programming-for-Data-Warehousing/tree/master/Chapter%202`.

Reading data extracts into SAS

SAS as a program provides a lot of different functionalities for handling datasets in any format. However, this leads to some complexity in coding, even when using SAS's native dataset formats. This section covers the following:

- SAS's native dataset format, `*.SAS7bdat`
- How to optimally use SAS's `WORK` directory
- How to specify and use `LIBNAME` for reading SAS data from a physical location
- How to read datasets in `*.SAS7bdat` format into SAS

This section will give you working knowledge of how to use `LIBNAME` to read in datasets in `*.SAS7bdat` format that are stored outside of the SAS environment.

Understanding SAS datasets

Up until now, when talking about data storage in SAS, we were talking mainly about reading data into the working memory of SAS. When a SAS session is started and data is read into SAS's working memory, it is said to be in the `WORK` directory. This directory will go away when the session ends, so data in the `WORK` directory is always lost at the end of a session.

Input/output (I/O) is fast in SAS when interacting with data that has already been placed in the WORK directory. However, getting data into the WORK directory is a challenge. As mentioned in *Chapter 1, Using SAS in a Data Mart, Data Lake, or Data Warehouse*, data analyzed in SAS was originally stored on punch cards. Later, datasets of various formats – such as *.csv or *.txt – could be read into SAS's working memory using the infile statement.

What was not covered yet in this book is information about data types specific to SAS, referred to as **SAS datasets**. Today, when a person refers to a SAS dataset, they usually mean a dataset in the format *.SAS7bdat. This is the format to which SAS automatically converts data when it is read into the WORK directory. If a dataset that is in *.txt or *.csv format originally is read into SAS then exported in **SAS format**, it will be in *.SAS7bdat format.

> **SAS file extensions**
>
> Although datasets in SAS are in *.SAS7bdat format, it is important to be aware of the other file extensions SAS uses for different SAS files. Important SAS file extensions are *.sas for code files, *.lst for output files, and *.log for log files. More SAS file extensions are listed here: https://v8doc.sas.com/sashtml/win/z1iles.htm.

Working with the WORK directory

The WORK directory is a part of SAS's memory that exists during the SAS session. Like a physical directory on a computer, it holds data files, but unlike a physical directory, it has no physical location. Running a data warehouse means becoming skilled at moving datasets in and out of the WORK directory during a SAS session while processing data, especially while performing **extract, transform, and load (ETL)** functions. That is because it is much faster for SAS analytics components to interact with datasets in the WORK directory compared with datasets physically stored on disk that need to be read into SAS.

When a new SAS session starts, there are no files in the WORK directory that were put there by the programmer. However, there are SAS files in WORK that are put there by the SAS system. The SAS system makes its own internal files (including datasets) when running **PROCs** and data steps, and these can be accessed by request from SAS's memory.

Let's look at what datasets are in my WORK directory as soon as I start a new session on my SAS University Edition account. We will use PROC DATASETS for this:

```
PROC DATASETS LIB = WORK;
QUIT;
RUN;
```

Notice the only option set is LIB = WORK, which specifies that I am interested in the contents of the library WORK. The results from the window follow:

Directory	
Libref	WORK
Engine	V9
Physical Name	/tmp/SAS_workABD0000009E8_localhost.localdomain/SAS_work9D0A000009E8_localhost.localdomain
Filename	/tmp/SAS_workABD0000009E8_localhost.localdomain/SAS_work9D0A000009E8_localhost.localdomain
Inode Number	145082
Access Permission	rwx------
Owner Name	sasdemo
File Size	4KB
File Size (bytes)	4096

#	Name	Member Type	File Size	Last Modified
1	REGSTRY	ITEMSTOR	32KB	09/26/2020 14:44:14
2	SASGOPT	CATALOG	12KB	09/26/2020 14:44:16
3	SASMAC1	CATALOG	208KB	09/26/2020 14:44:15
4	SASMAC2	CATALOG	20KB	09/26/2020 14:44:15
5	SASMAC3	CATALOG	20KB	09/26/2020 14:44:15
6	SASMAC4	CATALOG	20KB	09/26/2020 14:47:20
7	SASMAC5	CATALOG	20KB	09/26/2020 14:44:15
8	SASMAC6	CATALOG	20KB	09/26/2020 14:44:15
9	SASMAC7	CATALOG	20KB	09/26/2020 14:44:15
10	SASMAC8	CATALOG	20KB	09/26/2020 14:44:15
11	SASMAC9	CATALOG	20KB	09/26/2020 14:44:15
12	SASMACR	CATALOG	20KB	09/26/2020 14:44:17

Figure 2.1 – Results window after the PROC DATASETS command listing WORK directory contents

Under **Member Type**, SAS has only placed items of type **ITEMSTOR** or **CATALOG**. There are no datasets in **WORK** yet.

In *Chapter 1, Using SAS in a Data Mart, Data Lake, or Data Warehouse*, we created a SAS dataset by using a data step and doing data entry using code. Now, let's combine that code with the PROC DATASETS code. We will create a dataset in SAS's memory called THREEROWS, then run PROC DATASETS to see if it is indeed in WORK:

```
DATA THREEROWS;
    INFILE CARDS;
    INPUT _STATE SEX1 _AGE80;
    CARDS;
```

```
12 1 72
25 2 25
27 2 54
;
RUN;
PROC DATASETS LIB = WORK;
QUIT;
RUN;
```

Now when we look at the results, we will see the same thing as we saw earlier, with the addition of a file with a **Name** of THREEROWS, and a **Member Type** of DATA. This means that THREEROWS is a SAS dataset in SAS's memory – specifically, in the WORK directory. In order to export this SAS dataset to a physical location as a SAS dataset, meaning formatted as a *.SAS7bdat file, we will need to set LIBNAME.

Specifying LIBNAME

Specifying LIBNAME in SAS is a way of applying a label or **alias** that refers to a particular physical location outside of SAS where files are stored. Specifying LIBNAME allows SAS to access data files from this location, and to place files in this location.

The default directory when a SAS session starts is WORK. We already made a small SAS dataset in WORK called THREEROWS. Let's make a copy of this dataset in WORK using a data step. We will call the copy THREEROWS_COPY1:

```
DATA THREEROWS_COPY1;
    set THREEROWS;
RUN;
PROC DATASETS LIB = WORK;
QUIT;
RUN;
```

Notice that to copy the dataset, we used a very simple data step. The first line of the data step, DATA THREEROWS_COPY1, specifies the name of the SAS dataset to be output by the data step and the **library** (or directory) in which to store it. However, since WORK is the default library, we only specified the dataset name and not the library. The second line of the data step is the SET command, which specifies that the library and name of the dataset is being input into the data step. Again, since WORK is the default library, the library is not specified.

After running the preceding code, the results from the PROC DATASETS command show that there is now an extra dataset in your WORK directory called THREEROWS_COPY1. Now, let's consider saving a copy of the THREEROWS dataset to an external location outside of SAS.

First, we will use the LIBNAME statement to map the alias X to an external location using the following code. Note that there are rules for naming the alias, such as not starting with a number or containing special characters:

```
LIBNAME X "/folders/myfolders/X";
RUN;
```

Notice in the code that we start with the LIBNAME statement, then choose an alias – in this case, X. Then, we put our physical path in quotes to indicate to SAS where to go when mapping X to the physical location. You can replace /folders/myfolders/X in the code with the path to the directory you want to map to X. When we run that code and look in the log file, we see this confirmation:

```
NOTE: Libref X was successfully assigned as follows:
      Engine:         V9
      Physical Name: /folders/myfolders/X
```

Note that if you changed the path, it will be listed under **Physical Name** in the log.

Now that LIBNAME is specified, let's copy our small dataset, THREEROWS, into the new library, X. We will call the copy we save in X THREEROWS_COPY2. We will specify it by using the syntax of putting the LIBNAME alias, then a period, then the name of the dataset. This time, we will run PROC DATASETS on both the WORK library as well as our new X library to see what datasets are there after we run the code:

```
DATA X.THREEROWS_COPY2;
    set THREEROWS;
RUN;
PROC DATASETS LIB = WORK;
PROC DATASETS LIB = X;
QUIT;
RUN;
```

Now that we have done this, let's compare the second table of results from each of the PROC DATASETS commands we did for WORK and for X:

Contents of WORK

#	Name	Member Type	File Size	Last Modified
1	REGSTRY	ITEMSTOR	32KB	01/06/2020 10:04:14
2	SASGOPT	CATALOG	12KB	01/06/2020 10:04:14
3	SASMAC1	CATALOG	208KB	01/06/2020 10:04:14
4	SASMAC2	CATALOG	20KB	01/06/2020 10:04:14
5	SASMAC3	CATALOG	20KB	01/06/2020 10:04:14
6	SASMAC4	CATALOG	20KB	01/06/2020 11:37:23
7	SASMAC5	CATALOG	20KB	01/06/2020 10:04:14
8	SASMAC6	CATALOG	20KB	01/06/2020 10:04:14
9	SASMAC7	CATALOG	20KB	01/06/2020 10:04:14
10	SASMAC8	CATALOG	20KB	01/06/2020 10:04:14
11	SASMAC9	CATALOG	20KB	01/06/2020 10:04:14
12	SASMACR	CATALOG	20KB	01/06/2020 11:36:22
13	THREEROWS	DATA	128KB	01/06/2020 11:04:13
14	THREEROWS_COPY1	DATA	128KB	01/06/2020 11:20:20

Contents of X

#	Name	Member Type	File Size	Last Modified
1	THREEROWS_COPY2	DATA	128KB	01/06/2020 11:37:23

Figure 2.2 – Results windows after the PROC DATASETS command listing WORK and X directory contents

> **Note:**
>
> Even though you can have an alias longer than one letter in a LIBNAME statement, it is nice to keep it short, because you will be typing it a lot.

For any project, it is possible to set up multiple LIBNAME statements and use efficient data step processing to output different datasets into different physical libraries. LIBNAME statements also help you read SAS datasets into SAS.

Reading in SAS datasets

Earlier, we set up a LIBNAME statement and called it X. It referred to a library, also called a **directory**, outside of SAS. Let's imagine we stored a SAS dataset (a dataset in *.SAS7bdat format) in the directory to which the LIBNAME X points. We will place the chap2_1_sas.sas7bdat dataset in this directory, which is included in the datasets for this chapter.

Now, let's access this SAS dataset from SAS by setting `LIBNAME X` and then referring to the location of the dataset by first stating `LIBNAME`, then a period, then the dataset name. The following code block runs `PROC CONTENTS` on the dataset so that its contents can be displayed:

```
LIBNAME X "/folders/myfolders/X";
RUN;
PROC CONTENTS data=X.chap2_1_sas;
RUN;
```

As can be seen in the code, once a dataset is in `*.SAS7bdat` format, even if it is stored in an external location, it is easily called up in code by simply setting a `LIBNAME` statement and running commands on it by referring to it using the `LIBNAME` alias, followed by a period and the name of the dataset.

> **Note:**
>
> `LIBNAME` only needs to be mapped once per SAS session but may be remapped during the session. Therefore, if there are a lot of `LIBNAME` statements used in standard code, it is best to document what they mean so that communication is clear among the programming team.

This short demonstration seems to suggest that if you plan to analyze a lot of data in SAS, you should store the raw datasets in `*.SAS7bdat` format, because then referring to them will be easy in code, because you will simply need a `LIBNAME` statement. In addition, processing is faster in SAS when SAS is using a SAS dataset, where SAS does not need to use processing power to convert the dataset to SAS format in its memory.

However, there is a cost to transforming and storing datasets in SAS. As an example, I saved the `Chap2_1_sas.SAS7bdat` dataset as a `*.csv` for comparison (included in your files as `Chap2_1_CSV.csv`). As a `*.csv`, the dataset was only 4.2 MB, while the same dataset in `*.SAS7bdat` was over three times the size, at 13.5 MB. While the extra components SAS adds to SAS datasets can speed up processing in SAS, *converting raw datasets in a warehouse to SAS datasets is often not feasible due to storage limitations.*

Because SAS was aware that storing SAS datasets in `*.SAS7bdat` caused the datasets to take up a lot of space, SAS developed another data format intended for storing large datasets, called the XPT format. The XPT format was intended to shrink the size of `*.SAS7bdat` files so they would take up less space while being stored during transfer.

Using the SAS XPT format

Because the datasets in SAS's native dataset format, `*.SAS7bdat`, take up a lot of space, this makes it challenging to move big SAS datasets from one host environment to another. For this reason, SAS invented the XPT format. This section explains the following:

- The advantages, disadvantages, optimal uses, and features of storing datasets in XPT format

- How to create an XPT file from a `*.SAS7bdat` file, and how to convert the XPT file back to a `*.SAS7bdat` file

- The difference between using `PROC COPY` and `PROC CPORT/CIMPORT` in creating and extracting data from an XPT file

This section will provide you with a comprehensive understanding of the XPT file format so you can determine whether XPT could be helpful for archiving or transporting data in your data warehouse, data mart, or data lake.

Storing data in XPT format

Although the first worldwide release of SAS was in the 1970s, it wasn't until the 1980s that there was a need to transfer large data files from one system to another. This is because, in the 1970s, data was typically stored on punch cards that would be read into the computer each time analysis was to take place. As described by the US **Library of Congress (LOC** – see *Further reading* for citation), by the 1980s, when mainframes were the main data storage approach, SAS needed to develop a way of reducing the size of `*.SAS7bdat` files so they could be stored on static media, and then unpacked into a new electronic data storage system from static media. The purpose of this storage format was **transport**, which is why the file format was named XPT.

The US LOC explains that two versions of the XPT file format were developed, one called version 5, which was launched in 1989, and one called version 8, which was launched in 2012. They observe that even though version 8 was launched more recently, most file storage is actually in the version 5 format. This is because XPT has also been used extensively for archiving, and once an organization begins archiving in a certain format, they are unlikely to change it even if a new format is available.

> **Note:**
> The XPT format is also sometimes referred to as the XPORT format. The two terms mean the same thing. In documentation, this file format is also sometimes referred to as the `SAS_xport` format.

To be specific, the XPT format is a specification that was developed in particular for transporting data back and forth between IBM mainframes. An XPT file consists of records that are 80 bytes in length, with short records padded with ASCII nulls (hex 00) out to 80 bytes. Character data is stored in ASCII. The specification stores numeric formats as internal binary representations, rather than character representations. This was designed so that the accuracy of data points with many decimal places would not be lost from transfer to transfer. Per XPT specifications, integers are stored in **IBM-style integer format** and floating-point numbers are stored in **IBM-style double format**.

XPT files have unique header specifications. These header records appear at the beginning of the XPT file in ASCII. Positions in the headers without designated characters are **packed** with ASCII space characters (hex 20).

There is a primary file header made up of three 80-byte records:

- **The first header record** identifies the file as an XPT file format.
- **The second header record** names the version of SAS and the **operating system (OS)** used to create the file and includes the creation date.
- **The third header record** contains the last date of modification.

An XPT file can contain multiple datasets. For each dataset contained in an XPT file, there are 3 80-byte records:

- A member header record
- A descriptor header record
- A record identifying the version of SAS and OS used to create the file, along with the creation date

For each dataset in the XPT file, there is also a sequence of header records intended to describe each of the variables in a format called **namestr**. This is where SAS stores metadata about each variable – such as the name of the variable, the data type, and the size in bytes. The final header record is 80 bytes and indicates the location of the start of data observations in the data file. The values of these observations can then be placed into the variables described in the headers.

An example of where the XPT format is used can help illustrate its usefulness. Let's return to the dataset described in *Chapter 1*, *Using SAS in a Data Mart, Data Lake, or Data Warehouse*, the **Behavioral Risk Factor Surveillance System (BRFSS)**, which is a dataset arising from the annual health survey done by phone in the United States. The dataset is very large each year, with about 450,000 rows, and is made available for download on the internet. Many researchers use SAS to analyze this dataset, so it is logical to keep it in a SAS-native format. To facilitate transferring such a large dataset to SAS users over the internet, the dataset is made available in XPT format.

The 2018 BRFSS file has 437,436 rows and 275 columns. This is a good candidate dataset to be stored in the XPT format. This is because it is both long and wide; and because it is a survey, it would typically be loaded in memory deliberately prior to analyses (unlike a table in a production database, which would need to be constantly available).

Creating an XPT file

Let's practice making an XPT file out of a *.SAS7bdat file. Let's return to the *.SAS7bdat file included with this chapter, named Chap2_1_sas.SAS7bdat. We have already placed this file in the folder we mapped to X using the LIBNAME statement. Now, let's turn that file into an XPT file (containing only this dataset) that is saved in the folder mapped to X.

It is easier to do this operation in two steps. In the first step, we will create LIBNAME statements needed for the XPT conversion process, and copy the source *.SAS7bdat dataset we want to convert into XPT into one of the LIBNAME statements:

```
libname sasfile "/folders/myfolders/X";
libname xptfile xport "/folders/myfolders/X/Chap2_1_XPT.xpt";
data sasfile.temp;
    set X.chap2_1_sas;
RUN;
```

Let's consider this code line by line:

1. Remember, we have already mapped our LIBNAME X to an external folder outside of SAS that is holding our *.SAS7bdat dataset, Chap2_1_sas.SAS7bdat. Therefore, our first LIBNAME statement is part of the preparation for converting Chap2_1_sas.SAS7bdat to an XPT file. It maps whatever SAS datasets are in X to one LIBNAME. In our first line of code, that LIBNAME is sasfile. Therefore, from now on in our code, the LIBNAME sasfile refers to all the SAS datasets we could put in the library sasfile. This is why we will now need to refer to particular datasets in sasfile by using the syntax of LIBNAME followed by a period followed by the dataset name.

2. The second line is a LIBNAME statement that is special in that it includes the XPORT command. The LIBNAME alias is xptfile and this is followed by the XPORT command. Like the previous LIBNAME statement, this technically specifies the same directory as X but uses a different alias. Another important difference about this LIBNAME statement is that the item specified at the end of the line is not a directory, but actually a specific file. This is the name of the XPT file we are requesting SAS to make for us in the code. In the code, we have named this file Chap2_2_XPT.xpt.

3. Lines three and four of the code are data steps. The output dataset is sasfile.temp, and the input dataset is X.Chap2_1_sas. It is assumed by the code that Chap2_1_sas is a *.SAS7bdat file by the data step syntax. This data step copies our *.SAS7bdat dataset named Chap2_1_sas from the X library to a new dataset named temp in our sasfile library. Even though X and sasfile refer to the same location, the use of the naming is important in order to prevent confusion during the conversion of the XPT file.

In the second step, we will use PROC COPY to convert the *.SAS7bdat dataset into an XPT file:

```
PROC COPY in=sasfile out=xptfile memtype=data;
select temp;
RUN;
```

Let's look at this code line by line, too:

1. The first line with the PROC COPY command is easy to read now that we have created very intuitive LIBNAME statements. This line uses the in= option to tell SAS that it should read in the datasets mapped to the directory sasfile. This contains our *.SAS7bdat file called Chap2_1_sas. out= tells SAS that it should export the file it creates into the LIBNAME statement mapped to xptfile. Remember, this is a special LIBNAME statement that was mapped using the XPORT command. Therefore, SAS knows that when it is placing a dataset in the directory mapped to that LIBNAME statement, it should be in XPT format. Finally, memtype=data indicates to SAS that this is a data file.

2. The second line tells SAS to select the temp dataset from the LIBNAME sasfile to place in the XPT file. Remember, in the first step, we copied the source *.SAS7bdat dataset from X.Chap2_1_sas to sasfile.temp, and now select temp refers to selecting this *.SAS7bdat file. This is all followed by a run command.

Now, the XPT we created is in the directory mapped to X and is named Chap2_1_XPT.

Comparing PROC CPORT/CIMPORT to PROC COPY

We just converted a `*.SAS7bdat` file into an XPT file using `PROC COPY`. Another method is available in SAS for creating XPT files, which involves using `PROC CPORT` and `PROC CIMPORT`. J. Meimei Ma and Sandra Schlotzhauer explain in their SAS white paper the difference between these two methods (see the *Further reading* section for a citation).

Ma and Schlotzhauer recommend using the `PROC COPY` method rather than the `CPORT/CIMPORT` method for several reasons:

- First, when making an XPT file using the `PROC COPY` method, it will be more likely to be usable in any host environment, even those running older versions of SAS.

- Second, using the `PROC COPY` approach allows you to select specific datasets from a data library as part of the procedure (as is evident by the example code we used).

- Third, if you use `PROC CPORT` instead of `PROC COPY` to create the XPT file, you must use a compatible version of `PROC CIMPORT` to convert it back.

The main advantage of using the `CPORT/CIMPORT` engine is that it makes the resulting files even smaller than when we use `PROC COPY` (which uses the `XPORT` engine). To demonstrate this, in their white paper, Ma and Schlotzhauer created an XPT of the same data file using both the `XPORT` and the `CPORT/CIMPORT` engine. Using the `CPORT` engine, the resulting XPT file was only 27 KB, a little less than half the size of the corresponding XPT file from the `XPORT` engine, which was 52 KB.

> **Tips for storing data in XPT files:**
>
> Storing data in XPT files is only helpful if the main reason they are being stored is so they can be copied and transferred to other systems.
>
> Data accessed regularly as part of analytics (such as live data in a warehouse) should not be stored in XPT format.
>
> Backup files, archives, and extracts being prepared for transfer are optimal types of files to store in XPT format.
>
> If size is a constraint, consider using `PROC CPORT` to prepare the XPT files, as this engine makes smaller XPT files than the `XPORT` engine.
>
> If using `PROC CPORT` to prepare data for transfer, ensure that the receiving host system has the correct version of `PROC CIMPORT` to convert the data to `*.SAS7bdat` format.

Reading in XPTs using the XPORT engine

As described earlier, XPT is an excellent format for transporting files on static media, because they take up much less space than `*.SAS7bdat` files, especially when considering datasets with many rows and columns. Once the dataset is transported to the new host system, the XPT needs to be converted back to a `*.SAS7bdat` file in order for the host system's users to analyze the dataset in SAS.

We will use `PROC COPY` to convert the dataset we just placed in the XPT file named `temp` to a `*.SAS7bdat` file named `temp`:

```
libname sasfile "/folders/myfolders/X";
libname xptfile xport "/folders/myfolders/X/Chap2_1_XPT.xpt";
proc copy in=xptfile out=sasfile;
select temp;
RUN;
```

Notice that the code looks almost identical to the code we used to create the XPT file. The main difference is that the `in` option is now set to the `LIBNAME xptfile`, and the `out` option is set to the `LIBNAME sasfile`, indicating that we want to convert the XPT file to a `*.SAS7bdat` file.

You might want to observe the file sizes of the data files we are using in this chapter, named `Chap2_1_SAS` (in `*.SAS7bdat` format) and `Chap2_1_XPT` (in XPT format). This example dataset consists of 4 columns from the BRFSS 2018 file, which has 437,436 rows. In `*.SAS7bdat` format, this long, skinny dataset is 13,824 KB. Converting it to XPT reduces it by only about 150 KB to 13,672 KB. This is likely because the XPT file must contain the necessary headers, and this will always take up a certain amount of space in an XPT file. Therefore, the greatest gain in making XPT files is if the dataset being stored is both wide and long.

Working with other file formats

Up to now, in this chapter, we have talked about reading in SAS datasets in either XPT or `*.SAS7bdat` format. Now, we will examine reading in data from non-SAS formats:

- We will start by revisiting the `infile` statement, and examine options that can be set to help read in non-SAS data.
- Next, we will practice using the `infile` statement to read a `*.csv` file and a `*.txt` file into SAS.

- After this, we will examine PROC IMPORT, and experiment with how it can help us read in data without having us type a lengthy infile statement.

- Finally, we will discuss converting non-SAS data to SAS format for physical storage.

This section will provide an overview of approaches the analyst can take when reading data from non-SAS formats into SAS, and provide several examples.

Reading non-SAS data formats

With respect to the size of data files, you may have noticed that of the three formats for the Chap2_1_SAS, Chap2_1_XPT, and Chap2_1_CSV datasets, by far the smallest is the *.csv format, which is 4,327 KB. This is about one third the size of the *.SAS7bdat and XPT formats. This size difference between datasets in SAS and non-SAS formats is also seen in more typical files to store in XPT format. As an XPT, the BRFSS 2018 data file is 939,416 KB. However, as a *.csv file, it is less than half as big, at 357,668 KB. Seeing these numbers and considering the cost of storage space might lead you to ask why those running data warehouses should store their data in SAS formats in the first place. Might they take up less space when stored as *.csv? files

The answer is that storing data in non-SAS formats (such as *.csv) loses a lot of information SAS can use about the variables and datasets. Remember the many headers of information placed at the beginning of an XPT file? This is the type of information that is causing the XPT and *.SAS7bdat files to be so much bigger than the *.csv file. This is the same information that reduces I/O in SAS, allowing data to be manipulated more efficiently. Therefore, if the intended host system is SAS, then storing data in either *.SAS7bdat or XPT format is advantageous.

Still, most **production databases**, or databases that are currently conducting I/O as part of a business process, are in either **structure query language (SQL)** or some other database language that is not SAS. When receiving data extracts from a production database, those running SAS data warehouses usually receive the data in either *.csv or *.txt format (although some older database systems may provide extracts in *.dat or some other older format).

In *Chapter 1, Using SAS in a Data Mart, Data Lake, or Data Warehouse*, we looked at an example of using the infile statement to read a *.csv file into SAS. Let's revisit the infile statement now and use it to read in the dataset Chap2_1_CSV, which is in *.csv format:

- This dataset consists of four columns from the 2018 BRFSS dataset: _STATE, MARIJAN1, USEMRJN2, and RSNMRJN1.

- The variable _STATE contains a two-digit **Federal Information Processing Standards (FIPS)** code for state.

- MARIJAN1, USEMRJN2, and RSNMRJN1 are categorical variables that represent answers to questions on the survey about marijuana use.

- These variables are coded using a numeric code and can be missing, which is represented in SAS by a period.

The following code uses the INFILE statement to read Chap2_1_CSV.csv into a *.SAS7bdat dataset named Convert_CSV that will be placed in WORK:

```
data Convert_CSV;
%let _EFIERR_ = 0;
infile '/folders/myfolders/X/Chap2_1_CSV.csv' delimiter = ','
firstobs=2
    MISSOVER DSD lrecl=32767;
    informat    _STATE 2.;
    informat   MARIJAN1 2.;
    informat   USEMRJN2 2.;
    informat   RSNMRJN1 2.;
    format          _STATE 2.;
    format         MARIJAN1 2.;
    format         USEMRJN2 2.;
    format         RSNMRJN1 2.;
input
    _STATE
    MARIJAN1
    USEMRJN2
    RSNMRJN1
;
if _ERROR_ then call symputx('_EFIERR_',1);
RUN;
```

Let's consider the working of this code:

1. The code is essentially a data step that specifies the output dataset titled Convert_CSV. Because there is no LIBNAME statement and period before the name, SAS will automatically output Convert_CSV as a *.SAS7bdat file into the WORK directory.

2. The statements `%let _EFIERR_ = 0;` and `if _ERROR_ then call symputx('_EFIERR_',1);` provide error handling.

3. The `infile` statement is used in the data step in place of the `set` statement. Earlier, we mapped `LIBNAME X` to the directory where I placed our example dataset, which we are going to read in, named `Chap2_1_CSV.csv`. The path refers to the directory where I mapped `X`. The option `delimiter = ','` specifies that the format of the source file is **comma-separated values** (`*.csv`). The option `firstobs=2` indicates that the first observation is in the second row (as the first row has the column headings).

4. The option `MISSOVER` gives SAS instructions on how to handle processing each sequential data point by column and row, advancing left to right until the end of the row, and then moving to the next row, as each value is read into the dataset. `MISSOVER` tells SAS that if it comes to a row where it does not find values for all of the variables specified in the `INPUT` statement later in the code, it should fill in missing values for those variables.

5. The option `DSD` tells SAS the data is **delimiter-sensitive data**, meaning that the delimiter (in this case, a comma) should be observed to tell the columns apart. In a `*.csv` file, two commas together indicate that there is a blank field. By specifying `DSD`, SAS will ingest the delimited source data, seeing those two commas together as a blank variable.

6. The option `LRECL` specifies the logical record length. The maximum is 32,767, so typically, the `INFILE` option is specified as `LRECL = 32767`.

7. After the initial `INFILE` code, there are three sets of commands: `informat`, `format`, and `input`. `informat` specifies the names and source formats of the variables. In the case of the four variables in `Chap2_1_CSV`, all of them are two-digit integers, so the `informat` is set at 2 followed by a period. `Format` specifies the names and formats of the variables to output into the `*.SAS7bdat` dataset. These are also assigned the `2.` format because we want to keep them in the same format as the source format. Finally, the `INPUT` statement actually places each source variable into each formatted variable to define the new dataset.

SAS processing options for the INFILE statement

Here are four different options that can be specified in an `infile` statement that can control how SAS reads data:

`MISSOVER`: If SAS comes to a row where it does not find values for all of the variables specified in the `INPUT` statement, it will fill them in with missing values.

`STOPOVER`: If SAS comes to a row where it does not find values for all of the variables specified in the `INPUT` statement, it stops processing the data step. This is the option to choose if you are reading in a flat file that conforms to a given standard because `STOPOVER` will stop the file being read if it does not conform to that standard (as specified in code entered as part of the `INFILE` and `INPUT` statements).

`FLOWOVER`: If SAS comes to a row where it does not find values for all of the variables specified in the `INPUT` statement, it simply starts reading the next record in an attempt to assign values to the variables. This is the default behavior, and since this behavior is typically undesirable, it is important to specify one of the other options (usually `MISSOVER`) in the `INFILE` statement.

`TRUNCOVER`: If SAS comes to a row where it does not find values for all of the variables specified in the `INPUT` statement, it will try to assign whatever it is reading as the raw data value for the variable, even if the value is shorter than what is specified by the `INPUT` statement. Like `MISSOVER`, if SAS encounters the end of a record with variables with unassigned values, it will enter missing values for those observations.

Considering `infile` code for reading in a `*.csv` file, very similar code is used to read a file in `*.txt` format into `WORK`. We are assuming this file is tab-delimited, as many extracts from old databases are in this format. We will use the same dataset we have been working with, which has been converted to a `*.txt` tab-delimited format called `Chap2_1_TXT.txt`.

First, let's place that in the directory we have mapped to `LIBNAME X`. Next, let's run code to convert this file to a `*.SAS7bdat` dataset in `WORK`. The first four lines of code follow here:

```
data Convert_TXT;
%let _EFIERR_ = 0;
infile '/folders/myfolders/X/Chap2_1_TXT.txt' delimiter='09'x
firstobs=2
```

The rest of the code is identical to that of the code earlier, used to read in a `*.csv` file. Notice the changes from the previous code:

1. First, the output dataset is now called `Convert_TXT`.

2. Next, the input dataset has been changed to `Chap2_1_TXT.txt`.

3. Finally, the `delimiter` option has been set to `'09'x`, which means `tab-delimited`.

The `infile` statement is very useful, especially because each `format` and `informat` can be specified. However, sometimes this is too much detail and requires too much typing. In those cases, `PROC IMPORT` can be used instead.

Using PROC IMPORT

`PROC IMPORT` was added more recently to SAS to alleviate all the typing associated with the `INFILE` statement. Even though the example dataset we have been using only has four columns, this translates to at least 12 lines of `infile` code, because each variable requires an `informat`, a `format`, and an `input` line.

In `PROC IMPORT`, we are asking SAS to make guesses about these specifications by simply looking at the data in the first 20 rows of the file we are trying to import. Let's repeat what we did earlier to read `Chap2_1_CSV`, which is a `*.csv` file, into WORK, only this time, let's use `PROC IMPORT`:

```
PROC IMPORT datafile='/folders/myfolders/X/Chap2_1_CSV.csv'
out=Convert_CSV dbms=csv replace;
getnames=yes;
RUN;
```

The first line of code is very long and includes several options:

1. First, after the `PROC IMPORT` statement, the `datafile` option is set to point to the `*.csv` we want to read in.

2. Next, the `out=` option specifies the name of the `*.SAS7bdat` dataset we want to create in WORK as `Convert_CSV`.

3. After this, the `dbms=` option specifies `csv` as the file format.

4. Finally, the `replace` option tells SAS to automatically replace the dataset if there is already one in WORK with the name `Convert_CSV`.

5. A final option is set on its own line, which is `getnames=yes` followed by a semi-colon. This tells SAS to obtain the column names from the first row of the source dataset. If `getnames` is set to `no`, then SAS will name the variables sequentially as VAR1, VAR2, VAR3, and so on, and will read the header as an actual row of data. In general, it is good practice to set `getnames=yes` since Excel files generally have variable names in the first row.

In this case, it was not difficult for SAS to guess the information we normally would have put in the `informat`, `format`, and `input` lines had we used an `infile` statement. Therefore, the dataset was imported without any problems. However, there are cases where SAS guesses wrong.

Imagine a dataset with a variable with code in it that can either consist of a string of numbers or a string of letters. If the first 20 rows of the dataset have only numeric code for this variable, SAS will guess that the variable is numeric. SAS will continue reading sequentially until it encounters a character value in that variable, and then it will stop the import function and register an error. In those cases, the `infile` statement could be used to micromanage how SAS reads the dataset in.

The GUESSINGROWS option in PROC IMPORT

GUESSINGROWS is an option that can be set in PROC IMPORT. It specifies the number of rows SAS reads down the dataset in order to guess the correct code for `informat`, `format`, and `input` lines. This is SAS's method for trying to determine whether variables should be defined as numeric or characters.

GUESSINGROWS defaults to 20. It can technically be specified in a range of 1 to 2,147,483,547, but is of course limited by the rows of the dataset you are importing.

In a PROC IMPORT command, it is specified on its own line with a semi-colon at the end. Here's an example:

GUESSINGROWS = 100;

The purpose of PROC IMPORT is to save programming time. The problem with setting GUESSINGROWS to a high number is that it can greatly increase PROC IMPORT processing time. However, since the default is only 20, it can be helpful to expand this window to a slightly larger window of rows (such as 100, 250, or 500) so that SAS can make a better guess when automatically formulating code for `informat`, `format`, and `input` lines.

If we use PROC IMPORT and find that we cannot rely on SAS's guesses and we need to make infile code, there is an easy way to have PROC IMPORT help us with our task. PROC IMPORT automatically produces infile code when it runs, even if it guesses incorrectly with this code and encounters an error. The infile code produced by PROC IMPORT is automatically output into the log file. Here is an example of the top part of the log file from the code we just ran to import a *.csv file with a red box around the infile code:

```
1          OPTIONS NONOTES NOSTIMER NOSOURCE NOSYNTAXCHECK;
72
73         PROC IMPORT datafile='/folders/myfolders/X/Chap2_1_CSV.csv' out=Convert_CSV dbms=csv replace;
74         getnames=yes;
75         run;

NOTE: Unable to open parameter catalog: SASUSER.PARMS.PARMS.SLIST in update mode. Temporary parameter values will be saved to
WORK.PARMS.PARMS.SLIST.
76         /**********************************************************
77         *    PRODUCT:   SAS
78         *    VERSION:   9.4
79         *    CREATOR:   External File Interface
80         *    DATE:      26SEP20
81         *    DESC:      Generated SAS Datastep Code
82         *    TEMPLATE SOURCE:  (None Specified.)
83         ***********************************************************/
84         data WORK.CONVERT_CSV    ;
85         %let _EFIERR_ = 0; /* set the ERROR detection macro variable */
86         infile '/folders/myfolders/X/Chap2_1_CSV.csv' delimiter = ',' MISSOVER DSD lrecl=32767 firstobs=2
87            informat _STATE best32. ;
88            informat MARIJAN1 best32. ;
89            informat USEMRJN2 best32. ;
90            informat RSNMRJN1 best32. ;
91            format _STATE best12. ;
92            format MARIJAN1 best12. ;
93            format USEMRJN2 best12. ;
94            format RSNMRJN1 best12. ;
95         input
96                  _STATE
97                  MARIJAN1
98                  USEMRJN2
99                  RSNMRJN1
100        ;
101        if _ERROR_ then call symputx('_EFIERR_',1);  /* set ERROR detection macro variable */
102        run;

NOTE: The infile '/folders/myfolders/X/Chap2_1_CSV.csv' is:
      Filename=/folders/myfolders/X/Chap2_1_CSV.csv,
      Owner Name=root,Group Name=vboxsf,
      Access Permission=-rwxrwx---,
      Last Modified=07Jan2020:10:05:16,
      File Size (bytes)=4430173

NOTE: 437436 records were read from the infile '/folders/myfolders/X/Chap2_1_CSV.csv'.
      The minimum record length was 7.
      The maximum record length was 9.
NOTE: The data set WORK.CONVERT_CSV has 437436 observations and 4 variables.
NOTE: DATA statement used (Total process time):
      real time           0.57 seconds
      cpu time            0.52 seconds
```

Figure 2.3 – The top of the log file from the PROC IMPORT code run earlier

Notice how the infile code written by SAS's PROC IMPORT differs from the code we wrote earlier:

1. First, the code specified the WORK directory explicitly even though we did not do so in our code, stating the output dataset as WORK.CONVERT_CSV.

2. Next, notice that instead of using 2. for the informat and format specifications, for informat, SAS used best32., and for format, SAS used best12..

These formats are part of what is known as the BESTw. family of formats. The w in BESTw. stands for width and can take on a value between 1 and 32. BESTw. is the default numeric format SAS applies if no format is specified for a numeric variable.

Notice in the PROC IMPORT log file, SAS guesses best32. for informat. In other words, it guesses that the field is the maximum width it can be to make sure to read in data from the widest field possible. Next, it guesses format to be best12., because best12. is the default BESTw. format.

Features of the BESTw. format

Numeric format

The default for BESTw. is best12.

The maximum for BESTw. is best32.

BESTw. rounds the value.

If SAS can display at least one significant decimal in the decimal portion within the specified width, then it will produce a decimal result.

If it cannot, it will produce a result in scientific notation.

Regardless of the BESTw. format, SAS always stores the complete value of the variable.

To help us with our task, we can copy PROC IMPORT automatically-generated code from the log file and put it into a programming window. Now, we have informat, format, and input code for all of our variables – it just has errors in it. We can then edit our code, and write our custom informat and format code for each variable to correct the errors so the file imports properly.

Converting non-SAS data to SAS format

Generally, in a **data warehouse** or **data mart**, the data is stored in some dynamic structure, such as a production SQL database that is being actively queried by users. In those operations, the main work that is done by SAS is analysis, although SAS may also be involved in ETL functions that are part of the process of taking raw data and putting it into the data warehouse or data mart. In these cases, once non-SAS datasets are transformed into SAS datasets, they are immediately loaded into an SQL database for storage. They are not stored as SAS datasets in the host environment.

However, in the case of a **data lake**, where relatively raw datasets are stored without being actively connected in a warehouse environment, it is not unusual to see minimally-processed SAS datasets in storage. This is because the data lake users are using SAS, and having the relatively raw datasets in SAS format greatly improves their I/O as they join these datasets on the fly as part of their analyses.

A typical process for storing non-SAS data with SAS data in a data lake is depicted in the following diagram:

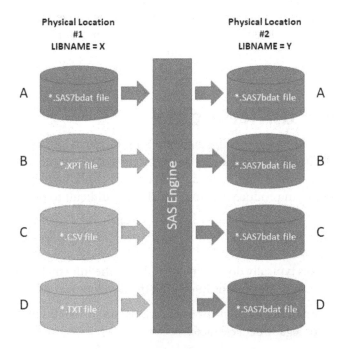

Figure 2.4 – Typical process for storing non-SAS data in a data lake

In the diagram, **Physical Location #1** (also called **LIBNAME = X**) refers to a secure location where raw datasets that are received from the source are stored. This location is only available to developers working on the data lake, as the source data has not been processed and may contain private information. The letters **A**, **B**, **C**, and **D** represent different source datasets in different source formats that are provided to the data lake. **Physical Location #2** (also called **LIBNAME = Y**) refers to a location that SAS users are approved to access to analyze the data.

For each of the datasets depicted in the diagram, although the code may simply tell SAS where **LIBNAME X** and **LIBNAME Y** are, and to read from **LIBNAME X** and write to **LIBNAME Y**, what is really happening is that each of the datasets is being processed through the SAS engine. Once the datasets are read into the SAS engine, they can be written to disk in several different formats. In the diagram, we choose `*.SAS7bdat` because we are using the scenario of a data lake where users are using SAS for analysis, making this format desirable.

The shape representing the SAS engine in the diagram demonstrates that because all the raw data needs to be ingested into SAS anyway in order to be converted, this provides an opportunity to edit the data. Even data in data lakes has been minimally edited to provide various functionality to users, and also to ensure the privacy of the data. We will talk more about this when discussing the transformation of variables.

Dealing with difficult data

Up until now, we have discussed reading data into SAS in typical SAS and non-SAS formats. Because SAS is able to handle such large datasets, it is common to receive data from **legacy** systems that were built many decades ago but continue to be used to store big data. These systems can produce data extracts that are difficult to read into SAS environments.

This section will cover the following:

- How legacy data is typically documented, and how the information in this documentation can be helpful for informing the development of `infile` code.

- How **file viewer** and **big data editor** applications can be helpful in developing `infile` code for legacy data extracts.

- How to adjust `infile` options, `format`, and `informat` to troubleshoot reading in difficult data.

- How to use `PROC PRINT` and `PROC CONTENTS` as part of troubleshooting reading in difficult data.

- How to use **file positions** to read in difficult data from a **fixed width** file.

- How to verify data files were successfully transferred.

This section will provide an overview of approaches that can be used for ingesting difficult data into SAS so they can be explored in SAS. From exploring the data in each field using SAS, the optimal `infile` code can be methodically built to read each variable in the dataset into SAS in the proper format.

Understanding legacy data

Receiving data from relational production databases to incorporate into a data warehouse today feels like a luxury compared to receiving extracts from legacy flat-file databases of the past. SQL programmers can produce data tables and views that serve up variables in formats needed by SAS data warehouse developers. Therefore, if SAS warehouse developers can communicate clearly with SQL programmers developing data extracts to be transferred to the SAS warehouse, the SAS developers can receive an extract that is relatively easy to read into SAS. Also, processing can be done on the SQL side to limit the columns and rows needed for transfer. This action can go a long way toward minimizing I/O when conducting ETL in the SAS warehouse environment on the data received.

It may be hard to believe, but even today, many of the old flat-file legacy systems are still in operation. These systems are still in common use in the health care, transportation, insurance, finance, and government sectors. In fact, customers of banks and airlines may interact with these legacy flat-file databases online without even realizing it. This is because new applications have been made to preprocess the data going in and out of these databases, which hide the underlying processes used by the **database management system (DBMS)** of the legacy database.

Ironically, these older databases are much more stable than today's new SQL databases. Also, although they are expensive to continue to manage, on balance, they continue to be very good at storing big data, and I/O continues to be fast. This is why so many industry sectors have not migrated to SQL. A new SQL environment might not hold all their data efficiently, might not ensure availability of the data, and might not be as stable. So while new databases built in the 2000s and later are typically in SQL or some sort of relational database structure, older ones that continue to be used – and continue to contain huge stores of historical and current data – remain in the flat-file structure.

These flat-file databases are the ones that are typically analyzed in SAS, because competing products such as **R** and **SPSS** could not handle all of that data. While these flat files hold a wealth of data, reading them into SAS can be especially challenging. First, these files typically are provided to the warehouse in `*.csv`, `*.txt`, or **fixed width** format, so they will need to be ingested into the SAS engine for conversion to a format usable in SAS. This means that an `infile` statement, complete with `informat`, `format`, `input`, and possibly other options will be needed.

The good news is that legacy flat-file databases tend to be well-documented. An example of such documentation is from the US **Centers for Medicare and Medicaid Services (CMS)** dataset, which is provided by the non-profit **National Council for Prescription Drug Programs (NCPDP)** (see *Further reading* for a link). An excerpt from this documentation follows in this table:

Name	Number	Definition	Format	Valid Values	Starting Position	Ending Position
Patient ID Qualifier	331-CX	Code qualifying the patient ID	X(02)	99	63	64
Patient ID	332-CY	ID assigned to the patient	X(20)	HICN	65	84
Date of Birth	304-C4	Date of birth of the patient	9(08)	Format = CCYYMMDD	85	92
Patient Gender Code	305-C5	Code indicating the gender of the patient	X(01)	0 = unspecified 1 = male 2 = female	93	93

Table 2.1 – Example of flat-file documentation

This documentation table has a lot of useful information for programmers who want to import this file into SAS:

- Notice the columns in the table titled **Starting Position** and **Ending Position**. Flat files were typically **fixed-width**, meaning that each column had a set width that it took up, regardless of whether or not it was populated. That way, if you know the widths of every source column, you can program these widths into `infile` code so that every column is wide enough to accommodate the columns you are reading in.

- For example, if you were reading in this file and it was in fixed-width format, you would expect in the 63rd and 64th data positions the code 99 for **Patient ID Qualifier**.

- The **Valid Values** column shows the values that are valid in the variable, so that it is possible to anticipate those in coding.

- Notice how sensitive variables such as **Patient ID Qualifier** and **Patient ID** are set to known values (**99** and **HICN**, respectively) to protect the patient's privacy.

- For the variable **Date of Birth**, the format of the date is given, and for the variable **Patient Gender Code**, the allowable codes and how to decode them are stated.

- The **Format** column is also useful, but is not intuitive in this example. Reading the documentation more carefully will allow the reader to understand the code used to express the format so the programmer can anticipate numeric versus character fields, and what format they should be in.

- Regardless of the format, however, the **length** of each variable is already specified and can be deduced by simply reviewing the starting and ending positions of the variable.

- The **Definition** column is useful to not only the analyst but the warehouse developer. It can help the developer troubleshoot data processing by being able to understand the potential contents of the field, and why logically there may be problems in processing.

In addition, in legacy databases, the **Number** column is usually more useful than the **Name** column for identifying variables in discussions with data providers. Typically, these databases did not store column names in headers; the first observation is a data row. Therefore, each analyst using the dataset could assign different variable names as they read the data into SAS using the `INFILE` statement. This created confusion in discussions. One solution was to identify the variable by its start position, but this could be confusing also, because when reading in data, SAS developers may have rearranged the columns to their liking. Another solution was to assign a unique code for the variable, which is **Number** in this dataset. This situation explains why in the **Name** column, the names of the variables have spaces in them, and could not be used literally as variable names in SAS.

Reading data with difficult formats

Legacy flat-file data is often challenging to read into SAS, but this challenge is reduced by the fact that most flat-file datasets that are analyzed regularly are well-documented, and this documentation can help inform the analyst developing `infile` code to read it into SAS. However, it is not unusual when working at an organization to receive a request to read a new flat file into SAS that has not been well-documented. These situations can be difficult for the programmer, as the datasets are often very long and wide, meaning that they can contain many different values, formats, and features that complicate reading code.

Every SAS developer working with reading big data into SAS will want to find their favorite **file viewer** for viewing big data files. The file viewer allows the analyst to view big data to try to troubleshoot why SAS might be encountering read errors. While small data files in `*.csv` and `*.txt` format are easy to open in **Notepad**, opening big data will cause Notepad to hang.

File viewers allow you to navigate and view the different data columns, but do not allow you to edit them. For that, you will need a **big data editor**. Those programs are also available for download. There are many big data file viewers and editors that are available for free, as well as more sophisticated ones that are available through paid application packages or suites.

Suggestions for a free file viewer and file editor

For a free file viewer for big data for Windows users, **Large Text File Viewer** is available for download online (`https://web.archive.org/web/20140908181354fw_/http://swiftgear.com/ltfviewer/features.html`). You can download a `*.zip` file and extract the contents into a folder. The files include the `*.exe`. Run the `*.exe` from the folder, then from within the application, open the large data file. You can view and navigate the data. The program quickly opens and navigates large data files. However, it cannot edit them. For that, you will need a file editor.

For a free file editor for big data for Windows users, **Large File Editor** is a program available for free as part of Liquid Studio's community edition license. This web page allows you to download the entire Liquid Studio: `https://www.liquid-technologies.com/trial-download`. You will need to install the application, and it will give you a choice between doing a 15-day free trial of their paid products, or only using the free community license version. Since the file editor is available under the community license, this is all that is necessary to run the large file editor component of the program. Because the program allows editing, it runs somewhat slower than a file viewer.

If you use macOS or Linux/Unix, you can get the functionality of both a **file viewer and a file editor** from **Vim** (which also works in Windows), available for free here: `https://www.vim.org/download.php`.

Let's imagine that the following example dataset, included in your chapter files as `Chap2_2_Difficult.csv`, comes from a legacy system:

ID	CODE1	CODE2	CODE3
1	820.00	RED	12
2	840.28	BLUE	34
3	E440.32	GREEN	56
4	F540.55	YELLOW	78
5	989.45	ORANGE	99

Table 2.2 – Example dataset

Notice that there are four columns:

1. The first column, ID, is an integer identifying the row.
2. CODE1 contains values that are numeric in some rows and alphanumeric in others.
3. CODE2 is entirely a character-only field.
4. CODE3 holds two-digit integers.

Now, imagine we did not know the data format. We could use the file viewer to look at the header. Even if there was no header, as this is a *.csv, we would be able to identify that there were four columns. The issue is that if this was a dataset with many rows, we would not be able to tell how wide each of the columns was, nor what type of data was in all the rows. Of course, if we could read all the data into SAS, we could use SAS to explore the answers to these questions.

When beginning to set up code to read in a data file in an unknown format, we can use an iterative process of creating a big `infile` statement, running it, looking at the results, and then going back and editing the statement to improve it. To start the `infile` statement, we should set the columns to a large width – in this case, we will choose 10 characters – and make all of the columns character format. Any type of data can be read into a SAS variable in character format. Let's place the dataset named `Chap2_2_Difficult.csv` into the directory mapped to LIBNAME X. Using the following code, we will read the dataset into WORK:

```
data Chap2_2_Difficult;
%let _EFIERR_ = 0;
infile '/folders/myfolders/X/Chap2_2_Difficult.csv' delimiter =
',' firstobs=2
    MISSOVER DSD lrecl=32767;
    informat    ID $10.;
    informat    CODE1 $10.;
```

```
        informat      CODE2 $10.;
        informat      CODE3 $10.;
        format            ID $10.;
        format            CODE1 $10.;
        format            CODE2 $10.;
        format            CODE3 $10.;
  input
        ID $
        CODE1 $
        CODE2 $
        CODE3 $
  ;
  if _ERROR_ then call symputx('_EFIERR_',1);
  RUN;
```

In the code, we are outputting the SAS dataset Chap2_2_Difficult into WORK and reading in the raw data file Chap2_2_Difficult.csv. Notice the different infile coding associated with character fields. Both the informat and format code includes $, such that informat and format of each character field is $10..

Now, we will want to look at the data in SAS that we just read in so that we can document what the fields contain by reviewing the data. One way to do this is to use PROC PRINT, but it is important to limit the printed rows to just a few. If the data is very wide, it might be helpful to limit the output to as few as five rows in the beginning. This can help you diagnose whether the initial choice of field width was wide enough to ingest all the data in the column. Even though our dataset is only five rows long, in the next PROC PRINT code, we will limit the printed rows to 5 with the obs option:

```
proc print data=Chap2_2_Difficult (obs=5);
  RUN;
```

Here is the output from PROC PRINT:

Obs	ID	CODE1	CODE2	CODE3
1	1	820	RED	12
2	2	840.28	BLUE	34
3	3	E440.32	GREEN	56
4	4	F540.55	YELLOW	78
5	5	989.45	ORANGE	99

Figure 2.5 – PROC PRINT output showing the full dataset read in with all columns in character format

Notice in the PROC PRINT output that the observation number, labeled **Obs**, is output as well. SAS keeps track of observation numbers of rows in a dataset using this internal variable. From viewing the data, it appears that the data has been imported properly. However, CODE3 is a particularly problematic variable because the column is clearly much wider than it needs to be, at a width of 10. Further, this two-digit code may need to be handled as a number, and it cannot be the way we read it in.

We can demonstrate this by trying to make a mean out of CODE3 using PROC MEANS:

```
proc means data=Chap2_2_Difficult;
var CODE3;
RUN;
```

However, our code does not run, and we get the following error in the log file:

```
ERROR: Variable CODE3 in list does not match type prescribed
for this list.
```

This shows that looks are deceiving, and it is necessary to investigate data types, formats, and other details of each column to make sure all the data is read in properly.

Let's change our code now:

- We can change format and informat for CODE3 to the format of 2. (and remove the $ from the input statement) in order to make it numeric.

- Also, now that we have viewed the data more carefully, we can edit the width of the other fields. If we had viewed CODE2 from only the first two rows of the dataset, we would have seen the entries RED and BLUE. This might have led us to believe that CODE2 had a width of four characters, so let's edit our code, imagining this is what we believed.

- Next, let's see what happens if we use a wide informat of $10. and use a much smaller format, $3., for CODE1.

Let's run our edited code:

```
data Chap2_2_Difficult;
%let _EFIERR_ = 0;
infile '/folders/myfolders/X/Chap2_2_Difficult.csv' delimiter =
',' firstobs=2
    MISSOVER DSD lrecl=32767;
    informat    ID 2.;
    informat    CODE1 $10.;
    informat    CODE2 $4.;
```

```
        informat    CODE3 2.;
        format              ID 2.;
        format              CODE1 $3.;
        format              CODE2 $4.;
        format              CODE3 2.;
input
    ID
    CODE1 $
    CODE2 $
    CODE3 $
;
if _ERROR_ then call symputx('_EFIERR_',1);
RUN;
```

Now, to examine the difference in how the data read in compared to our previous code, we will rerun our PROC PRINT code, as well as running PROC CONTENTS:

```
PROC PRINT data=Chap2_2_Difficult (obs=5);
PROC CONTENTS data=chap2_2_Difficult;
RUN;
```

Our PROC PRINT output looks different this time:

Obs	ID	CODE1	CODE2	CODE3
1	1	820	RED	12
2	2	840	BLUE	34
3	3	E44	GREE	56
4	4	F54	YELL	78
5	5	989	ORAN	99

Figure 2.6 – Output from PROC PRINT showing CODE1 and CODE2 cut off, and CODE3 reformatted as numeric

Let's first examine the results:

- For CODE2, we can see that because we designed the width of the variable as four characters, that the values in rows three through five are cut off. Whenever the analyst sees this, they know they need to work on making this variable wide enough to accommodate all the data in the source variable.

- Next, we can see this time that CODE3 is right-justified instead of left-justified, which is a hint from SAS that this time, it was read in as a numeric variable.

- The PROC PRINT output also shows that CODE1 is only displaying three characters.

Let's continue to consider CODE1 while looking at the last table in the PROC CONTENTS output:

Alphabetic List of Variables and Attributes					
#	Variable	Type	Len	Format	Informat
2	CODE1	Num	8	7.	7.
3	CODE2	Char	6	$6.	$6.
4	CODE3	Num	8	2.	2.
1	ID	Num	8	2.	2.

Figure 2.7 – PROC CONTENTS output showing different format and informat for CODE1, too small of a width for CODE2, and the numeric format of CODE3

Notice that PROC CONTENTS reports format and informat of all the variables. We can see that even though we read in 10 characters in informat for CODE1, the format is only three characters wide, so that is all that is being displayed.

It is also easy to get an error if you misunderstand data types. Only the first two rows of CODE1 are numeric. If this field is designated numeric, and code is run to read it in, the following error will register in the log:

```
NOTE: Invalid data for CODE1 in line 4 3-9.
  RULE:        ----+----1----+----2----+----3----+----4----+----5--
--+----6----+----7----+----8----+----9----+----0
  4           3,E440.32,GREEN,56 18
  ID=3 CODE1=. CODE2=GREEN CODE3=56  _ERROR_=1  _N_=3
```

This error has four lines:

1. The first announces the row and positions of the invalid data.

2. Next, there is a ruler that wraps around to help the analyst match up the data following the ruler with the column positions.

3. The third line simply lists the data values SAS is trying to read in. In the final line, SAS shows how it links up the stated variable names and the values coming in.

4. _ERROR_=1 indicates an error code for _N_=3, which indicates observation three.

Some legacy files, however, have errors that are very hard to detect. For example, let's say all the periods in the CODE1 field were accidentally stored as commas in a *.txt comma-delimited file. The dataset included with this chapter that is in this format is called Chap2_3_error.txt. Let's place it in the directory mapped to LIBNAME X. The following code should be perfectly tailored to read in the dataset if the periods in CODE1 have not been turned into commas. However, now that there are extra delimiters, SAS will become confused:

```
data Chap2_3_Error;
%let _EFIERR_ = 0;
infile '/folders/myfolders/X/Chap2_3_Error.txt' delimiter = ','
firstobs=2
    MISSOVER DSD lrecl=32767;
    informat    ID 2.;
    informat    CODE1 7.;
    informat    CODE2 $6.;
    informat    CODE3 2.;
    format          ID 2.;
    format          CODE1 7.;
    format          CODE2 $6.;
    format          CODE3 2.;
input
    ID
    CODE1
    CODE2 $
    CODE3
;
if _ERROR_ then call symputx('_EFIERR_',1);
RUN;
PROC PRINT data=Chap2_3_Error (obs=5);
RUN;
```

In this code, we are outputting the dataset named Chap2_3_Error in WORK, and we are reading in the file Chap2_3_Error.txt, which is the comma-separated file with extra commas as errors. Even though the informat, format, and input statements are correct with respect to data types and widths, the PROC PRINT output shows that the extra commas have shifted the columns:

Obs	ID	CODE1	CODE2	CODE3
1	1	820	RED	12
2	2	840	28	.
3	3	.	32	.
4	4	.	55	.
5	5	989	45	.

Figure 2.8 – PROC PRINT output showing extra commas causing shifting columns

As can be seen by this short exercise, reading undocumented large data files into SAS starts with a tedious game of trial and error. Working to perfect `infile` code while inspecting data in the file viewer or even editing it in a data editor is often a large task when encountering a new dataset. In addition, a large collection of formats can be applied to data, and decisions need to be made about which ones to use.

Unfortunately, sometimes due to various constraints, it is necessary to actually skip reading in certain columns that need to be dealt with later. This can be done by specifying read positions in the `infile` command.

Specifying data locations in a fixed-width file

The example shown earlier of the documentation of the NCPDP dataset revealed that the NCPDP dataset is a **fixed-width file**. Fixed-width files are common from legacy databases. In these files, there are no delimiters. Instead, there is a map of characters in every row. Every row (or **record**) must be a certain amount of characters, and if there are character **positions** where there is no character value, a space is used as a filler. In your files for this chapter, there is a dataset called `Chap2_4_Fixed.txt`, which is an example of a fixed-width file, also shown here:

```
1 820     RED     12
2 840.28 BLUE     34
3 E440.32GREEN    56
4 F540.55YELLOW   78
5 989.45 ORANGE   99
```

Let's look at some features of this dataset:

- Notice this is the same dataset we have been working with, which has four columns – ID, CODE1, CODE2, and CODE3 – and five rows.

- However, this dataset has no headers. This is because legacy databases generally did not use headers in exports, as mentioned earlier.

- Since this is a fixed-width dataset, this dataset also has no delimiters. Notice how in the fourth row, the value in CODE1 is immediately followed by the value for CODE2 with no delimiter in between.

- Notice how due to the fixed-width format, the ID column appears to be two spaces, because there is a space after each of the entries. From this, we can deduce that ID takes up two character positions in the source dataset (regardless of whether or not they both are populated), and its values are in positions one to two.

- In fact, from reviewing the file, it is possible to count the length of each record. It appears that each record has a fixed width of 18 characters.

Imagine we were having trouble reading this file into SAS because we were getting errors, and therefore, we were viewing it in a file viewer. File viewers have tools that allow the user to click on the data file and know the position of the mouse. Using those tools, we could determine the start position of each of the columns. For example, we see that positions 17 and 18 are taken up by CODE3, so the start position for CODE3 would be 17, and the end position would be 18.

Or imagine we wanted to focus on studying CODE2 so we could develop a format for each variable as we build our infile code. We want to read it in as a character so we do not have any trouble with data type mismatches, and we want to make sure the column is wide enough to input all of the data. More importantly, if we are getting errors from other columns, we want to tell SAS to skip the other columns and only read in the values in CODE2 so we can study those.

Let's add the dataset Chap2_4_Fixed.txt to the directory mapped to X. Then we will use the following code to read in only CODE2 from this source dataset:

```
data Chap2_4_Fixed;
infile '/folders/myfolders/X/Chap2_4_Fixed.txt';
length          CODE2 $7;
informat        CODE2 $7.;
input              CODE2 1-7    @10;
RUN;
```

Let's look at each line of code separately:

1. To begin, the data step specifies outputting SAS dataset `Chap2_4_Fixed` into `WORK`.

2. The `infile` statement refers to the raw dataset `Chap2_4_Fixed.txt` placed in `X`.

3. This time, instead of using the `format` statement, we are using the `length` statement. This defines the variable, `CODE2`, as having a length of seven. The `$` indicates it is a character field. A `length` statement would be used for each variable intended to be in the output dataset.

4. The `informat` statement defines reading in variable `CODE2` as a seven-character field.

5. The `input` statement tells SAS to put into the newly created variable `CODE2` the contents of the source file starting at position 10 (indicated by `@10`). Because we tell SAS that `CODE2` is seven characters long and we are inputting into character positions one to seven, SAS will start at position 10 in the source file and read the next seven characters into `CODE2` positions one through seven in the new dataset we are inputting into `WORK` named `Chap2_4_Fixed`.

The resulting dataset only has `CODE2` in it. This would theoretically allow an analyst to examine this one column to better understand its contents, and develop optimal `infile` code.

Troubleshooting reading data after transport

As stated earlier, when running a data warehouse in SAS, the warehouse developers typically receive raw data files in `*.txt`, `*.csv`, or sometimes XPT format on static media or through a file server. Those who provide the data have already made agreements with the warehouse as to the specifications of what data they are providing, how often, and how it will be transported. Once these policies are set up, there is always the day-to-day work of transferring this data regularly from the source system to the SAS warehouse. People involved in these regular transfers have a natural propensity to work collaboratively, as they are stewards of the same big datasets.

As part of this working relationship, the person transferring data from the source system should provide some metadata about each data file to the person receiving the transfer at the SAS warehouse. If the source data is in SAS, the person at the source system who is preparing the data for transfer could simply keep a record of PROC CONTENTS output for each dataset being transferred, and provide this record to the receiving person (for example, in an emailed *.PDF). This would provide enough information about the SAS dataset to allow the receiving person to assure themselves that the data described in the PROC CONTENTS output matches the actual files they received once they load them in to SAS.

When transferring a dataset that is not in SAS format, it is trickier to verify the transfer. It becomes clear that errors in reading the data into SAS need to be separated from errors inherent in the transferred data file. Before the data is read into SAS, the following can be verified between the source and the receiver:

- The source file and the received file have the same file metadata (for example, the last update date).
- The source file and the received file are the same size.

The next step is to read the file into SAS to verify more details about the file, but this is also where the SAS user can inadvertently introduce errors that do not arise from the source data. Before verifying any details, it is important to ensure that the entire dataset is adequately read into SAS and there are no errors in the log file.

For transfers that are part of routine ETL processes, typically, there are no problems reading each new file because the code has been developed and debugged. However, on occasion, problems reading routinely transferred files arise due to unanticipated changes in the structure of the source data. If these structure changes are the source of the error, this can be determined by opening the data in a file viewer and finding the changes using guidance from error reports in SAS's log file. Also, it is possible to see in the file viewer whether the data file appears corrupt, and this is the source of the error.

If the dataset reads into SAS without errors, the following items can be matched up between the received dataset and the transferred dataset with the source provider:

- The number of rows in the dataset
- The number of columns in the dataset
- Specified values in specified rows

From time to time, there is an error when preparing the transfer file, or an error is introduced during transfer, that causes less than the total file to be transferred. In those cases, these troubleshooting techniques will identify that one of these types of errors occurred, and the source file either needs to be regenerated, or it was generated properly and it needs to be retransferred.

Summary

This chapter described many strategies and considerations to be made when developing code to read big data into SAS. SAS native data file formats `*.SAS7bdat` and `XPT` contain metadata that makes reading them into SAS easy and accurate. However, the downside is that these datasets can take up a lot of storage space.

`LIBNAME` statements in SAS point to external locations for reading and storing data. Using `LIBNAME` statements, the SAS user can convert `*.csv` and `*.txt` files to `*.SAS7bdat` datasets for storage, and can also convert them to XPT. While storing data in `*.csv` and `*.txt` format can conserve space, the downside is that oftentimes specialized `infile` code is necessary for reading this data in so that it is properly formatted in SAS. While the automation involved in `PROC IMPORT` can help with this, when regularly transferring large raw data files, usually, long and detailed `infile` code must ultimately be developed and maintained.

Because of SAS's ability to handle big data analysis, it is not unusual for SAS data warehouses to host datasets from large legacy flat-file systems that are still in operation today in many industries. While valuable, data from these systems can have characteristics that make them difficult to read into SAS. This chapter provided guidance on how to troubleshoot reading in legacy data with difficult formats, including how to handle reading in fixed-width data.

In the next chapter, we will look more at `PROC CONTENTS` and other PROCs and strategies for understanding and managing data. We will go over how to use SAS labels and formats, and talk about different approaches to viewing datasets in SAS.

Questions

1. What is the XPT format for?

2. What does the `GUESSINGROWS` option in `PROC IMPORT` do?

3. How can `PROC IMPORT` help the analyst develop `infile` code, even if the procedure hits an error?

4.   Why is data from fixed-width datasets challenging to read into SAS without documentation?

5.   Imagine you have a *.csv file that you want to convert to a *.SAS7bdat dataset to transfer to a coworker using static media. How would you use LIBNAME statements in your code?

6.   Imagine your company assembles a large dataset once per year using SAS and analyzes it in SAS for their annual report. The company rarely uses the dataset once the report is generated, and therefore, it can be kept in archives. What is the optimal format for storing these yearly datasets? State your rationale.

7.   Imagine you are a developer at a SAS data warehouse. A data provider for your warehouse asks you in what format you would like to receive the data. You have the choice between *.csv, *.txt, or *.SAS7bdat. What should you choose and why?

8.   A data provider tells you that the *.SAS7bdat dataset you requested is too large to fit on static media. What options could you suggest and why?

Further reading

- US Library of Congress describes preservation of data in digital formats: https://www.loc.gov/preservation/digital/formats/fdd/fdd000464.shtml

- SAS white paper about PROC CPORT and PROC CIMPORT *Tips and Techniques For Moving Between Operating Environments* by J. Meimei Ma and Sandra Schlotzhauer – available here: https://support.sas.com/resources/papers/proceedings/proceedings/sugi23/Begtutor/p56.pdf

- Documentation from US CMS dataset from the NCPDP: https://www.cms.gov/Medicare/Billing/ElectronicBillingEDITrans/downloads/NCPDPflatfile.pdf

3
Helpful PROCs for Managing Data

The purpose of this chapter is to introduce a few helpful PROCs and data step maneuvers that can be particularly useful when managing a **Statistical Analysis System (SAS)** data warehouse. First, the chapter will describe how to use PROC CONTENTS to understand what is in a dataset. Next, the use of **SAS formats** and **SAS labels** to attach **metadata** to variables and values will be covered. The chapter will demonstrate examples of applying labels, **user-defined formats**, and **SAS native formats** to variables. We will also go over what happens when leaders of a SAS data warehouse choose to use SAS labels and formats in the warehouse. If labels and formats are used, the data will display differently in SAS windows and certain PROCs, which may or may not be desired. Also, the choice to maintain labels and formats adds a specific type of **management** layer to the data warehouse, which is described along with **alternatives to using labels and formats**. Finally, this chapter introduces different ways to use PROC PRINT, PROC SQL, and other features in SAS to view raw datasets.

On completing this chapter, you will understand what SAS labels and formats are, how to make them and apply them to datasets, and how they are used in reporting. You will also learn about keeping documentation on SAS datasets, different ways to view big SAS datasets, and the syntax for arithmetic operators in SAS.

In summary, this chapter covers the following main topics:

- Using PROC CONTENTS for exploring datasets
- Creating and using *codebooks* for determining text to use in SAS labels and formats
- Creating and applying labels, user-defined formats, and SAS native formats to change how data displays in SAS
- Using PROC PRINT and PROC SQL to view data
- Viewing data through SAS windows

Technical requirements

The *.SAS7bdat demonstration dataset used in this chapter can be found on GitHub: https://github.com/PacktPublishing/Mastering-SAS-Programming-for-Data-Warehousing/blob/master/Chapter%203/Data/chap3_1.sas7bdat.

The code bundle for this chapter is available on GitHub here: https://github.com/PacktPublishing/Mastering-SAS-Programming-for-Data-Warehousing/tree/master/Chapter%203.

PROCs for understanding data

SAS warehouse developers become accustomed to receiving datasets from other systems, then performing **extract, transform, and load (ETL)** procedures on these datasets to load them into the warehouse. But in order to set up the ETL procedures, an analyst has to explore and understand the data. Values in each column need to be examined, as well as other characteristics of the dataset, in order for ETL code to be developed.

This section introduces several helpful approaches for exploring these datasets. First, we talk about ways to use PROC CONTENTS to understand new datasets. Next, the role of codebooks in providing documentation about the variables is discussed. Variables can be annotated with SAS labels, and levels of categorical variables can be annotated with user-defined SAS formats. In addition, native SAS formats can be applied to improve the display of data. Different strategies for applying labels and formats are demonstrated.

Using PROC CONTENTS to understand data

The most basic PROC for understanding datasets is PROC CONTENTS. When PROC
CONTENTS is used on a dataset, it provides output that lists the names of the variables
in the dataset, along with other information.

Let's see what the default PROC CONTENTS output looks like. First, we will map
LIBNAME X to a known folder, as follows:

```
LIBNAME X "/folders/myfolders/X";
RUN;
```

In the code, replace /folders/myfolders/X with the location in which you want
to store the datasets for this chapter. Then, place the SAS dataset provided, named
Chap3_1.SAS7bdat, in the location mapped to the LIBNAME X. After that,
run the following code:

```
PROC CONTENTS data=X.chap3_1;
RUN;
```

PROC CONTENTS produces three tables in the default results. Let's start by looking at the
first table in the results, as follows:

The CONTENTS Procedure

Data Set Name	X.CHAP3_1	Observations	38901
Member Type	DATA	Variables	4
Engine	V9	Indexes	0
Created	01/21/2020 15:07:06	Observation Length	32
Last Modified	01/21/2020 15:07:06	Deleted Observations	0
Protection		Compressed	NO
Data Set Type		Sorted	NO
Label			
Data Representation	SOLARIS_X86_64, LINUX_X86_64, ALPHA_TRU64, LINUX_IA64		
Encoding	utf-8 Unicode (UTF-8)		

Figure 3.1 – First table of output from PROC CONTENTS run on the Chap3_1.SAS7bdat dataset

Notice that the table has four columns, detailed as follows:

- The first and third columns are labels, while the second and fourth columns are the
  values pertaining to our dataset.

- In the first two columns, an important value to notice in data warehousing is **Data Set Name**, which specifies LIBNAME and the dataset name in the second column. The reason why it is important to pay attention to this particular value when doing warehousing is that you are often dealing with multiple datasets, such as monthly files going back several years. Looking under **Data Set Name** in PROC CONTENTS helps ensure that you are thinking about the correct dataset when reading the rest of the output.

- In the third and fourth columns, important values for data warehousing are **Observations** and **Variables**. When transferring a dataset to the warehouse, the data provider should tell whoever is receiving the data the number of observations and the number of variables in the dataset. Then, when the dataset is read into the SAS warehouse, PROC CONTENTS can be used to reveal this information and help troubleshoot whether or not the full dataset was properly transferred to the SAS environment. In the case of this example, the dataset has 38,901 observations and four variables.

We will skip looking at the second table in the PROC CONTENTS results and will focus on the third table, displayed here:

Alphabetic List of Variables and Attributes					
#	Variable	Type	Len	Format	Informat
2	SEX1	Num	8	2.	2.
3	_AGE80	Num	8	2.	2.
4	_BMI5	Num	8	4.1	2.2
1	_STATE	Num	8	2.	2.

Figure 3.2 – Third table of output from PROC CONTENTS run on Chap3_1.SAS7bdat dataset

Let's review some of the important information in the third PROC CONTENTS table, as follows:

- **Column number**: This number is under the # character and shows the order of the columns in the dataset. _STATE is the first column as it is labeled 1, and the last column is _BMI5, labeled 4.

- **Variable name**: The variable name is listed under **Variable**. The default PROC CONTENTS will sort the variables alphabetically. As can be seen from the output, the underscore sorts after alpha characters. Also, the table is titled **Alphabetic List of Variables and Attributes**.

- **Type**: PROC CONTENTS gives the data type for each variable. The Chap3_1 dataset only has numeric data, so all are listed as Num.

- **Format and informat**: As described in *Chapter 2, Reading Big Data into SAS*, INFORMAT indicates the data format used when the data was read into SAS, and FORMAT indicates the data format used when the data was placed into the SAS dataset.

In datasets with a lot of variables, PROC CONTENTS output can get long. Because the default sort is in alphabetical order, it is helpful to learn the details of alphabetical sort order in SAS.

Sorting order in SAS:

The smallest-to-largest comparison sequence for numeric variables in SAS is as follows:

SAS missing values (usually displayed as a period)

Negative numeric values

Zero

Positive numeric values

For sorting character variables, either the **Extended Binary Coded Decimal Exchange Code (EBCDIC)** order or the **American Standard Code for Information Exchange (ASCII)** order is used, depending on the operating system, as explained here:

EBCDIC order: Used in **Conversational Monitor System (CMS)** and OS/390

ASCII order: Used in Macintosh, **MS-DOS** (which stands for **Microsoft Disk Operating System**), OpenVMS (where **VMS** stands for **Virtual Memory System**), OS/2, **PC DOS** (which stands for **Personal Computer Disk Operating System**), Unix and its derivatives, and Windows.

The EBCDIC and ASCII orders are different in the following ways:

EBCDIC: Lowercase letters are sorted before uppercase letters. Uppercase letters are sorted before digits. Some special characters interrupt the alphabetic sequences, and the blank space is the smallest displayable character.

ASCII: Digits are sorted before uppercase letters. Uppercase letters are sorted before lowercase letters. As with EBCDIC, the blank space is the smallest displayable character.

A link to a web page with all SAS sort sequences available is listed under the *Further reading* section.

To print variables in the third table from `PROC CONTENTS` in the order in which they appear in the dataset (as shown under # in *Figure 3.3*), the `VARNUM` option should be used, as illustrated here:

```
PROC CONTENTS data=X.chap3_1 VARNUM;
RUN;
```

Here is the output:

Variables in Creation Order					
#	Variable	Type	Len	Format	Informat
1	_STATE	Num	8	2.	2.
2	SEX1	Num	8	2.	2.
3	_AGE80	Num	8	2.	2.
4	_BMI5	Num	8	4.1	2.2

Figure 3.3 – Third table of output from PROC CONTENTS run on the
Chap3_1.SAS7bdat dataset with results sorted by column number

As shown in the output, the variables are now in the order of the # column instead of being listed alphabetically. Also, notice the table is now titled **Variables in Creation Order**.

> **Note:**
> In data warehousing, data is typically transferred to the warehouse in a standardized format from each data provider. Next, the warehouse adds columns through **ETL processing**. For this reason, running PROC CONTENTS on a transformed dataset and using the VARNUM option effectively lists the transformed columns in the order they were created. This can help with troubleshooting ETL code.

The `PROC CONTENTS` results we've been reviewing are typical for what is seen when a non-SAS dataset is read into SAS. SAS provides features that can be used to add metadata to the variables. These include adding **SAS labels** to describe variables, and adding user-defined **SAS formats** to categorical variables to describe each level. If these SAS functions are used, this information will be added to the `PROC CONTENTS` results (and can also be used in other PROCs).

Let's review how this works by adding labels and formats to our `Chap3_1.SAS7bdat` SAS dataset and rerunning `PROC CONTENTS`. Before we do that, however, we have to check our documentation to make sure we have the information we need about the variables in our dataset, so we will look at the **codebook**.

Documenting SAS data with codebooks

The four variables in the `Chap3_1.SAS7bdat` example dataset are named `_STATE`, `SEX1`, `_AGE80`, and `_BMI5`. These are four variables taken from the **2018 Behavioral Risk Factor Surveillance System (BRFSS)** dataset, which is an annual phone survey about health done in the US every year. In the technical requirements for this chapter, you can download the **codebook** for this dataset, which provides information about each variable.

If we look in the codebook for `_STATE`, we will find information about that variable. `_STATE` is a categorical variable referring to the state in the US where the survey respondent resides. The values stored for the states are two-digit numeric codes called **Federal Information Processing Standard (FIPS)** codes.

In the codebook, for categorical variables, you will also find the frequencies for each level of the categorical variable. The `Chap3_1.*SAS7bdat` dataset only includes records from three states: Florida (FIPS code 12), Massachusetts (FIPS code 25), and Minnesota (FIPS code 27). The codebook entry for `_STATE` with the information for these three states is reproduced here:

```
                        LLCP 2018 Codebook Report
                 Overall version data weighted with _LLCPWT
                 Behavioral Risk Factor Surveillance System
```

```
Label: State FIPS Code
Section Name: Record Identification
Section Number: 0
Question Number: 1
Column: 1-2
Type of Variable: Num
SAS Variable Name: _STATE
Question Prologue:
Question: State FIPS Code
```

Value	Value Label	Frequency	Percentage	Weighted Percentage
12	Florida	15,242	3.48	6.67
25	Massachusetts	6,669	1.52	2.15
27	Minnesota	16,990	3.88	1.68

Figure 3.4 – BRFSS codebook entry for _STATE with information about three states included in the Chap3_1 dataset

As you may have noticed, the native BRFSS dataset appears to be in SAS. We can tell this from hints we can see in the codebook, detailed as follows:

- First, we see in the upper left the term **Label**. When using a codebook for a SAS dataset, the word **Label** can indicate that this is the actual SAS label that was applied to the variable.

- Other useful information about the variable is in the upper left, including **SAS Variable Name**, as well as the survey question that was asked to elicit this answer.

In the table in the lower part of the codebook screenshot, information for the three states in the `Chap3_1.SAS7bdat` dataset is listed. Notice that under **Value** is the actual numeric FIPS code stored in the variable, and the **Value Label** decodes the FIPS code with the state name. In addition, the frequency of records for each state and other information about each categorical level is printed in the codebook.

Automatic generation of codebooks from SAS

The purpose of the BRFSS codebook is to document a SAS dataset. It is typical to make a codebook to document SAS datasets, especially surveys. For this reason, SAS users often wonder if there is a function in SAS that can automatically produce a codebook from a dataset.

The short answer is that there is no automatic function, but programmers have developed **macros**—or custom SAS routines that can be run with inputs—that can help. One method described in a SAS white paper by Louise Hadden is a four-step process for creating a codebook that includes using Excel. Naoko Stearns and John Gerlach describe a different approach in their SAS white paper, which demonstrates ways to use macros to develop and output a final codebook report. Links to these white papers are available under the *Further reading* section.

Before applying SAS labels and SAS formats to variables, it is important to have all the information needed about them from the codebook. We now have the information we need for _STATE to apply labels and formats to that variable in our `Chap3_1.SAS7bdat` dataset. We will need to look up the information for the other three variables, which are `SEX1`, `_AGE80`, and `_BMI5`. We will continue to do this as we add labels and formats to all the variables in `Chap3_1`.

Using labels for variables

Labels in SAS refer to a specific SAS function that attaches an alias in the form of a string of characters to a variable, for the purpose of describing that variable. As we saw earlier from the codebook, the label that was intended to be applied to the _STATE variable is State FIPS Code. After looking in the codebook, we can summarize the labels we need to apply to the four variables in our Chap3_1.SAS7bdat dataset, as follows:

Variable	Label
_STATE	State FIPS Code
SEX1	Sex at Birth
_AGE80	Imputed age value collapsed above 80
_BMI5	Body Mass Index (BMI)

Table 3.1 – Text for variable labels from the BRFSS codebook

Now that we know which labels to put on each of the variables, we can add them to the SAS dataset with a data step. The following code inputs the Chap3_1.SAS7bdat dataset from LIBNAME mapped to X, attaches labels to the variables, and outputs the Chap3_1_Labels SAS dataset to the same directory:

```
data X.Chap3_1_Labels;
    set X.Chap3_1;
    label _STATE = "State FIPS Code"
    label SEX1 = "Sex at Birth"
    label _AGE80 = "Imputed age value collapsed above 80"
    label _BMI5 = "Body Mass Index (BMI)"
    ;
RUN;
```

Let's look more closely at the features of this code, as follows:

- **Data and set statements**: Notice that the data and set statements alone would only copy the dataset. It is after these statements that label statements indicate to SAS that labels must be attached to the dataset before making a copy of it.

- **Format of label statements**: Each label statement is on its own line and has the syntax of label followed by the variable name, then an equals sign and the label text in quotes. A label statement in SAS can be up to 256 characters long and can include spaces.

- **Semicolon for the label statements**: Notice that, unlike most SAS code that has a semicolon at the end of each line, the set of `label` statements has only one semicolon at the end. Typically, this is placed alone on the last line after the last `label` statement.

Now, when we run `PROC CONTENTS` on the new dataset with the `Chap3_1_Labels` labels we made, we see more information in the third table of the output, as follows:

```
proc contents data=X.Chap3_1_Labels;
RUN;
```

This would give us the following output:

Alphabetic List of Variables and Attributes						
#	Variable	Type	Len	Format	Informat	Label
2	SEX1	Num	8	2.	2.	Sex at Birth label
3	_AGE80	Num	8	2.	2.	Imputed age value collapsed above 80 label
4	_BMI5	Num	8	4.1	2.2	Body Mass Index (BMI)
1	_STATE	Num	8	2.	2.	State FIPS Code label

Figure 3.5 – Output from PROC CONTENTS with labels

The variable labels not only prints out in the output for `PROC CONTENTS`, but can also be seen in the output from other PROCs, which you will be able to see later in the next section.

Adding user-defined formats to categorical variables

User-defined SAS formats are similar to SAS labels in that their purpose is to place a label on a value. The difference is that SAS labels are for placing a label on a variable to describe what the variable means, while user-defined SAS formats are for placing labels on levels of categorical variables to explain what each level means.

In our `Chap3_1_Labels` example dataset with the labels, we have only three state values in the `_STATE` variable, which are the values for the states of Florida, Massachusetts, and Minnesota. They are represented by the codes 12, 25, and 27, respectively. The `_STATE` variable is an example of where we could apply user-defined SAS formats to interpret the codes of 12, 25, and 27 into the actual state names.

Adding user-defined formats to SAS datasets takes place in two steps, as follows:

1. **Creating user-defined formats in PROC FORMAT**: First, using PROC FORMAT, the user needs to create the mappings of what the codes mean to the text strings, explaining the levels of the variable.

2. **Applying these formats to variables in a PROC or data step**: In this second step, a data step or PROC is used to apply each set of user-defined formats to the appropriate variables in the dataset. As an example, numeric values with formats can then be displayed with label (character) values.

Let's do the first step by making a format for _STATE for the Chap3_1_Labels dataset. We will name the user-defined format we are making as state_f, as illustrated in the following code snippet:

```
PROC FORMAT;
    value state_f
            12 = "Florida"
            25 = "Massachusetts"
            27 = "Minnesota"

    ;
RUN;
```

Let's review the features of this code, as follows:

- **PROC FORMAT statement**: Notice that the PROC FORMAT statement is alone on a line, followed by a semicolon. As with the label statements, there is no semicolon at the end of each line. Since we are not attaching formats to a dataset in this step, there are no references to datasets in the code.

- **Naming the user-defined format**: The value state_f line of code tells SAS that we are going to define the values in the state_f format in the following lines. This is where we tell SAS that we want this format to be named state_f.

- **Decoding the values**: Notice that the value state_f line of code does not have a semicolon, and neither do the lines of code below this that link the codes with the text strings, using an equals sign. Each numeric code is stated alone on a line with an equals sign, followed by the text label for that level in quotes. There is one semicolon at the end that ends the code statements.

When the code is run, there is no output. However, the log file will report the following:

```
NOTE: Format STATE_F has been output.
```

> **Character limits in PROC FORMAT:**
>
> The name of a format for a character variable can be up to 31 characters in length, and for a numeric field can be up to 32 characters in length.
>
> The text label for each level of the categorical variable can be up to 32,767 characters, but what is seen on the output by SAS can be limited to the first 8 or 16 characters.

Let's look up the other categorical variable, SEX1, in the codebook and make a format for this variable. In the codebook, we find that for SEX1, 1 = **Male**, 2 = **Female**, 7 = **Don't Know**, and 9 = **Refused**. In our data warehouse, we will want to group together values 7 and 9 under the label *unknown*. Therefore, we will make a sex_f format that does this.

We will add the code to make the format sex_f onto the PROC FORMAT code we started so that we can see how to create one long PROC FORMAT code that creates all of our formats, as follows:

```
PROC FORMAT;
      value state_f
              12 = "Florida"
              25 = "Massachusetts"
              27 = "Minnesota"
        ;
      value sex_f
              1 = "Male"
              2 = "Female"
              7 = "Unknown"
              9 = "Unknown"
        ;
RUN;
```

Let's examine this code, as follows:

- Notice how each format statement starts with value followed by the user-defined format name, then lists the mapping for each code to the text label. The semicolon comes at the very end of the list of all the value mappings for each format (not at the end of each line, as with most SAS code).

- Also, notice how we made a slight modification from the codebook when creating the SEX1 format. Since in our data warehouse we want to group together values 7 and 9 under *unknown*, we created the same text value of Unknown for the label for values 7 and 9. We can see later how this strategy impacts the way SAS formats output when analyzing data with formats applied.

Format lists tend to fall into one of two categories: those that are used on multiple variables throughout the data warehouse, and those that are only used on one or a few variables in the warehouse. In the case of a survey, there are often several questions with the answers of 1 = **Yes**, 2 = **No**, and 9 = **Don't Know**. In this case, one format could be created named YNDK_f and applied to several variables of the same format. But the _STATE variable provides a perfect example of a format that would only be applied to variables storing state FIPS codes. Because other state-coding systems exist, care would have to be taken to apply the formats for the correct coding system to the correct state variables.

Naming conventions for formats:

Most SAS formats in a data warehouse are only applied to a few different variables throughout the warehouse. For example, the user-defined format we made called state_f would be used on all of the variables for state stored as FIPS codes in the warehouse, which would likely be few.

In those cases, it is tempting to give the format the same name as the variable from the dataset. In our example, it would be tempting to give the format for _STATE the same name: _STATE. However, this is problematic for a number of reasons. First, as described earlier, this format is for state FIPS codes, and the name of the format does not help with this distinction. Secondly, another variable for state FIPS code may exist in the warehouse under a different name. Or, worse, there could be another variable named _STATE from a different dataset in the warehouse where _STATE is not coded with an FIPS code.

Therefore, it is helpful to always name SAS formats with the _f suffix. This tells the user that we are talking about a SAS format, not a variable. Other rules can be added to naming conventions for formats, such as including the system in the name. For example, we could have state_FIPS_f as the name of a format for state that decodes the FIPS code.

Now that we have created the state_f and sex_f formats, our next step is to apply these formats to our Chap3_1_Labels dataset. Although formats can be added in PROCs to enhance output, we will attach these formats to the variables in a data step, as follows:

```
data X.Chap3_1_Formats;
      set X.Chap3_1_Labels;
      format      _STATE state_f.
                  SEX1 sex_f.

      ;
RUN;
```

Let's review the code, as follows:

- The code creates a new `Chap3_1_Formats` dataset in the directory mapped to `LIBNAME X` and reads in the dataset we just created with labels, called `Chap3_1_Labels`, in the `set` statement.

- To connect the formats to the variables, the `format` command is used. Notice that there is only one semicolon after the `format` command at the very end, and on each line, the variable is stated, followed by the name of the format, with a period at the end (such as `SEX1`, followed by `sex_f`).

Again, this code is a data step that essentially copies the dataset and, in doing so, adds the formats. For this reason, it produces no output, but we can verify by looking in the log file that the code ran.

In order to verify that the formats were attached, we could print the top five observations using `PROC PRINT`, as follows:

```
PROC PRINT data=X.Chap3_1_Formats (obs=5);
RUN;
```

Because we set `obs=5` in our `PROC PRINT` statement, we will be able to see the values in the first five rows of `Chap3_1_Formats`. Because the native dataset was sorted by `_STATE`, the records for Florida (code `12`) are at the top and are the ones that show up on the output from `PROC PRINT`, as illustrated in the following screenshot:

Obs	_STATE	SEX1	_AGE80	_BMI5
1	Florida	Female	76	31.8
2	Florida	Male	72	35.3
3	Florida	Female	46	26.5
4	Florida	Female	51	33.3
5	Florida	Male	53	41.6

Figure 3.6 – Output from PROC PRINT with formats applied to the _STATE and SEX1 variables

As can be seen from the output, now, the values in `_STATE` and `SEX1` no longer appear as codes but instead appear as their labeled formats. The actual data values did not change. Also, if we run `PROC CONTENTS` as follows, we will see in the contents output that the formats have been applied:

```
PROC CONTENTS data=X.Chap3_1_Formats;
RUN;
```

Now, we can look at the third table of the PROC CONTENTS output and see that the formats have been applied, as follows:

Alphabetic List of Variables and Attributes					
#	Variable	Type	Len	Format	Informat	Label
2	SEX1	Num	8	SEX_F.	2.	Sex at Birth label
3	_AGE80	Num	8	2.	2.	Imputed age value collapsed above 80 label
4	_BMI5	Num	8	4.1	2.2	Body Mass Index (BMI)
1	_STATE	Num	8	STATE_F.	2.	State FIPS Code label

Figure 3.7 – PROC CONTENTS output showing that the sex_f and state_f formats have been applied

As can be seen in the output, under the Format column, for SEX1 and _STATE, the applied SEX_F and STATE_F formats are now being reported.

> **Different kinds of native SAS formats:**
>
> A link to a SAS support page that lists all the native SAS formats is available under the *Further reading* section. SAS native formats fall into the following four categories:
>
> **Character**: Character formats attach characters' data values to display to variables in character format.
>
> **Date and Time**: Date and time formats are applied to numeric variables that can be interpreted to be date or time variables so that they display according to a particular date or time format.
>
> **ISO 8601**: These formats instruct SAS to display date, time, and datetime values using the **International Organization for Standardization (ISO)** 8601 standard. This format is generally used in the pharmaceutical industry.
>
> **Numeric**: These formats instruct SAS to display numeric variables as numeric data in a particular format.
>
> SAS native formats have generic names that involve inserting a variable value—for example, the BESTw. format is the format to which SAS assigns numeric variables when using PROC IMPORT to read in a non-SAS dataset. As described in *Chapter 2, Reading Big Data into SAS*, PROC IMPORT assigns the BEST12. format as the default format of the output of numeric variables into the imported dataset. In that case, the w in BESTw. is filled in with 12.

Up to now in this chapter, we have discussed using PROC FORMAT to create user-defined formats for categorical variables. SAS also has a set of native formats that can be applied to variables to control their display (such as to make a numeric date display in a way that spells out the name of the month). We used very basic versions of these native formats earlier when reading in datasets in *Chapter 2, Reading Big Data into SAS*.

The other variables in the Chap3_1_Formats dataset, _AGE80 and _BMI5, are numeric and continuous. As can be seen from the PROC CONTENTS output in *Figure 3.7*, the more basic format of 2. is applied to _AGE80, and the format 4.1 is applied to _BMI5. There are more sophisticated native formats that also can be applied to numeric variables. The next section will talk about using SAS formats with numeric variables.

Using native SAS formats with numeric variables

SAS has a lot of native formats that can be applied to numeric variables. This section will give a few examples, but they are by no means exhaustive. SAS data warehouse developers will need to decide on the importance of applying formats to SAS data in their warehouses, because formats mainly impact the display of data. They do not impact the underlying data values. Therefore, if data is not displayed using SAS components, then applying formats may have less value in the warehouse.

A common situation where SAS data warehouse developers need to apply formats has to do with variables reporting currency. Imagine a COST field that holds values of the cost of several products in dollars and cents but is stored as a decimal number, and therefore does not automatically display as dollars and cents. It may be difficult for the user to interpret that variable as having to do with currency.

Let's do an exercise where we create a COST field in the Format_Money dataset, and then apply a format so that it displays as currency. We will start by creating the Format_Money dataset with one column called COST and two values that should be formatted as currency, as illustrated in the following code snippet:

```
data Format_Money;
    INFILE CARDS;
    INPUT COST;
    CARDS;
123.45
678.90
;
RUN;
```

The `Format_Money` dataset is in the `WORK` directory and has one column, `COST`, with two rows. Each row has one numeric value: `123.45` and `678.90`. The resulting dataset looks like this:

COST
123.45
678.9

Table 3.2 – The format_Money dataset before applying formats

Now, we will use a data step to add the `DOLLARw.d` generic native SAS format, which is explained here:

- The word `DOLLAR` in the name of the format indicates that it formats numeric variables as dollar currency.

- The `w.d` part of the name indicates that the user will need to determine the value of `w` (or the entire width of the resulting field display) and `d`, which is the number of units to display to the right of the decimal.

Looking at the variables, we already know that we will want `d` to be 2, as we will want two places to the right of the decimal to be displayed. But in order to calculate `w`, we will need to add up all the positions that will be in the display, as follows:

1. The first position will be a dollar sign.

2. The second, third, and fourth positions will be the dollar values.

3. The fifth position will be the decimal point.

4. The sixth and seventh positions will be two places after the decimal.

This means we will want the entire field to be seven-character positions; so, for us, `w` will equal seven. Therefore, the format we will want to use is `DOLLAR7.2`.

We will now use a data step to apply the `DOLLAR7.2` format to the `COST` variable. We will output the same dataset we read in, which is the dataset we just made, named `Format_Money`, as illustrated in the following code snippet:

```
data Format_Money;
    set Format_Money;
    format COST DOLLAR7.2;
RUN;
```

After we run this code, we see that the COST field in Format_Money will now display with the dollar sign and both places to the right of the decimal, as illustrated here:

COST
$123.45
$678.90

Table 3.3 – The format_Money dataset after applying the DOLLAR7.2 format

Another common problem can be seen with large integers, as the default display of these integers does not have a comma separator every three digits. In SAS, this display problem is often solved by using the COMMAw.d format.

We will look at an example. First, we will use code to create a dataset in WORK called Format_Comma that has two rows and one column called BIGNUMBER, as follows:

```
data Format_Comma;
    INFILE CARDS;
    INPUT BIGNUMBER;
    CARDS;
123123
456456
;
RUN;
```

The two values we inputted are 123123 and 456456. These are hard to read without a comma every three digits, as illustrated here:

BIGNUMBER
123123
456456

Table 3.4 – The format_Comma dataset before applying a format

As with the DOLLARw.d currency format, with COMMAw.d we will have to fill in d, which is the number of places to the right of the decimal, and w, which is the total width of the display field. We already know d will be 0 because we don't have any decimal places in the native dataset. But this means that w will have to be the same width as the native field, plus one position to make space for the comma. In this case, since each of the numbers is six digits, adding one space for the comma equals seven. Hence, we will use the format COMMA7.0.

Let's do data step code to apply this format to the BIGNUMBER variable in the Format_Comma dataset, as follows:

```
data Format_Comma;
    set Format_Comma;
    format BIGNUMBER COMMA7.0;
RUN;
```

After running the code, when looking at the Format_Comma dataset, we now see the comma included, as can be seen here:

BIGNUMBER
123,123
456,456

Table 3.5 – The format_Comma dataset after applying the COMMA7.0 format

> **Note:**
>
> In a SAS data warehouse, decisions need to be made as to whether reporting functions will be run through SAS or whether they will be done in another application.
>
> **If reporting is to be done using SAS components**, there are many benefits to applying both formats and labels, as they can be used quite adeptly in the reporting process to enhance the display and quality of reports.
>
> **But if reporting is to be done in another application**, SAS labels and formats are generally useless as they do not translate to other applications. In those cases, other approaches should be taken to enhance the display of data on reports.

Applying user-defined formats to continuous variables

This section will provide an example of applying a user-defined format to a continuous variable in our example dataset we have been editing, Chap3_1_Formats. The _BMI5 variable refers to **Body Mass Index (BMI)**, which is a number used for determining a person's weight status.

According to the US **Centers for Disease Control and Prevention (CDC)**, this integer can be classified into the following four categories:

BMI Value	Classification
<18.5	Underweight
18.5 to <25.0	Normal
25.0 to <30.0	Overweight
30.0 or higher	Obese

Table 3.6 – BMI classifications

Even though BMI is stored as a continuous variable, we can use a format to have SAS collapse the various BMI values into classifications that can be used in output. First, we will start by using PROC FORMAT to make a format we will call BMI_f, as follows:

```
PROC FORMAT;
    value BMI_f
            LOW - <18.5 = "Underweight"
            18.5 - <25 = "Normal"
            25 - <30 = "Overweight"
            30 - HIGH = "Obese"
            . = "Unknown"
        ;
RUN;
```

Notice some features of how we created the BMI_f format with our code, as follows:

- **LOW and HIGH**: The LOW and HIGH statements were used to indicate the minimum and maximum. LOW and HIGH are reserved keywords in SAS. This way, all the data was classified without needing to set hard boundaries in the code.

- **Use of – and <:** The dash was used to indicate a range of numbers, and < was used on one of the boundaries to ensure that all values were classified in the range of numbers (such as in 18.5 - <25).

- **Classifying missing values**: The period in SAS indicates a missing value in a numeric field. Therefore, making a format for a period as done in the code using . = "Unknown" will cause blank records to be classified as Unknown.

Now, we will apply this format to our most recent dataset we edited in the LIBNAME mapped to X, which is Chap3_1_Formats, as follows:

```
data X.Chap3_1_Formats;
    set X.Chap3_1_Formats;
    format      _BMI5 BMI_f.

    ;
RUN;
```

To see how our data looks now, we can rerun PROC PRINT on the dataset as we did before, with an option (obs = 5) limiting our output to the top five records. With all the formats we have applied including BMI_f, the top five records now look like this:

Obs	_STATE	SEX1	_AGE80	_BMI5
1	Florida	Female	76	Obese
2	Florida	Male	72	Obese
3	Florida	Female	46	Overweight
4	Florida	Female	51	Obese
5	Florida	Male	53	Obese

Figure 3.8 – PROC CONTENTS output showing all formats applied, including BMI_f

On one hand, it is helpful to see these codes decoded on this output. It is immediately obvious that the first five records are from Florida, and are for three women and two men. It is also obvious that in the first five records, four of the individuals are obese and one is overweight.

But what if we wanted to know the exact numbers underlying the formats? This would be especially important in the _BMI5 column, as we may be interested in viewing the actual BMI value, not the classification. An analyst may wonder if they should actually use SAS to create a new variable holding the BMI classification instead, and not use the format function to simply achieve a display that does not actually aggregate the underlying data values.

This is why the decision to use SAS labels and formats for attaching information to data in a SAS data warehouse needs to be undertaken with care. Where you will see SAS labels and formats displayed in data warehouse output is when a SAS component that can make use of the SAS labels and formats is involved in generating the output. The next section will provide an example of when labels and formats in displays can be helpful in SAS processing.

Using labels and formats in processing

Up to now, we have applied labels and formats, but we have not seen how these labels and formats are handled in PROCs (with the exception of PROC CONTENTS, where the attached labels and formats are printed on the output). In this section, we use PROC FREQ to demonstrate how labels and formats can improve the display of data. First, we observe how labels and formats can provide clarification on PROC FREQ output. Next, we consider what would be necessary to maintain updated labels and formats on all variables in a SAS data warehouse. Practical challenges to maintaining labels and formats are discussed.

Unlike files that are readable in many programs, such as tables stored in *.csv format, SAS format and label files are only usable in the SAS application. Therefore, if the SAS data warehouse intends to use another application for data visualization—such as R or Tableau—the SAS format and label files can't be used. The information contained within them—variable labels, and definitions of levels of categorical variables—will need to be extracted and reprogrammed into the visualization application. In that case, alternatives to SAS labels and formats can be used, and these are described in the *Alternatives to using labels and formats in a warehouse setting* section.

Using PROCs with labels and formats

Some PROCs in SAS, such as PROC TABULATE, are designed specifically for reporting functions. Other PROCs, such as PROC LOGISTIC, are designed to do a statistical maneuver, such as creating a logistic regression model. In SAS, labels and formats are heavily used with PROCs that are aimed at doing reporting, such as PROC TABULATE. However, analysts will notice that labels and formats, when applied to the underlying SAS datasets, will also display on output from statistical functions such as PROC LOGISTIC.

One of the most basic statistical functions in SAS is running one- and two-way frequencies, which is done with PROC FREQ. Let's run a two-way frequency on our Chap3_1_Formats dataset in LIBNAME X, which has all the formats we just applied, as follows:

```
PROC FREQ data=X.chap3_1_formats;
    Tables SEX1 * _BMI5;
RUN;
```

Notice the syntax of PROC FREQ, described here:

1. **First line**: The line starts by calling PROC FREQ and specifying the dataset. Options can also be added to this line.

2. **Second line**: The Tables command is used to request a two-way frequency between SEX1 and _BMI5. Had only one variable been listed, we would have been requesting a one-way frequency. Options can be added to the Tables command as well.

The output is somewhat complicated, but for now, let's focus on where our labels and formats appear on the output, rather than on the actual numbers reported. The output is shown here:

The FREQ Procedure

Frequency Percent Row Pct Col Pct	Table of SEX1 by _BMI5				
	_BMI5(Body Mass Index (BMI))				
SEX1(Sex at Birth label)	Underweight	Normal	Overweight	Obese	Total
Male	205	4340	7262	5440	17247
	0.58	12.22	20.44	15.31	48.55
	1.19	25.16	42.11	31.54	
	31.98	38.65	55.92	50.98	
Female	434	6879	5706	5215	18234
	1.22	19.36	16.06	14.68	51.32
	2.38	37.73	31.29	28.60	
	67.71	61.26	43.94	48.88	
Unknown	2	11	18	15	46
	0.01	0.03	0.05	0.04	0.13
	4.35	23.91	39.13	32.61	
	0.31	0.10	0.14	0.14	
Total	641	11230	12986	10670	35527
	1.80	31.61	36.55	30.03	100.00
Frequency Missing = 3374					

Figure 3.9 – PROC FREQ output showing the use of labels and
formats for the SEX1 and _BMI5 variables

The two-way frequency output is a table that has the BMI categories across the top as column headings, and the sex categories as row headings. The labels and formats we applied play the following roles in this output:

- **Defining variables**: Notice that the column heading actually says `_BMI5(Body Mass Index (BMI))` and the row heading says `SEX1(Sex at Birth label)`.

- **Grouping classifications**: Earlier, we defined the `sex_f` format to code both seven and nine as *unknown*. In the output, we see these two classifications are grouped together under **Unknown**.

- **Grouping a continuous variable**: In the dataset, `_BMI5` is still stored as a continuous variable. But now that the `BMI_f` format has been applied, it is handled as a categorical variable in this `PROC FREQ` code, which groups the ranges together and labels them according to the text strings we supplied in the format.

Formats do not need to be applied to the underlying dataset in order to be used in PROCs on the fly, so long as the formats are already loaded. By now, in editing `Chap3_1_Formats`, we have loaded all the formats. But our original SAS dataset, `Chap3_1`, does not have these formats applied. Also, `Chap3_1` does not have any labels applied. If we run the following code using the `Chap3_1` dataset and applying the loaded formats while also specifying labels on the fly, we will get almost the same output as when we did `PROC FREQ` with the labels and formats applied to the underlying dataset:

```
PROC FREQ data=X.chap3_1;
    Tables SEX1 * _BMI5;
    format       SEX1 sex_f.
                 _BMI5 BMI_f.
    ;
    label        SEX1 = "Sex at Birth"
                 _BMI5 = "Body Mass Index (BMI)"
    ;
RUN;
```

Let's look at this code, as follows:

- Notice that the first two lines of code are identical to the code we used before with `Chap3_1_Formats` except for specifying the dataset without labels and formats, which is `Chap3_1`.

- The next two lines apply the `sex_f` and `BMI_f` formats to the `SEX1` and `_BMI5` variables, respectively, and are followed by a semicolon.

- The next two lines after that specify the label text for SEX1 and _BMI5, followed by a semicolon.

This demonstration using PROC FREQ shows how labels and formats applied to underlying data can help make output easier to understand and read. It also shows that there is a choice when applying labels and formats to data, described as follows:

1. One option is that they can be applied to the underlying dataset stored in a data warehouse or data lake.

2. Another option is to apply the formats and labels on the fly when running a PROC, and not attach them to the underlying dataset.

3. A third option is to not use labels and formats at all, and instead deal with metadata differently.

We will consider these different choices in the next section, where we cover maintaining labels and formats in a SAS data warehouse.

Maintaining labels and formats

If your data warehouse, data mart, or data lake chooses to maintain labels and formats for your SAS datasets, you will need to develop a strategy that maintains updated labels and format files for datasets as they are processed into the data environment. The strategy will likely be different for labels as compared to formats due to differences in the code syntax used to establish and apply them.

A typical scenario in a data warehouse is to receive monthly files from a data source that need to undergo ETL processing into the warehouse format. It is during this ETL processing that labels and formats will be applied. This requires taking into consideration the processing order for the variables.

Let's take the example of the BRFSS data, which came with a codebook specifying labels. As we read in the four native variables in Chap3_1, we were able to apply the labels we read from the codebook. But let's say we are processing multiple years of BRFSS data, so we want to add a variable called FileYr to indicate that the year of the file is 2018. We will do that to our already processed dataset with labels and formats, named Chap3_1_Formats, in the following code, and run PROC CONTENTS afterward to see how our new variable looks in the output:

```
data X.Chap3_1_Formats;
    set X.Chap3_1_Formats;
    FileYr = 2018;
```

```
RUN;
PROC CONTENTS data=X.Chap3_1_Formats;
RUN;
```

Notice that to create the `FileYr` variable, we simply added a line to our data step code and set `FileYr` to equal 2018. But now, on our `PROC CONTENTS` output, `FileYr` looks very different than the other variables, as can be seen in the following screenshot:

Alphabetic List of Variables and Attributes						
#	Variable	Type	Len	Format	Informat	Label
5	FileYr	Num	8			
2	SEX1	Num	8	SEX_F.	2.	Sex at Birth label
3	_AGE80	Num	8	2.	2.	Imputed age value collapsed above 80 label
4	_BMI5	Num	8	BMI_F.	2.2	Body Mass Index (BMI)
1	_STATE	Num	8	STATE_F.	2.	State FIPS Code label

Figure 3.10 – PROC CONTENTS output showing that the new FileYr variable has no label

As we can see, `FileYr` has no label. This has a few implications, detailed as follows:

- First, it is now unclear what the added `FileYr` variable actually means. The data warehouse could make it a policy that in code that creates variables, labels must be created along with the variables and attached before the end of the code. This would ensure that metadata is maintained on all datasets as variables are added during data processing.

- But what also has to be considered is that multiple files of the same format (such as monthly files) would need to have the same labels attached to variables if they were ever to be appended together. In other words, a policy must be made to ensure that the correct labels are affixed to the correct variables at the correct stage in processing.

One way this can be facilitated is with **no-run code filled with label statements**. This code cannot be run. Instead, it is seen as a *cheat sheet* to use when creating labels. If this label code is placed where all data processing staff can access it, they can use it to apply labels and standardize their code. Here is an example of no-run label code from our demonstration:

```
label FileYr = "File Year"
label SEX1 = "Sex at Birth"
label _AGE80 = "Imputed age value collapsed above 80"
label _BMI5 = "Body Mass Index (BMI)"
label _STATE = "State FIPS Code"
```

Notice how the code is sorted in order of variable name so that a programmer can easily locate the label statement desired. For example, a programmer adding `FileYr` to the dataset could then copy the label statement for `FileYr` from this no-run code and paste it into the data step in their SAS code after creating the variable.

Maintaining user-defined SAS formats, however, requires a totally different process than maintaining SAS labels, as explained here:

- First, unlike with labels, standalone code such as the code we used for creating formats can exist independently.

- Then, when doing ETL processing, the first step in processing could be to run all the format files to add all the formats to SAS memory.

To explain why this is necessary, we will do a demonstration. Let's go back to our edited dataset, `Chap3_1_Formats`, and try to apply an unloaded format called `unloaded_format_f` to the `_STATE` variable, as follows:

```
data X.Chap3_1_Formats;
    set X.Chap3_1_Formats;
    format        _STATE unloaded_format_f.
    ;
RUN;
```

The code runs and executes the data step, but registers the following error in the log:

```
NOTE 484-185: Format UNLOADED_FORMAT_F was not found or could
not be loaded.
```

Looking at the data, we can see that the `_STATE` variable still appears as a FIPS code. Imagine an error such as this happened during a long ETL processing routine. It may not be identified during processing because the data step executed successfully. This is why it is important to first load all the formats you are planning to apply before beginning ETL processing.

While it easier to maintain separate SAS format files than label files, the issue of policy still exists. Data warehouse leaders will need to designate who maintains the official format files, implement a process for identifying which formats to use on which variables, and maintain oversight of ETL such that formats are applied properly during processing. It will be necessary to keep track of the data documentation from the data provider that provides the source of information for labels and formats as well.

> **Formats and catalogs:**
>
> When formats are saved in SAS, they are stored in a catalog, which is a term that means a collection of formats. The default catalog is in the WORK directory and is called WORK.FORMATS. PROC CATALOG can be used to list the contents of the format catalog. This catalog can be searched, and metadata about the formats can be accessed. See the *Further reading* section for two white papers that provide advanced guidance to format management.

Alternatives to using labels and formats in a warehouse setting

Traditionally, SAS shops have carefully maintained label and format files not only to improve data display but also to document the data. However, as the internet grew and big datasets became more common, it became clear that the amount of documentation allowed in labels and formats was not enough. The maintenance of **data dictionaries**, codebooks, original data collection instruments, and other curation files became necessary to keep all warehouse staff and users on the same page.

As it turns out, this type of documentation is simple to keep in commonly used programs such as Microsoft Excel and Microsoft Word. An excellent example is the US **Military Health System Data Repository** (**MDR**) data dictionary, which is in Excel (a link to the web page is available under the *Further reading* section).

Keeping data curation documentation outside of SAS not only provides the opportunity for extra documentation beyond what can be contained in formats and labels but also allows non-SAS users to benefit from the documentation. If the only place that variable names are decoded is in a no-run SAS label file, non-SAS users will not be able to access it.

As described before, if reporting is not being done in SAS, then SAS labels and formats cannot be used to improve data display because they are not useful to other applications. Therefore, deciding whether or not to maintain labels and formats in SAS in today's data warehouse environment is a big decision. Since other data curation files will inevitably be needed, a data leader will need to weigh the pros and cons of keeping up label and format files to improve data display, or use some other approach that relies on existing data curation files (such as data dictionaries already in Excel). The considerable amount of extra effort that is required for maintaining policies around label and format files can be worth it if SAS is being used for reporting, but this may be wasted effort if other reporting approaches are used or if existing data curation files can be leveraged to achieve the same outcomes as formats and labels would.

Viewing data in SAS

When preparing ETL code, it is important to be able to view raw data. However, this can be tricky because typically, a programmer wants to view only certain parts of the dataset, such as certain columns, or rows with certain values. This section first describes what SAS users have traditionally done to view data, which is to use PROC PRINT. Options that can be set on PROC PRINT to control the data being viewed are discussed. Next, we consider using a newer procedure for managing data, PROC SQL. PROC SQL performs the same tasks as PROC PRINT but uses **Structured Query Language (SQL)** syntax, so may be easier for SQL programmers to learn. Because we are using criteria to filter data, and these require us to use **arithmetic operators**, this section also covers how to use arithmetic operators in SAS. Finally, this section covers different ways of viewing SAS datasets through windows in SAS.

Using PROC PRINT to view data

In this book, we have already used PROC PRINT to view the top few rows of data in our datasets. PROC PRINT has traditionally been used in SAS to view data, although now, there are other alternatives, such as PROC SQL. We have already used PROC PRINT with the obs option to limit the data displayed to the top five observations. Let's explore other ways PROC PRINT can be used to view data.

Imagine that when we edited the data to make the Chap3_1_Formats dataset, we were concerned that the males in the dataset (SEX1 = 1) were not having their BMI formats applied properly. What we could do is first sort the dataset by SEX1 using PROC SORT, and then look at the first 10 observations using the obs = 10 option. Next, we will do PROC PRINT, but this time, we will include both the BY and VAR options, as illustrated in the following code snippet:

```
PROC SORT data=X.chap3_1_formats;
    BY SEX1;
PROC PRINT data=X.chap3_1_formats (obs = 10);
    BY SEX1;
    VAR _BMI5 SEX1;
run;
```

Here are some features to point out about the code:

- **Leading PROC SORT**: Because we are planning to use the BY statement in PROC PRINT to sort the output by SEX1, we will need to run PROC SORT first and sort the SEX1 variable.

- **Use of BY variable**: The same BY variable used in PROC SORT, which is SEX1, is then used again in PROC PRINT to sort the output by SEX1.

- **Use of VAR statement:** In PROC PRINT, the VAR statement lists the variables to output. We only chose to output two variables from our dataset, _BMI5 and SEX1.

Let's view the output, as follows:

Sex at Birth label=Male

Obs	_BMI5	SEX1
1	Obese	Male
2	Obese	Male
3	Normal	Male
4	Obese	Male
5	Overweight	Male
6	Obese	Male
7	Obese	Male
8	Normal	Male
9	Overweight	Male
10	Overweight	Male

Figure 3.11 – PROC PRINT output sorted by SEX1

We should not be surprised to see the output. It has Sex at Birth label = Male as the heading, because this is the first classification in the sorted SEX1 variable that is encountered by SAS as it goes through the sorted dataset to display the top 10 rows in the response to use, specifying obs = 10. The two columns we put in the VAR statement, _BMI5 and SEX1, are the only ones displayed. The data is displayed with formats on.

Note:

As you may remember from *Chapter 1, Using SAS in a Data Mart, Data Lake, or Data Warehouse*, many PROCs rely on data being sorted in a certain way. Often, when a BY command is included in the PROC, the dataset will need to be sorted by the variables included in the BY command prior to running the PROC.

The PROC PRINT output suggests that the male BMI scores are being classified in the format, which was our original scenario, motivating us to run PROC PRINT.

But what if we were interested in knowing whether values of SEX1 at the bottom of the sort were getting classified? Then, we could add a descending option to our initial PROC SORT to have it sort SEX1 from high to low, as illustrated in the following code snippet:

```
PROC SORT data=X.chap3_1_formats;
    BY descending SEX1;
```

Notice how the descending option is added in the BY statement before the variable name. If the same PROC PRINT code is run after this PROC SORT code, the output will be from the bottom of the SEX1 sort, as follows:

Sex at Birth label=Unknown

Obs	_BMI5	SEX1
1	Obese	Unknown
2	Obese	Unknown
3	Unknown	Unknown
4	Unknown	Unknown
5	Unknown	Unknown
6	Overweight	Unknown
7	Unknown	Unknown
8	Normal	Unknown
9	Unknown	Unknown
10	Unknown	Unknown

Figure 3.12 – PROC PRINT output sorted by SEX1 descending

We can see from the output that the last value in the dataset in SEX1 is classified as Unknown because of the format we applied. Using descending is a quick way to look at the observations at the bottom of a dataset.

But what if we wanted to look at values in the middle of the dataset? We could sort by SEX1, and then move our window of 10 observations down further. We can see through earlier log files that the Chap3_1_Formats dataset has a total of 38,901 rows. If we use the additional option of firstobs and set that to 20000, and then set obs to 20010, we will be able to see the 10 records in the rows 20,000 to 20,010. The code can be seen in the following snippet:

```
PROC SORT data=X.chap3_1_formats;
    BY SEX1;
PROC PRINT data=X.chap3_1_formats (firstobs = 20000 obs =
20010);
    BY SEX1;
    VAR _BMI5 SEX1;
RUN;
```

Notice in the code where the firstobs and obs options are set to 20000 and 20010, respectively. The code produces this output:

Sex at Birth label=Female

Obs	_BMI5	SEX1
20000	Normal	Female
20001	Unknown	Female
20002	Obese	Female
20003	Normal	Female
20004	Overweight	Female
20005	Normal	Female
20006	Normal	Female
20007	Obese	Female
20008	Overweight	Female
20009	Normal	Female
20010	Obese	Female

Figure 3.13 – PROC PRINT output from observations 20000 through 20010

Traditionally, SAS programmers used PROC PRINT with options to view data as part of troubleshooting ETL. However, when SQL became popular in the early 2000s, SAS's PROC SQL also became more popular to use, especially for viewing data.

Using PROC SQL to view data

SQL is a style of data management language that comes in different **flavors**, depending upon who designed the version of SQL. Perhaps the most popular flavors of SQL are **Oracle** and **Microsoft**, but open source **MySQL** and other SQLs exist and are in active use.

While these flavors of SQL are all slightly different, they are similar in that they use a small set of basic commands to query and update datasets. There is a similar syntax between all flavors of SQL. For this reason, as more SQL programmers were growing their careers, they came to expect SAS to behave like SQL. In addition, data warehouses needed experienced programmers, and as SQL was so popular for production databases, many SQL programmers were finding themselves with jobs that included SAS, and encountered data steps in SAS environments. Those individuals tend to prefer PROC SQL because the syntax is similar to other SQLs, and the commands are very basic and aimed specifically at data management (rather than analysis). However, in reality, the output that is created from PROC SQL could be created from using other SAS PROCs or data steps.

To demonstrate this, we will reproduce the same PROC PRINT output using PROC SQL commands. Let's start by using PROC SQL to display the _BMI5 and SEX1 variables for the first 10 observations of the Chap3_1_Formats dataset, as follows:

```
PROC SORT data=X.chap3_1_formats;
    BY SEX1;
PROC SQL;
    select _BMI5, SEX1
    from X.Chap3_1_Formats (Obs = 10);
quit;
```

Let's review this code, as follows:

1. First, the code sorts the data by SEX1, which is similar to what we did when we used PROC PRINT.

2. Next, we start using PROC SQL code. All of the PROC SQL code will go between the PROC SQL command and the ending quit command.

3. The syntax for SQL starts with a select statement, which specifies the variables we want to select to view. As before, we specify _BMI5 and SEX1.

4. Next, the syntax for SQL expects us to declare in a from statement which dataset to use. It is in the from statement that we specify the Chap3_1_Formats dataset in the X library, and we add the option (obs=10) so that it will print the first 10 observations only.

Body Mass Index (BMI)	Sex at Birth label
Obese	Male
Obese	Male
Normal	Male
Obese	Male
Overweight	Male
Obese	Male
Obese	Male
Normal	Male
Overweight	Male
Overweight	Male

Figure 3.14 – PROC SQL output for the same query run before in PROC PRINT

As can be seen with the output, there are some minor differences between the PROC SQL and PROC PRINT output. The heading on the PROC PRINT output is not on the PROC SQL output, and the observation numbers are not listed on the PROC SQL output. However, the data values retrieved are the same.

> **Error with the order by clause in PROC SQL:**
>
> Programmers familiar with SQL will be aware that typically, an order by clause is used in SQL programming to do sorting, and it therefore seems that we should not have to use PROC SORT. In SQL, the order by clause usually comes after the from statement and can sort the dataset by variables in either ascending or descending order.
>
> SAS has built the order by clause into PROC SQL, but unfortunately, it has an error. This error makes it so PROC SQL may ignore an order by clause. For this reason, it is advisable to use PROC SORT to order data when using PROC SQL. More information from SAS is available about the error on SAS's support page (see link under the *Further reading* section).

Previously, we reran our PROC PRINT with the data sorted by SEX1 in descending order to look at the opposite end of the dataset. Here, we can do the same thing with PROC SQL, which is to sort the dataset in descending order with PROC SORT, then rerun the preceding PROC SQL code. If we do that, we will get results in the same format as we did in PROC PRINT earlier, as illustrated here:

Body Mass Index (BMI)	Sex at Birth label
Overweight	Unknown
Unknown	Unknown
Unknown	Unknown
Unknown	Unknown
Unknown	Unknown
Unknown	Unknown
Obese	Unknown
Obese	Unknown
Unknown	Unknown
Normal	Unknown

Figure 3.15 – PROC SQL output for the same query run before in PROC PRINT
sorted by SEX1 descending

This time, the data printing out does not look exactly the same as the data printing out from PROC PRINT. We could troubleshoot why this is if we had the observation number. In PROC PRINT, the observations are printed by default on the output, but that does not happen automatically in PROC SQL.

A simple way to see the observation number in PROC SQL is to simply add it as a variable in the dataset, as follows:

```
data X.Chap3_1_Formats;
    set X.Chap3_1_Formats;
    rownumber = _N_;
RUN;
```

In this data step, we added the rownumber variable to the Chap3_1_Formats dataset. We set rownumber = _N_, as _N_ is a code in SAS for observation number. This data step code copies the observation number from SAS's memory into the dataset as an actual variable named rownumber.

Now, when we rerun our code, we can include the rownumber variable for troubleshooting why we are getting different query results, as illustrated in the following code snippet:

```
PROC SORT data=X.chap3_1_formats;
    BY descending SEX1;
PROC SQL;
```

```
    select rownumber, _BMI5, SEX1
    from X.Chap3_1_Formats (Obs = 10)
    order by SEX1;
quit;
```

Now, we can compare the observation numbers between the PROC SQL output and the earlier PROC PRINT output, as follows:

rownumber	Body Mass Index (BMI)	Sex at Birth label
6	Overweight	Unknown
10	Unknown	Unknown
3	Unknown	Unknown
7	Unknown	Unknown
9	Unknown	Unknown
5	Unknown	Unknown
1	Obese	Unknown
2	Obese	Unknown
4	Unknown	Unknown
8	Normal	Unknown

Figure 3.16 – PROC SQL output for the same query run before in PROC PRINT
sorted by SEX1 descending with row numbers for troubleshooting

Our experiment reveals that the same records were being pulled by PROC SQL as were being pulled by PROC PRINT, but in the output, they are not sorted in the same order. This situation demonstrates the utility of adding a rownumber variable while preparing ETL code in PROC SQL.

In PROC PRINT, we did a query that showed us observations 20000 through 20010. It is possible to use firstobs and obs to restrict which observations we are being shown. We can use these options in PROC SQL by adding them to the from clause, as follows:

```
PROC SORT data=X.chap3_1_formats;
    BY SEX1;
PROC SQL;
    select rownumber, _BMI5, SEX1
    from X.Chap3_1_Formats (firstobs = 20000 Obs = 20010);
quit;
```

However, when we run the code, we do not get the records we expect, as can be seen here:

rownumber	Body Mass Index (BMI)	Sex at Birth label
1909	Normal	Female
1910	Unknown	Female
1911	Obese	Female
1912	Normal	Female
1913	Overweight	Female
1914	Normal	Female
1915	Normal	Female
1916	Obese	Female
1917	Overweight	Female
1918	Normal	Female
1919	Obese	Female

Figure 3.17 – PROC SQL output using firstobs= and obs=

SAS is fundamentally different from SQL in a few ways, as explained here:

- One of the important ways that SAS is different from SQL is that in SAS, there are many variables, system files, and other items that SAS uses *behind the scenes* (such as observation number) that are not obvious, and the programmer must apprise themselves of these resources in order to access and use them in coding.

- In contrast, SQL's functions generally rely on inputs directly from the underlying data values, so the same type of information tends to be stored more concretely in the SQL dataset.

Part of the reason for this difference is that SAS's history starts at a time when keeping this processing work behind the scenes was a way of creating efficiency. Now, some of these features may feel more like bugs to the modern programmer.

So, when working with PROC SQL, it can be easier to program if variables are created and queried. For example, if using PROC SQL, it may be advantageous to not use SAS formats, and instead create another variable with the format label in it. In our scenario, SEX1 could still be stored as a code, but another variable would exist, called something such as SEX1_DESC, that would include the format descriptor as a character variable. Of course, this situation could force a PROC SQL programmer to create a lot of extra variables in a big dataset that would not be necessary if not using PROC SQL. When using PROC SQL, these tradeoffs have to be considered.

Since we already created the `rownumber` variable, in order for us to query the records for observations `20000` through `20010`, it is best if we actually set our query criteria on the `rownumber` variable using a `where` clause, as follows:

```
PROC SQL;
    select rownumber, _BMI5, SEX1
    from X.Chap3_1_Formats
    where rownumber ge 20000 and rownumber le 20010;
quit;
```

In SQL, the `where` clause comes after the `from` clause and specifies criteria that need to be satisfied in order for the records to be selected for the query. Notice we used the code `ge` to stand for *greater than or equal* and `le` to stand for *less than or equal*, and we used the word `and` to connect the criteria.

The `where` clause in `PROC SQL` is very helpful when applying criteria to queries. If criteria are complex in a query, many SAS programmers turn to the `where` clause in `PROC SQL` rather than struggle with setting criteria in data steps or the PROCs.

Using arithmetic operators in SAS

In the previous code, we used `ge` and `le` as arithmetic operators in SAS. A link to a SAS web page that provides extensive information on arithmetic operators in SAS is available under the *Further reading* section.

Commonly used operators are summarized in the following table:

Operator (Choices separated by comma)	What it means	Example in applying criteria
`"="`, `"eq"`	Is equal to	`where _BMI5 = 30`
`"^="`, `"NOT ="`, `"^ eq"`, `"NOT eq"`, `"ne"`	Is not equal to	`where _BMI5 ne 30`
`"<"`, `"lt"`	Is less than	`where _BMI5 < 30`
`"NOT <"`, `"NOT lt"`	Is not less than	`where _BMI5 NOT lt 30`
`"<="`, `"le"`	Is less than or equal to	`where _BMI5 <= 30`
`"NOT <="`, `"NOT le"`	Is not less than or equal to	`where _BMI5 NOT <= 30`
`">"`, `"gt"`	Is greater than	`where _BMI5 gt 30`
`"NOT >"`, `"NOT gt"`	Is not greater than	`where _BMI5 NOT > 30`
`">="`, `"ge"`	Is greater than or equal to	`where _BMI5 >= 30`
`"NOT >="`, `"NOT ge"`	Is not greater than or equal to	`where _BMI5 NOT >= 30`

Operator (Choices separated by comma)	What it means	Example in applying criteria
`"&", "and"`	And	where (_BMI5 > 18.5) & (_BMI5 < 30)
`"|", "or"`	Or	where (_BMI5 = 25) or (_BMI5 = 30)
`"= ."`	Is missing	where _BMI5 = .
`"NOT = ."`	Is not missing	where _BMI5 NOT = .

Table 3.7 – Summary of commonly used arithmetic operators in SAS

As can be seen from the preceding table, many operations have multiple ways of expressing the operation in code. SAS programmers tend to prefer certain styles of expressing arithmetic operators, so when working on a programming team, it is important to have a working knowledge of the multiple ways to express different operations.

Viewing data through SAS windows

When SAS is run on a mainframe system using Command Prompts instead of windows, there are no automatic displays of data from SAS. Users must call PROC PRINT, PROC SQL, or some other PROC to view the raw data in the dataset through SAS. The only other way to view the raw data when using a big dataset on such a system is to use a **file viewer** or **file editor** application designed for opening big datasets, as described in *Chapter 2, Reading Big Data into SAS*.

However, once PC SAS was developed, datasets could then be viewed in windows within SAS. PC SAS has an explorer pane on the left that provides a visual image of the mapped LIBNAMES including WORK, displaying them as folders. Clicking into these folders can locate the SAS datasets mapped into those LIBNAMES, and double-clicking the dataset in the folder can open it in a window to the screen.

The ability for PC SAS to do this represents a great improvement in functionality from the Command Prompt environment. However, when dealing with big data, PC SAS can easily run out of memory and crash if a SAS dataset is opened this way.

SAS University Edition is a free, online version of SAS available for anyone to use. SAS University Edition does not have all of the functionality that PC SAS can have (depending on the components included in the build), and it runs on a server in the cloud rather than locally. But SAS University Edition uses a different interface that makes viewing data very easy, as can be seen in the following screenshot:

Code tab

Output Data tab

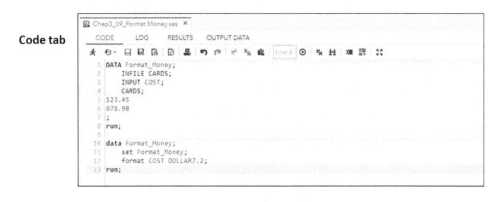

Figure 3.18 – CODE and OUTPUT DATA tabs in SAS University Edition

The preceding screenshot shows the code we ran to create and format a currency variable. The code is created and run in the **CODE** tab. After the code is run, the **OUTPUT DATA** tab becomes available. To view the dataset, the user only needs to navigate to the output data tab; the data is already on display, and controls are available to allow easy navigation of the data display.

Summary

This chapter provided an overview of some PROCs and data step approaches that can be helpful for exploring, understanding, and documenting datasets in SAS. First, we talked about what we can learn about a SAS dataset from PROC CONTENTS. Next, we practiced applying SAS labels and formats to document our data and change the look of our output.

While labels and formats improved the display of data in some positive ways, we found that adding them also introduced limitations. Some data warehouses embrace SAS labels and formats, and a description of how these are managed in such an environment was provided. Other data warehouses use other methods to document and report their data, and these alternatives were discussed.

Finally, the chapter covered ways to view data in SAS using the traditional `PROC PRINT` and the newer `PROC SQL` methods. Since we were using criteria, we went over arithmetic operators in SAS. Finally, we talked about how SAS data can be viewed through windows in PC SAS and SAS University Edition.

In this chapter, you were given an overview of SAS labels and formats. You were shown how to make SAS format files and how to apply SAS labels and formats to datasets. You were shown different examples of how to keep documentation about SAS datasets. You were also given different options for approaches to viewing big SAS datasets, and were introduced to the syntax used for arithmetic operators in SAS.

When maintaining a data warehouse, a considerable amount of effort is devoted to studying and documenting new datasets. The PROCs and data step approaches described in this chapter can help explore new datasets in order to develop documentation. Labels and formats can help with this documentation. PROCs such as `PROC CONTENTS`, `PROC PRINT`, and `PROC SQL` can also be used to provide valuable information about the data.

Once new datasets are studied and documented, it is time to start designing how the dataset will *fit* into the warehouse. This is all managed through the warehouse's ETL code. The next chapter will describe how to manage ETL in SAS. First, a description will be given of how to set up the analytic environment. Next, the chapter will review issues related to transforming variables, and how to design and document the variables that will be loaded into the warehouse.

Questions

1. Which option is added to `PROC CONTENTS` to the variables to print in the order in which they are in the dataset, not in alphabetical order?

2. What does `ge 10 and le 20` mean?

3. Why would a programmer want to attach user-defined formats to a categorical variable with five categories coded 1, 2, 3, 4, and 5?

4. Why might a SAS warehouse choose not to use labels and formats?

5. Imagine a new programmer is hired at a SAS warehouse. Should they be told to use `PROC PRINT` or `PROC SQL` to view data?

6. In this chapter, we used a user-defined format to make a continuous variable, BMI, display in categories. What is the main problem with doing this?

7. How is knowing the observation number of a row in `PROC PRINT` helpful?

Further reading

- Web page listing all SAS sort sequences: `https://v8doc.sas.com/sashtml/proc/z1epts.htm`

- SAS white paper by Louise Hadden about making codebooks—available at `https://www.lexjansen.com/pharmasug/2017/QT/PharmaSUG-2017-QT07.pdf`

- SAS white paper by Naoko Stearns and John Gerlach about making codebooks—available at `https://support.sas.com/resources/papers/proceedings/proceedings/sugi23/Appdevel/p9.pdf`

- SAS support page listing all native SAS formats: `http://support.sas.com/documentation/cdl/en/lrdict/64316/HTML/default/viewer.htm#a001263753.htm`

- US CDC adult BMI classifications—available at `https://www.cdc.gov/obesity/adult/defining.html`

- SAS white paper by John Ladds about advanced format management—available at `https://support.sas.com/resources/papers/proceedings12/048-2012.pdf`

- SAS white paper by Jack Shoemaker about advanced format management—available at `https://support.sas.com/resources/papers/proceedings/proceedings/sugi27/p056-27.pdf`

- US MDR data dictionary: `https://www.health.mil/Military-Health-Topics/Technology/Support-Areas/MDR-M2-ICD-Functional-References-and-Specification-Documents`

- SAS support information about error with `order by` clause in `PROC SQL`: `http://support.sas.com/kb/57/042.html`

- Web page listing SAS operators in expressions: `https://documentation.sas.com/?docsetId=lrcon&docsetTarget=p00iah2thp63bmn11t20esag14lh.htm&docsetVersion=9.4&locale=en`

4
Managing ETL in SAS

This chapter covers issues and topics that need to be considered and addressed in the management of **extract, transform, and load (ETL)** functions in a SAS data warehouse. First, we discuss the different storage environments needed within the SAS data warehouse, and how staff user groups must be designated to correspond to these different storage environments so that access to them is internally controlled. Second, managing the storage of documentation for source datasets and recommendations for dataset naming conventions are covered.

Next, we describe SAS arrays, and a demonstration explaining how they are used in data steps (when performing transformation) is provided. The use of arrays can impact variable naming conventions in SAS, which are discussed with regard to using arrays in transformation code, as well as with respect to maintaining data in an evolving warehouse environment. Modular code is described and a set of code-naming conventions is recommended along with suggestions for a certain style and format of SAS code. Finally, considerations regarding, and recommendations for, format and label policies, data transfer policies, and other data stewardship policies are presented.

In this chapter, we will discuss the following main topics:

- Storage areas and corresponding user groups
- Naming conventions
- Storing and updating documentation
- SAS arrays and how they work with variable names
- Data transfer policies
- Other policies needed to manage ETL

Technical requirements

The dataset in `*.SAS7bdat` format used as a demonstration in this chapter is available online from GitHub: `https://github.com/PacktPublishing/Mastering-SAS-Programming-for-Data-Warehousing/blob/master/Chapter%204/Data/chap4_1.sas7bdat`.

The code bundle for this chapter is available on GitHub here: `https://github.com/PacktPublishing/Mastering-SAS-Programming-for-Data-Warehousing/tree/master/Chapter%204`.

Setting up an analytic environment

An **analytic environment** conceptually refers to a technological environment in which analytics takes place. This can be as simple as a designated directory on a hard drive for storing files associated with analyzing a small survey dataset. In SAS data warehousing, the analytic environment encompasses the parts of the data warehouse in which analytics take place. While this is not a very specific definition, it can be better understood when thinking of permissions associated with user groups. Analysts and developers who build and maintain the SAS warehouse need access to the parts of the analytic environment they must use in order to perform their functions. Users who are outside the analytic environment could be considered *end users* of the SAS data warehouse. These include readers of reports from the SAS data warehouse or users of a web dashboard placed on data from a SAS data warehouse.

Setting up and running a SAS data warehouse requires a lot of administration. Data with different levels of privacy and confidentiality considerations must be stored in different places in the warehouse and controlled under different policies. It is important to limit access to highly secure data to those who need it, and this generally means giving different members of staff different levels of access, depending on their roles. This section describes how to plan for an analytic environment administratively in the SAS data warehouse, data mart, or data lake.

Designating storage and user groups

Running a data warehouse is a little like running a physical warehouse in one notable way—there is a primary focus on managing storage. Just like a warehouse (for a chain of popular retail stores) has to use careful logistics to ensure that goods are moved in and out of the warehouse appropriately and are stored in appropriate and documented places while in the warehouse, the data warehouse has to have parallel policies about moving data in and out of the warehouse, and regarding how data is stored when in the warehouse.

End users of SAS data warehouses, data marts, and data lakes access the data in these systems in a variety of ways. End user access to data warehouses and data marts is often controlled through an application, such as **IBM's data warehousing application COGNOS**. When this happens, authentication of the application can be where permission levels for end users of the data warehouse or data mart are controlled. The area of the warehouse to which end users have access only contains data appropriate for end users. As they use the application, they will only be allowed to access data for which they have permissions, as controlled by the application. Of course, end user interfaces that are meant for the public, such as those available on the World Wide Web can be deployed to easily limit end user access to data available via the web frontend.

On the other hand, data lakes often do not have sophisticated interfaces. Typically, a SAS data lake consists of a set of minimally processed SAS datasets sitting on a secure server. In this case, end users of the data require much more sophistication and may need to go through more vetting before being granted access to the server containing the data lake.

Up to now, we have talked about *permission levels* for end users of the data warehouse, mart, or lake. It is important to observe that there are usually many staff involved in the maintenance of a data warehouse, mart, or lake. User groups within the data system on the processing side need to be designated to steward the data properly. How the staff are grouped into user groups has to do with the functions they serve; for example, staff who process raw data in the warehouse would be in one group, and staff who work on designing reports would be in another group. Since these groups relate to the function of the staff in these groups, these functional groups will also relate directly to the type of storage access needed for each group. It is very likely that those who process raw data in the warehouse will need access to different data storage areas than those who build reports on top of processed data. The following table summarizes suggested storage areas in a data warehouse and who would have access to them:

Purpose of storage area	Characteristics of storage area	What is in the storage area?	Who should have access to the storage area?
To store raw data sent from data providers	Highly secure May be stored in static media and not be live online Few users have access	Raw data extracts from data providers in the native form in which the extracts arrive	Individuals involved in ETL processing from raw data At least one senior programmer and one leader should have access
To store processed datasets in a data lake	Secure but has minimal risk data Online and live Many users have access	Processed SAS datasets that can be accessed by users in SAS	Users who combine datasets from the data lake into their own analytic datasets
To archive old data	Highly secure Stored offline in static media	Raw data extracts from data providers no longer needed Previously used files and datasets in the data warehouse that are no longer being used	At least one senior programmer and one leader should have access

Purpose of storage area	Characteristics of storage area	What is in the storage area?	Who should have access to the storage area?
To store dataset documentation	May be split into security levels where some are in the public domain and some are private for approved data users only

Components may be online in the form of a reference website | Documentation from data sources—documentation generated by the staff at the data warehouse | Public domain files should be accessible by many users and should serve as a marketing tool for the warehouse

Other files can be made accessible to approved users through a password-protected web portal |

Table 4.1 – Proposed storage areas in a data warehouse, mart, or lake, and recommended staff user access

As shown in the preceding table, there are four types of storage areas to which various data warehouse staff may need access:

- Storage for raw datasets from the data provider
- Storage for processed datasets in a data lake
- Storage for archived data
- Storage for data documentation

As you can see in the table, storage for raw extracts directly from data providers and storage for archived data should be handled very sensitively. The storage areas must be *highly secure*. Examples include storing the data on static media kept in a safe or storing the data on a server that is operational but not connected to the internet or any other servers. Certainly, highly secure storage locations that are accessible over the internet can be set up. However, simply exposing the data store to the internet itself is a risk. In cases where exposing data to the internet is deemed to be too much of a risk, data can be *easily secured* by being stored in a storage location that is not connected to the internet. This is reasonable to contemplate, as accessing raw data from data providers and accessing archival data are tasks that are not performed frequently.

In a warehouse, often, the staff involved in developing and running ETL code are the same ones who are involved in archiving. Therefore, there tends to be a small group of programmers who need access to both these highly secure storage areas. Nevertheless, this group should remain as a small, elite subset of the programming team, and should include at least one leader with the authority to make policies in relation to these storage areas. This leader does not need to be a regular programmer, but should be adept enough at SAS to be able to use SAS to access and examine the data in each storage area to execute oversight.

In a data lake, minimally processed datasets are made available on a server accessed by a group of approved users who have been granted access through **data stewardship policies**. Although this storage area needs to be secured, it generally contains data that is de-identified or otherwise designed so that a breach would not have a high impact. Therefore, it does not have to be as highly secured as the previously described storage areas. Using an application to control access to a data lake is also an option. Importantly, staff involved in ETL will not only need access to the highly secure storage areas, but will also need access to the data lake environment to load the minimally processed datasets at the end of the ETL process.

Table 4.1 does not cover all the different types of storage environments that might be needed in a data warehouse, but it does provide a list of commonly needed ones. If you are running a SQL data warehouse but using SAS for ETL, an additional storage area may be needed, and this is known as a **staging area**. This is an area where raw data is staged or reshaped into the correct relational structure before being loaded into a relational database. This staging area would be where those conducting ETL would have access. Their access would be controlled based on their designated tasks in the warehouse:

- Those doing ETL, *but not needing access to the raw data*, would only be given access to the staging area, and not to the storage area with the raw data from data providers.

- Those doing ETL, *but not loading data into the final storage area*, would only be given access to the staging area, and not to the storage area with the loaded data.

Managing documentation storage

The last storage area mentioned in the table is an official area to store dataset documentation. Regardless of the data system, it is important to have official documentation from the data sources that can be available to all users. If some of this documentation is proprietary and cannot be shared, warehouse staff themselves need to generate *documentation of the native variables* being provided to users so that users understand the meaning and source of the native variables. Also, warehouse staff need to provide *documentation of transformed variables* being served to users so they can understand them when using them in their queries, analyses, and reports.

There are different ways to manage the provision of this documentation. A common method is through a **web portal**, which can be available for the public or limited in access through online authentication. In a data lake, this may be accessed as data on a server, or documentation may be stored with the datasets or on a separate **documentation server**. Besides, programming resources, such as code with SAS labels and formats, can be placed on the documentation server.

> **Note**
>
> Warehouse leaders should be proud of their documentation, as it provides insight into not only the value of the datasets they are making available, but the high-quality level of their management. Well-documented warehouses should promote this feature in marketing materials as a strength to attract users.

Setting naming conventions for datasets

As implicated by the many storage areas for warehouse data, once data has entered the warehouse, it may undergo a variety of transformations. These are the typical transformation steps in a SAS data warehouse:

1. **Read in raw data**: Raw data tables from the data provider are loaded into the storage area designated for these datasets. If they are not already in SAS format, basic transformation code may be run on them to convert them to `*.SAS7bdat` format with the data otherwise unprocessed.

2. **Run modular transformation code**: Multiple small and modular code files need to be run in a particular order during transformation. Each code file reads in a version of the dataset from disk, conducts transformations in WORK, and writes out the next version of the dataset to disk.

3. **Publish or load data**: In a data lake, the final dataset from the transformation is published to the data lake. In a data warehouse or data mart, the transformed data is loaded into the system.

These steps are roughly depicted in the following diagram:

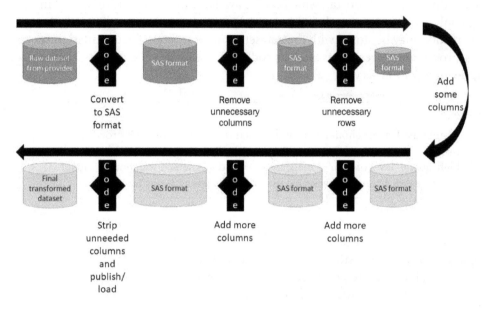

Figure 4.1 – Typical transformation steps during ETL processing

As indicated in the preceding diagram, during early steps in transformation, unneeded columns from the raw dataset are removed to improve processing time. Unneeded rows are removed as well, although effective communication between the data warehouse and the data provider tends to minimize the numbers of rows erroneously transferred that need to be removed.

Next, as shown in the diagram, multiple transformed columns are added to the dataset. As a rule, no variables from the native dataset are edited during transformation. Instead, the variable may be copied to a new, transformed variable, and edited there.

> **The reason for many modular code files versus having one big transformation code file:**
>
> Having modular code that only includes a few transformations between writing versions of the dataset to disk allows the programmer the opportunity for a **manual rollback**. This means that if a problem with a transformed variable is found in the dataset, it is easy to roll back to the transformation step where the variable was added and troubleshoot what went wrong. The shapes marked **Code** in the diagram indicate an intermediate transformation step where data is written to disk. In the last transformation step, unneeded native and transformed variables are removed, and the final dataset is either published or loaded into the destination data system.

At each of these transformation steps, naming conventions for datasets are needed. This table recommends a set of naming conventions that could be used in a SAS data warehouse for these datasets at different steps:

Type of dataset	Naming convention	Comments
Raw data from data provider	Keep datasets native name	This can be changed if there are duplicate names or issues with confusing native names
Intermediary datasets kept in WORK during ETL processing	Name each iteration with a suffix of underscore and successive letters	Example: BRFSS_a, BRFSS_b, BRFSS_c
Datasets written to disk during ETL processing that are not final datasets	Name of the final dataset written to disk with the suffix _v plus an increasing number	Example: BRFSS_v1, BRFSS_v2, BRFSS_v3
Final dataset with all added transformed columns	Name of the final dataset written to disk with the suffix _v plus a number	Example: Last ETL dataset named BRFSS_v12
Final dataset stripped of unneeded columns to be loaded into the data lake or warehouse	Date incorporated into the name of the published dataset	Example: A dataset published on December 15, 2019 could be named 191215_BRFSS

Table 4.2 – Recommended dataset naming conventions for a SAS data warehouse

As shown in the table, it is best to keep the raw datasets from the data provider named the same way as when they were received from the provider. However, there may be situations in which a different policy is made in order to ensure extracts are stored properly and do not have duplicate or confusing names. Also, there may be a reason to rename these raw files in order to leverage automation in SAS that relies on certain naming conventions, which will be discussed at length in *Chapter 8, Using Macros to Automate ETL in SAS*.

As shown in *Figure 4.1*, each modular code file does one or a few transformation steps before writing another version of the dataset to disk. During the transformation code, datasets are read from and written to WORK. As shown in the preceding table, it is suggested that these datasets be named with suffixes with increasing letters.

In *Chapter 1, Using SAS in a Data Mart, Lake, or Warehouse*, we reviewed the **Behavioral Risk Factor Surveillance System (BRFSS)**—the annual health survey done by phone in the United States. We have been using extracts from this dataset in our coding examples. Imagine we read in the 2018 version of the BRFSS in its original, raw form:

1. The first version of the dataset in WORK could be called BRFSS_a.

2. Imagine we use transformation code to remove unneeded columns from BRFSS_a. The next dataset we write to WORK should be called BRFSS_b.

3. Since there are only expected to be a few transformations in each of the modular code files, it is unlikely we will run out of letters for suffixes before the end of the code.

As indicated in the diagram, in each successive transformation code, the last version of the transformed dataset is read from physical disk into WORK, the code transforms it in WORK, and then the next transformed version is written to disk. In a literal sense, these datasets will be read in from, and written to, physical locations that the programmer has mapped to LIBNAMES. Many programmers find it easiest to map one LIBNAME and use only that location for reading and writing datasets during transformation.

In the table, it is recommended that each time a dataset is written to disk between transformation code, it should be labeled with an increasing version number. In our example, imagine after transforming our dataset to BRFSS_c, we want to write it to disk. Our first version written to disk would be BRFSS_v1. The next successive transformation code would read BRFSS_v1 into WORK as BRFSS_a, potentially transform it a few times, and then write it to disk as BRFSS_v2.

At the end of the transformation, the dataset written to disk will have some final version number attached to it. Over time, this final number may vary as the transformation code is groomed, maintained, and edited to keep up with changes in the underlying source datasets or in user demands for other transformed and native variables. In a data lake, there is usually a final transformation step and the publishable dataset is given a specific name to indicate its status. As shown in the table, it is often named with the date of publication so as to not confuse users who have been working with historical data that may have been recently refreshed with edited or newly transformed variables. The naming of the final dataset is not an issue in a data mart or warehouse, where the final load step copies the rows from every table to be loaded into the relational structure of the data system.

In addition to planning and setting policy for user group permissions, data storage, and dataset naming conventions, data system leaders need to set other policies as part of ETL code development. These include other naming policies, such as for SAS arrays, variables, and code.

Planning for data transformation

During transformation, modular code is run to develop transformed variables based on native variables. When the same operation is to be performed on a group of similarly formatted variables, a *SAS array* is used in the data step. When managing a SAS data warehouse, policy needs to be set in relation to the naming of arrays, variables, and code. Furthermore, code style and format must be specified.

This section provides examples and background about SAS arrays, and covers the topics of setting naming and style conventions for SAS arrays, variables, and code.

Understanding arrays in SAS

An **array** is a type of object in SAS that we have not discussed in this book until now. An array holds a series of values or elements in it and can be designated as either a *numeric or character array*. The main ways arrays are leveraged in data processing is when they are used in do loops occurring within data steps. In SAS data warehousing, analysts often set up arrays that include lists of variable names. Then, they combine a do loop with this array in a data step. This way, the same data step code executes on all the variables in the array.

Even though these ideas are complex, we can understand them if we take the concepts one at a time:

1. First, let's choose a physical location and place the SAS dataset Chap4_1 for this chapter in that location.

2. Now, let's map that location to LIBNAME X, read the dataset into SAS's WORK directory with a data step, and run PROC CONTENTS on the file with the VARNUM option, so that the variables print out in the order they are in in the dataset:

    ```
    LIBNAME X "/folders/myfolders/X";
    data Chap4_1;
         set X.Chap4_1;
    proc contents data=X.Chap4_1 VARNUM;
    RUN;
    ```

3. If you replace /folders/myfolders/X in the code with the location of the directory where you have stored the data, then the code will map LIBNAME X to that directory.

4. Let's view the third table from the PROC CONTENTS results:

Variables in Creation Order					
#	Variable	Type	Len	Format	Informat
1	_STATE	Num	8	2.	2.
2	SEX1	Num	8	2.	2.
3	_AGE80	Num	8	2.	2.
4	_BMI5	Num	8	4.1	BEST32.
5	MARIJAN1	Num	8	2.	2.
6	USEMRJN2	Num	8	2.	2.
7	RSNMRJN1	Num	8	2.	2.

Figure 4.2 – PROC CONTENTS results show information about the
seven variables in the Chap4_1 dataset in order of variable number

This dataset represents seven variables from the BRFSS 2018 dataset. This is an annual anonymous health survey done by phone in the United States. Each annual dataset has between *400,000 and 500,000 records*. Although it is done in all states, represented by the _STATE variable, only the states coded as 12 **(Florida)**, 25 **(Massachusetts)**, and 27 **(Minnesota)** are included in the Chap4_1 dataset for demonstration. This dataset therefore only contains 38,901 records.

Because we used the VARNUM option in our PROC CONTENTS code, the variables printed in the order in which they were placed in the dataset. _STATE, the first variable, is a code for the state and, as described earlier, will only take on a value of 12, 25, or 27 in the Chap4_1 dataset. The variable SEX1 is a numeric code for the respondent's sex, and _AGE80 is the respondent's age in years, with ages over 80 coded as 80. _BMI5 is a continuous variable representing **Body Mass Index (BMI)**, a measure of obesity. The last three variables, MARIJAN1, USEMRJN2, and RSMRJN1, represent numeric codes for answers to three relatively new survey questions added about using marijuana for both medical and recreational purposes.

To demonstrate how arrays are used, we are going to set up a scenario using the Chap4_1 dataset. We are going to concentrate on the number of records coded as missing for the three marijuana variables, MARIJAN1, USEMRJN2, and RSNMRJN1. In order to learn the number of records with missing values for each of these three variables, we need to run a one-way frequency using PROC FREQ on each variable:

```
PROC FREQ data=X.chap4_1;
  tables MARIJAN1;
  tables USEMRJN2;
  tables RSNMRJN1;
RUN;
```

Notice how in the first line, PROC FREQ is called, and the LIBNAME and dataset X.chap4_1 are specified. Then, three different one-way frequencies are run using the tables command. One tables command is used for each variable and stated alone on each line, with a semi-colon at the end of the line.

The information about missing numbers prints out at the end of the PROC FREQ output, which is summarized here:

The FREQ Procedure

MARIJAN1	Frequency	Percent	Cumulative Frequency	Cumulative Percent
1	295	1.05	295	1.05
2	234	0.83	529	1.88
3	114	0.40	643	2.28
⋮	⋮	⋮	⋮	⋮
99	106	0.38	28167	100.00

Frequency Missing = 10734

USEMRJN2	Frequency	Percent	Cumulative Frequency	Cumulative Percent
1	1642	77.13	1642	77.13
2	158	7.42	1800	84.55
3	8	0.38	1808	84.92
4	214	10.05	2022	94.97
5	41	1.93	2063	96.90
6	47	2.21	2110	99.11
7	11	0.52	2121	99.62
9	8	0.38	2129	100.00

Frequency Missing = 36772

RSNMRJN1	Frequency	Percent	Cumulative Frequency	Cumulative Percent
1	586	27.55	586	27.55
2	809	38.03	1395	65.59
3	709	33.33	2104	98.92
7	11	0.52	2115	99.44
9	12	0.56	2127	100.00

Frequency Missing = 36774

Figure 4.3 – Frequency of missing values in native BRFSS 2018 marijuana variables MARIJAN1, USEMRJN2, and RSNMRJN1

Looking at the first set of output for MARIJAN1, the end of the frequency table indicates that **Frequency Missing = 10734**. The question associated with MARIJAN1 is, *During the past 30 days, on how many days did you use marijuana or cannabis?*. Therefore, in terms of answers, values of 1 to 30 refer to days. However, the value of 77 means **Don't know/ not sure**, 88 means **None**, and 99 means **Refused**. This question was only asked in some states; if it was not asked, it was left blank (indicated by a period in SAS). These blanks are what are being counted in **Frequency Missing = 10734**.

The other two questions, USEMRJN2 and RSNMRJN1, ask about the primary method whereby the respondent uses marijuana, and the reason the respondent uses marijuana. Respondents are asked to choose from a list of answers, with 7 being coded as **Don't know/not sure**, 9 as **Refused**, and blank indicating the question was not asked. As can be seen by the output, USEMRJN2 has **Frequency Missing = 36772**, and RSNMRJN1 has **Frequency Missing = 36,774**.

Because the frequency missing is being printed below the frequency table, these records are not included in the denominators of the **Percent** and **Cumulative Percent** columns. Imagine that you wanted to change the missing value to 99 in all three of these variables so that they were no longer missing and were counted in the percentage columns in PROC FREQ. You could do three different operations in a data step to recode each variable. But let's say you wanted to recode missing to 99 on all three variables in one data step. Then, using an array that included the names of all three variables would be a reasonable approach.

Let's first consider the code we would use to edit just one variable. Imagine we wanted to create a new dataset in WORK called Chap4_1_RecodeMissing that was a copy of our original dataset Chap4_1 in the library X, but we were going to just edit one of the variables we just reviewed, MARIJAN1. The code would look like this:

```
data Chap4_1_RecodeMissing;
    set X.Chap4_1;
    if MARIJAN1 = . then MARIJAN1 = 99;
RUN;
```

In the data step code, we tell SAS that if MARIJAN1 is coded as missing (as represented by a period), it should replace all the values with 99 before outputting the dataset Chap4_1_RecodeMissing. Of course, we would not really do this type of operation during ETL, because during ETL, we are not allowed to edit the native variables. We are just doing this now for demonstration purposes.

Now, let's do this operation on all three marijuana variables at once using an array. However, to ensure that we can easily troubleshoot our code and do not make any mistakes, we will build the code in two steps. In the first step, we will simply include an array in the data step code. Then, in the next step, we will add a do loop that uses the array.

An array cannot be created as a standalone object, the way a format can be created from PROC FORMAT, or a dataset can be created from a data step. It must be created from within a data step. In the following code, we use a data step to copy our dataset X.Chap4_1 to Chap4_1_RecodeMissing, which is in WORK. We also create an array named marvars_list in the data step:

```
data Chap4_1_RecodeMissing;
    set X.Chap4_1;
    array marvars_list(3) MARIJAN1 USEMRJN2 RSNMRJN1;
RUN;
```

Let's examine the array code more carefully:

- **Begin with the array command**: When you declare an array, you start on a new programming line within the data step and start with the array command.

- **Name the array**: After the array command, you declare the name of the array. In the code, we have named the array marvars_list.

- **Include the number of elements in the array in parentheses**: Immediately after the name marvars_list is (3). This indicates that we are going to add three elements, or variable names, to the array.

- **Include the list of variable names as elements in the array**: The three variables we want to edit, MARIJAN1, USEMRJN2, and RSNMRJN1, are included in the array. Note that if these variables do not exist in the dataset, then SAS will create them.

Naming conventions for arrays

When doing ETL in SAS, arrays are used to such an extent in data step language that they can become easily confused with variable names. Imagine a medical dataset with a set of diagnosis variables named DIAG1, DIAG2, DIAG3, DIAG4, and DIAG5. Creating an array of these five variables named DIAG would then be confusing when referenced in the rest of the code, because DIAG looks so similar to the variable names, and it is not clear it is an array.

One solution is to make the rule that the suffix _list must be added to the end of each array name. In the scenario regarding the diagnosis variables, naming the array DIAG_list would clear up confusion and indicate that DIAG_list was an array and not a variable.

If we run the preceding code, the dataset X.Chap4_1 will be copied into WORK as dataset Chap4_1_RecodeMissing, and the array we developed named marvars_list will be created. This array, however, was not used in this code. That is because we are building this code one step at a time. In our next step, we will add a do loop to the data step code that uses this array, marvars_list.

Let's now add this do loop to our code. This is where we will include the recoding instructions to SAS we reviewed earlier when looking at just the first variable in the array, MARIJAN1. The purpose of the do loop is to instruct SAS to go through each of the marijuana variables stored in the array in the order we stored the variables, and if the variable has a missing value, SAS is instructed to update the value to 99:

```
data Chap4_1_RecodeMissing;
    set X.Chap4_1;
    array marvars_list(3) MARIJAN1 USEMRJN2 RSNMRJN1;
    do i = 1 to 3;
            if marvars_list[i] = . then marvars_list[i] = 99;
    end;
    drop i;
RUN;
```

Let's examine the new code added to the data step we did previously:

- **Beginning the do loop**: The first line of the do loop says do i = 1 to 3, followed by a semi-colon. The variable i is designated by the programmer; any variable can be used. However, it is a convention to use i, j, k, and advancing letters as variables in SAS loops. Saying i = 1 to 3 is telling SAS to create a variable called i, and first, set it equal to 1. Then, go through and do all the code below in the loop before the end command, assuming that i = 1. After doing the code in the loop and hitting the end command, update i to equal the next number, which is 2, and go through the loop again. Because we said do i = 1 to 3, the last time SAS will go through the loop is when i = 3. This time, when SAS hits the end command, it will stop processing the loop.

- **Editing the data using the array**: The next line says `if marvars_list[i]` `= . then marvars_list[i] = 99`, followed by a semi-colon. Imagine we were updating only the first variable in the array, `MARIJAN1`. Then we could write code shown earlier that said `if MARIJAN1 = . then MARIJAN1 = 99`, and that would update all the missing values in `MARIJAN1` to `99`. But since we put `MARIJAN1` along with two other variables into the `marvars_list` array, we now want SAS to go use the array, and go through each element in the array in order, replacing the variable name in the code with that element. Since `MARIJAN1` is the first element in `marvars_list`, on the first loop when `i = 1`, SAS will read the array as `marvars_list[1]`.

  Notice that brackets (not parentheses) are used around the variable `i` after stating the array. Therefore, when executing the first loop, SAS sees the `i` in the brackets as being a `1`. This means SAS will search in the `marvars_list` array for element 1, which is `MARIJAN1`, and replace `marvars_list[i]` in the data step with `MARIJAN1`. On the second loop, SAS will see `i` as equaling 2, and will see `marvars_list[i]` as `marvars_list[2]`, and will therefore replace the variable in the code with the second element in the `marvars_list` array, which is `USEMRJN2`. Because the loop was designated as `do i = 1 to 3`, after SAS sees `marvars_list[i]` as `marvars_list[3]` for a loop, replacing the variable in the code with the third and final element, `RSNMRJN1`, it will stop, because 3 was designated as the upper boundary of the loop. Arrays are a short-hand method alternative to these three repeated statements.

- **Ending the loop**: SAS knows when to finish executing the current loop and go back around to start executing the next loop because it hits the end line. Imagine there was another way to edit all three variables the same way, such as changing all values of 9 in each of the variables to 0. This code could be added before the `end` line, and it would execute in the loop on all three variables as the loop goes around all three times. However, everything added after the `end` line is outside the loop, and will only be executed once.

- **Dropping the i variable**: Because it is a convention to use the letter `i` in loops, if you do not drop the `i` variable created for this loop, it might accidentally be used in a future loop. Therefore, the `drop i` command is very helpful to run at the end of a loop so the loop variable is then cleared and you can safely use the same variable, `i`, for the next loop.

The code we just ran copied our `Chap4_1` dataset into `Chap4_1_RecodeMissing`, and, using an array and a `do` loop, recoded all the missing values in the three marijuana variables to `99`. We will now rerun `PROC FREQ` on those variables to see how different it looks now that the values of missing are coded `99` in those variables:

```
proc freq data=Chap4_1_RecodeMissing;
tables MARIJAN1;
tables USEMRJN2;
tables RSNMRJN1;
RUN;
```

Here is the output from `PROC FREQ`:

The FREQ Procedure

MARIJAN1	Frequency	Percent	Cumulative Frequency	Cumulative Percent
1	295	0.76	295	0.76
2	234	0.60	529	1.36
3	114	0.29	643	1.65
99	10840	27.87	38901	100.00

USEMRJN2	Frequency	Percent	Cumulative Frequency	Cumulative Percent
1	1642	4.22	1642	4.22
2	158	0.41	1800	4.63
3	8	0.02	1808	4.65
4	214	0.55	2022	5.20
5	41	0.11	2063	5.30
6	47	0.12	2110	5.42
7	11	0.03	2121	5.45
9	8	0.02	2129	5.47
99	36772	94.53	38901	100.00

RSNMRJN1	Frequency	Percent	Cumulative Frequency	Cumulative Percent
1	586	1.51	586	1.51
2	809	2.08	1395	3.59
3	709	1.82	2104	5.41
7	11	0.03	2115	5.44
9	12	0.03	2127	5.47
99	36774	94.53	38901	100.00

Figure 4.4 – PROC FREQ on marijuana variables in BRFSS 2018 after replacing missing values with 99

First, it is now obvious that there are no more missing values in all three variables as **Frequency Missing** is not stated below the tables. The value of 99 had already been used in the first variable, MARIJAN1, but had not been used in the other two. Now, in the first table for MARIJAN1, we see the frequency on the 99 line increase substantially to include the missing values we recoded. In the second two tables, we see that 99 appears with the frequency of rows that matches the frequencies that were previously coded as missing. In addition, we observe that this maneuver has caused our percentages to have these records now included in the denominator.

This example used an array that contained three elements. Normally, a list of only three elements does not warrant the extra work of array processing. Array processing is very common in ETL when many more fields are involved. Imagine data for health insurance claims with up to 20 diagnostic codes in variables named DIAG1 through DIAG20. When 20 variables that are in an identical format need to be edited identically, it is an excellent scenario for using a SAS array to help with editing. Not only does it minimize coding, but it gives instructions to SAS's engine in the most efficient way. Therefore, code with arrays generally takes a shorter time to run than code that accomplishes the same tasks without arrays.

Imagine, instead of three marijuana variables, that we had 20 marijuana variables that we wanted to include in an array and edit similarly in one data step. It could take a long time to type out that array. After declaring array marvars_list(20), the programmer would have to type out the name of all 20 variables. This could be a very onerous and error-prone task, especially if there are challenging variable names to retype accurately. Therefore, SAS has designed a shortcut that programmers can use to avoid having to type multiple variable names into an array. If the variables included in the array are named strategically, then the shortcut can be used.

Let's demonstrate this feature of SAS arrays by renaming our marijuana variables first according to the naming approach that can work with the SAS shortcut that will eliminate our having to type the name of all the variables in the array. To do this, we need to pick a character string, and then name each variable we want to put in the array with that string plus an integer that keeps incrementing for each variable. In our case, the logical character string to select would be *mar*, and then the integers we would choose are *1*, *2*, and *3*. In other words, we will use a data step and the RENAME command to rename the three marijuana variables from MARIJAN1, USEMRJN2, and RSNMRJN1 to mar1, mar2, and mar3:

```
data Chap4_1_Renamed;
    set X.Chap4_1 (RENAME=(
        MARIJAN1 = mar1
        USEMRJN2 = mar2
```

```
          RSNMRJN1 = mar3));
RUN;
```

Please observe the following features of the code:

- The code copies the Chap4_1 dataset from X into a dataset in WORK called Chap4_1_Renamed.

- Notice how the RENAME command is inserted between the dataset name and the semi-colon on the same line as the set command. It is possible to format the RENAME command as one long line. For readability, we separated this long command, and have placed the recode of each of the three variables being recoded on three separate lines.

- Notice how the parentheses are used. There is an open parenthesis before the RENAME command, and another open parenthesis after the = after the rename command. On each line, the existing name of the variable is stated on the left, and linked with an equal sign to the new variable name on the right. There are two closed parentheses at the end of the line before the semi-colon.

Essentially, the code makes a new dataset in WORK named Chap4_1_Renamed. This dataset has the marijuana variables now named mar1, mar2, and mar3. We can verify this by running PROC CONTENTS with VARNUM as we did before, only this time on our dataset with renamed variables called Chap4_1_Renamed:

```
proc contents data = Chap4_1_Renamed VARNUM;
RUN;
```

Here is the output from PROC CONTENTS:

Variables in Creation Order					
#	Variable	Type	Len	Format	Informat
1	_STATE	Num	8	2.	2.
2	SEX1	Num	8	2.	2.
3	_AGE80	Num	8	2.	2.
4	_BMI5	Num	8	4.1	BEST32.
5	mar1	Num	8	2.	2.
6	mar2	Num	8	2.	2.
7	mar3	Num	8	2.	2.

Figure 4.5 – PROC CONTENTS showing marijuana variables renamed as mar1, mar2, and mar3

Now that the marijuana variables have been renamed according to this format, we can rewrite our SAS data step with the array and the do loop with less typing, because we can specify the variables that we want processed in the array using a range. We will do that in the following code, which creates the `Chap4_1_RecodeMissing2` dataset:

```
data Chap4_1_RecodeMissing2;
    set Chap4_1_Renamed;
    array marvars_list(3) mar1-mar3;
    do i = 1 to 3;
        if marvars_list[i] = . then marvars_list[i] = 99;
    end;
    drop i;
RUN;
```

Notice in the preceding code, aside from a different name for the output dataset, that the only difference from the previous code is in how we state the array. Instead of typing the name of each of the three marijuana variables, we can now specify the variables using the range `mar1-mar3`.

Running the code will produce the `Chap4_1_RecodeMissing2` dataset in `WORK`, and running the following `PROC FREQ` code will produce the same output as we achieved using the loop to recode the marijuana variables before they were renamed:

```
proc freq data=Chap4_1_RecodeMissing2;
tables mar1;
tables mar2;
tables mar3;
RUN;
```

On the one hand, renaming the marijuana variables as `mar1`, `mar2`, and `mar3` made it easier to call the array, and made the array processing more efficient. But, on the other hand, now our variables are named `mar1`, `mar2`, and `mar3`, and these are not very useful names for data processing. For example, `mar2` used to be named `USEMRJN2`; this is the answer to a question about the type of marijuana the respondent used. Therefore, having the variable named `USEMRJN2` is helpful to the person doing data processing, because this name provides a hint as to what the variable means.

There can be other issues with renaming variables to make it easier for them to be used in arrays. The native coding for the variable `RSNMRJN1` (renamed `mar3`) is in six levels (including missing). `RSNMRJN1` is a variable holding the response to a question asking the respondent about the reason they are using marijuana.

Imagine there were multiple variables coded that way in the dataset that represented the answers to questions about reasons for using a certain medication or treatment. If RSNMRJN1 was to be used in an array recoding all of these variables, it would need to be renamed in any case, even if it had already been handled in the array described and renamed mar3. In other words, the programmer would have to go through and give mar3 and the other variables in the next array another set of names in order to make it easy to state a range of names in an array command in a data step. That effort itself might not be worth the gain in efficiency conferred by this strategy to array processing.

The reason why arrays needed to be discussed in this section before we move on to discussing naming conventions for variables is that *this feature of arrays needs to be considered when designing names for variables in SAS*. In situations where it is intuitive to name a group of similarly coded variables names that would work an array range, such as the example given with diagnosis variables DIAG1 through DIAG20, this convention should definitely be followed. However, when managing a data warehouse that can include multiple years of data, and therefore multiple versions of the same variables evolving through time, other considerations are taken into account when setting up naming conventions that may be prioritized more highly than staging data for easy array processing.

Setting naming conventions for variables

When developing naming conventions for SAS variables in a data warehouse, it is worthwhile reviewing how the SAS application sees named items. When the term **SAS name** is used when talking about the SAS application, it refers to either elements used in the application to which the user has applied names (such as variables and datasets), or elements used by SAS to which SAS has applied a name. The elements we have talked about so far in this book that can have a SAS name include variables, SAS datasets, formats, PROCs, and arrays, but other elements we have not talked about can also have SAS names (more information is available under the *Further reading* section). Regardless of the element named and how it gets named, the SAS name has to follow certain rules.

Here are some of the rules that all SAS names must follow that are helpful to know when designing variable names (as well as the names of other elements in SAS):

- The allowable length of the SAS name depends on the element, but most either permit a maximum of 32 characters or a minimum of 8 characters.
- Names of SAS datasets, variables, and arrays can be up to 32 characters. Names of formats can be up to 8 characters. Names of variable labels can be up to 256 characters.
- The first character of the name must either be a letter or an underscore. A SAS name cannot start with a number.

- SAS names cannot contain blanks. Generally, an underscore is used in place of where a blank would go.

- Special characters (except the underscore) cannot be used in SAS names for the elements we have discussed so far.

- SAS has reserved some names that the user cannot use. For example, SAS reserves the name **work** for the WORK directory.

Traditional naming conventions for SAS versus SQL variables

When SAS was invented, variable names could be a maximum of 8 characters, and all programming was done in uppercase letters. This means that variable names were often in all caps and were shortened versions of what they meant (for example, a variable from a survey where the respondent was asked what type of marijuana was used might be named USEMRJN2).

Given that processing time and effort were at a premium, naming variables strategically was important to reduce processing time. Therefore, variables tended to be named in ranges that were usable by arrays (for example, DIAG1 through DIAG20). The unfortunate side effect of these naming limitations was that SAS variable names were unintuitive. This situation enhanced the necessity to use SAS labels and formats to make sense of the data while processing it, not just while analyzing it.

As described in *Chapter 1, Using SAS in a Data Mart, Lake, or Warehouse*, **structured query language (SQL)**—the primary relational database language currently used—was gaining prominence in the early 2000s. SQL handles the naming of variables and tables differently. The rise of SQL brought influence from SQL programmers as to naming in SAS environments in a number of ways:

SQL users prefer camel case. This is where capital letters are used alongside lowercase letters in a name to increase readability (for example, a dataset named RecodeMissing).

SQL users use consistent suffixes (usually placed after an underscore) in a name. The purpose of this naming convention is to indicate the type of object that is named so this could be deduced from simply looking at its name. Two examples include naming a table listing products product_tbl, and naming the variable that holds the identification number of the product product_ID. All tables in the database would be named with a _tbl suffix, and all ID variables would be named with an _ID suffix.

Both of these new SQL conventions served to make variable names longer. Now, data loaded into a SAS data warehouse from a SQL source may come with native variables that have longer names containing many underscores. Therefore, when running a SAS data warehouse, native names of datasets and variables need to be acknowledged along with other factors and considered when developing naming convention policy.

Even though SAS now allows longer variable names, due to earlier limitations, many SAS datasets have variables with names that are eight characters or less. There are still good reasons to keep the variable names short:

- First, it is easier to type a short name than a longer name, so programming can go faster.

- Second, in today's data warehouse, so much metadata needs to be maintained outside of programming that it may not matter that the variables have short, unintuitive names.

- It may be easier to simply use a data dictionary while programming to navigate a sparsely-named dataset.

As SQL programmers tend to use a different set of naming conventions for datasets and variables, those running a modern SAS data warehouse often have to contend with data originally housed in SQL with names more appropriate for SQL than SAS. The following table recommends naming conventions for a SAS data warehouse that attempts to leverage the most useful components of naming conventions arising from both SQL and SAS: Table 4.3 – Recommended SAS variable naming conventions:

Circumstance	Variable naming convention
Serving up a copy of the variable from the native dataset	Keep the name in its native form, or change the name according to policy
Transforming an edited copy of a native variable	Add successive numbers as suffixes
Creating an ordinal variable from a continuous variable	Add _GRP or GRP as a suffix
Creating a set of two-state flags (binary indicator variables) from levels of an ordinal variable	Choose a reference group, and add suffix numbers accordingly
Creating a set of two-state flags (binary indicator variables) from levels of a nominal variable	Choose a reference group and add an indicator variable for all other groups. Name the variable after what it means when the variable equals one.
Creating novel groupings introduced by the data warehouse	Establish a short character mnemonic indicating the specific variable Establish a short character suffix to indicate the type of novel grouping Use numeric suffixes for different versions of the same variable

Table 4.3 – Recommended SAS variable naming conventions

This table covers suggested naming conventions for different circumstances. Imagine we wanted to serve up the native variable from BRFSS called SEX1 (respondent's sex) in our data warehouse without editing it. We could just continue to call the variable SEX1 in the data warehouse. However, let's say we wanted to edit SEX1. Maybe we wanted to change which code was used to indicate missing in SEX1. We would create a new transformed variable and edit it. According to the table, we would then call this new version of the variable SEX2.

As mentioned earlier, the BRFSS variable _AGE80 indicates the age of the respondent. According to the table, if I wanted to make a grouping variable for _AGE80, I could call it _AGE80_GRP. If we want to create several transformed variables based on _AGE80 with age grouped three different ways, we could use the variable names _AGE80_GRP1, _AGE80_GRP2, and _AGE80_GRP3 for the transformed variables.

Imagine I generated the transformed variable _AGE80_GRP1, which had five ordinal levels. I may want to create a set of **two-state flags**, also called **indicator variables**, for this categorical variable. When this occurs, one level of the ordinal variable is chosen as the reference level, and an indicator variable coded as either 1 or 0 is created for each of the other levels. According to the table, in this case, if I chose the youngest level of age (AGE1) as the reference group, for the other levels two through five, I would create the indicator variables AGE2, AGE3, AGE4, and AGE5.

However, imagine I instead have a nominal categorical variable, such as USEMRJN1. The question for this variable asked, *During the past 30 days, which one of the following ways did you use marijuana the most often?*, and the potential answers provided were as follows:

- 1 = **Smoke it**
- 2 = **Eat it**
- 3 = **Drink it**
- 4 = **Vaporize it**
- 5 = **Dab it**
- 6 = **Use it some other way**
- 7 = **Don't know/not sure**
- 9 = **Refused**

To create a set of indicator variables for this variable, according to the table, each variable should be named according to the state indicated by 1. If you were to create an indicator variable to equal 1 for all of the respondents who answered 3 = **Drink it** (and 0 for everyone else), the advice in the table indicates you should name the variable DRINK. If, instead, you choose to make the variable equal 1 for respondents who did not report drinking it, the variable should be called NODRINK. The level designated as the reference level needs to be made explicit in documentation because, unlike with the ordinal variable, this will not be apparent from viewing the names of the other indicator variables in the set.

Finally, the table recommends setting up a system to handle sets of variables representing novel groupings introduced by the warehouse. Let's consider an example using the *Charlson co-morbidity index* (also called the **Charlson score**). The Charlson score is a way of calculating a score for the level of illness based on an individual reporting their health conditions.

There are different ways this can be calculated. Imagine that developers in a SAS data warehouse with health data wanted to calculate Charlson scores in their data using different equations. They could designate a short character mnemonic of CHAR to indicate that the variable is from the set of Charlson scores, and a suffix of SCR to indicate that the variable is a score. Then, they could name each of the variables arising from different equations with successive numbers: CHARSCR1, CHARSCR2, CHARSCR3, and so on.

Setting naming conventions and style for code

As mentioned earlier, transformation codes are modular, and they do very few operations per code file. Of course, following this rule will mean that you ultimately generate many short code files that need to be run in a particular order to generate the intended dataset at the end. Therefore, naming conventions for code can benefit from prepending numbers to the names of code files.

Although SAS names cannot start with a number, the names of SAS code files typically start with numbers. The number designates the order in which the code file will run. Imagine you observed SAS code files with the following names in order on a server:

- 100_Read in data
- 105_Remove unneeded columns
- 110_Remove unneeded rows
- 115_Add age groupings
- 120_Add BMI groupings

Notice the format of the names. Each starts with a three-digit number followed by an underscore. The numbers and underscore are followed by a short description of what the code does using camel case and spaces in the name. Intuitively, from reading the names, you would imagine that these are five files of modular transformation code. The first one, with the prefix of 100, probably reads in the raw data from the server and converts it to SAS, writing it to disk before the next transformation step. The second file, with a name starting with 105, likely reads in the SAS file transformed in the last code, removes the unneeded columns, and writes the result to disk.

By leaving five units between numbers prepended to each filename in the naming convention, we are leaving ourselves an opportunity to add another transformation step in between two transformation steps. For example, if it was found that I needed to add another variable to the dataset before I could add the BMI groupings in 120, I could develop a file of code with 118 prepended. This code would read in the dataset written out by 115 and write to disk the dataset that will then be used by 120 for the transformations. By adhering to the naming convention, I would ensure that all the code continued to be run in order.

The following table summarizes suggested naming conventions for SAS code in a data warehouse:

Prefix numbers	Operations	Comment
000 through 099	No-run documentation code, macro code that can be called from other programs, and PROC FORMAT code	Keeping these code files under the 000's helps them stand out and can remind the person doing ETL to run the format code before the transformation code.
100 through 199	Transformation code	Each code file contains data steps or other code that achieves transformation (such as PROC SQL). The first code file reads in the raw data and writes it to disk as a SAS dataset named with suffix _v1. Each successive code file reads in a version of the SAS dataset output by the last successive code (for example, BRFSS_v1) and outputs a transformed version to disk (for example, BRFSS_v2).

Prefix numbers	Operations	Comment
200 through 899	Customizable	Depending on the data warehouse, these can be appropriated for different uses.
900 through 999	Exploratory and research code	This is code saved as part of researching the data in order to develop ETL. Saving this is helpful in that it documents why certain programming decisions were made during ETL.

Table 4.4 – Suggested system for determining prefixes in named SAS code in a data warehouse.

The table recommends reserving code name prefixes 000 through 099 for code that is either not intended to be run (such as no-run label code, described in *Chapter 2, Reading Big Data into SAS*), or is intended to be run prior to transformation code (such as PROC FORMAT files). Although 100 through 199 should be reserved for transformation code, longer ranges, such as those in the 200s and 300s, can be appropriated to transformation code if necessary.

If a warehouse has a set of code for specialized functions, the 700s or 800s could be designated for these files. It is recommended, however, that the 900 through 999 range contains code that was used as part of exploratory research that was done to develop ETL code. This code does not need to be functional, in the sense that it does not need to be able to be run on the fly. However, it should be saved for posterity, because it informs the development of the transformation code present in the 100s and possibly longer ranges.

In addition to setting up naming conventions for code, it is important to emphasize a common readable style and format for code files (as demonstrated in the code presented in this book). These considerations include the following:

- **Keep lines of code short**: Long commands should be broken up into separate lines to improve readability.

- **Use tabs to enhance readability**: Tabs should be used to help the programmer understand the code intuitively. For example, in do loops, the do command that launches the loop should line up vertically with the end command that closes the loop, and the steps within the loop should either be in line vertically or be indented one tab-space to indicate that they are within the loop.

- **Add blank lines between code blocks**: This helps the programmer see the breaks between different blocks of code.

- **Use consistent case**: If you choose to use all caps when calling PROCs, then use all caps whenever calling PROCs throughout the code.

Up to now, we have discussed policies that control data storage in the data warehouse, and policies that control naming conventions. The next section covers other policies that need to be developed when running a SAS data warehouse, data mart, or data lake.

Developing policy

SAS data warehouses require policies for smooth management, and this section focuses on some high-priority policies relating to ETL that need to be developed to ensure that the data warehouse functions responsibly. First, how SAS format and label policies can be developed and implemented is described. Next, a spotlight is placed on procedures transferring data into the data warehouse from external sources and transferring data from inside the warehouse to external recipients.

Because both of these activities increase the risk for the data warehouse, policies that can ensure these tasks are done with reduced risk are recommended. Finally, other policies that are optimal for a SAS data warehouse to implement are suggested.

Setting format and label policies

In *Chapter 3*, *Helpful PROCs for Managing Data*, we went over how SAS labels can be applied to variables in SAS datasets as text strings that annotate the variables, explaining what they are. We also went over how we can use PROC FORMAT to develop user-defined formats to annotate levels of categorical variables (such as having the 1 in SEX1 be decoded into what the 1 means, which is **Male**).

Attaching labels to variables is done with the label command in a data step. In contrast, attaching user-defined formats to variables must be done in multiple steps. As we reviewed in the previous chapter, first, PROC FORMAT code must be run to create a format in SAS. Next, the programmer must run a data step or some other code to instruct SAS to attach the saved formats to the designated variables. In addition, we went over how SAS has a multitude of native formats that can be applied using data steps or used to format output in PROCs.

The fact that SAS labels and formats are best used when SAS is used for reporting was emphasized, as SAS labels and formats cannot be used in other applications. If reporting is done in SAS, then SAS labels and formats can be very useful and even necessary. If SAS labels and formats are used, then the development and maintenance of SAS labels and formats should be covered by policy.

Before setting policy in relation to SAS label and format development and maintenance, it is important to recognize which staff in the data system would likely have the information about the correct values to put in SAS labels and formats. SAS programmers who develop and manage the early ETL steps that read raw data from the provider, as well as programmers who complete the final ETL steps that involve loading and publishing the data, tend to be the programmers who know the dataset documentation the best. These individuals should be in charge of not only setting SAS label and format policy, but also enforcing it.

As described in the previous chapter, a method for setting policy around labels can involve maintaining a set of no-run label code made available in dataset documentation that simply connects the variable names with their annotations in a `label` command that is not part of any particular data step. Programmers can simply access this code file and copy these statements from it into data steps in their transformation code.

If this approach is taken, two policies could be set:

- **Native variable label policy**: This policy says that each time a native variable is read in from an external dataset, a label is applied before writing the next version of the dataset to disk. The label wording should be based on documentation from the data source. No-run label code should be updated to contain the native variable and correct label wording.

- **Transformed variable label policy**: This policy says that for each new variable created in transformation code, a label needs to be attached before writing the next version of the dataset to disk. In addition, the no-run label code file should be updated to contain the new variable and label.

Setting format policies around SAS native formats could be done similarly to setting SAS label policies. For example, a SAS data warehouse containing many currency variables might set a policy to use a certain SAS native format for these variables (such as `DOLLARw.d`). As with enforcing label policy, one way to help enforce this native SAS format policy would be to create no-run code that attaches appropriate native formats to variables that could be copied into data step code.

To set SAS user-defined format policies, we need to acknowledge that user-defined formats are developed and attached to variables in two steps. Therefore, we need to set policies around the process of developing the `PROC FORMAT` files that should be run prior to transformation, as well as set policies around actually attaching the formats during processing. Again, those who conduct high-level ETL in the warehouse, such as reading raw data from the provider, or loading final datasets into the warehouse, should be the ones in charge of setting and enforcing SAS format policy. That way, they are able to easily adjust and revise the policy as necessary, and can be involved in implementing revisions as the underlying data and needs of the data system evolve.

As with the label policy for native variables, a staff member in the warehouse familiar with the original documentation of native variables should be the one to develop PROC FORMAT code to attached user-defined formats, as they will be the one most likely to know how the levels should be annotated. With respect to user-defined formats for transformed variables, these would be more easily crafted by the programmers developing the transformed variables. In any case, policy needs to be made about what PROC FORMAT files are updated, and who updates them. Further, policy needs to be established in relation to running these PROC FORMAT files during ETL, as well as making them available in documentation.

SAS labels and formats, when used in a SAS data warehouse, need to be standardized and enforced by the programming staff. However, how SAS labels and formats impact the end user experience depends on the data system. If the system is a data lake, and the users are relatively sophisticated and are actually using SAS to analyze data, making labels and format code available to users may be helpful. It may also be helpful to provide code to users that allows them to strip the native labels and formats from the data you are providing.

The following code provides an example of how to strip labels and formats from a dataset. First, we will make a small dataset in WORK named Format_Money with one column, COST, and two rows, each with a value representing the cost of an item in US dollars. We will apply a label and format to the COST variable, and show the output from PROC CONTENTS:

```
data Format_Money;
      INFILE CARDS;
      INPUT COST;
      format COST DOLLAR7.2;
      Label COST = "Cost of item";
      CARDS;
123.45
678.90
;
RUN;
PROC CONTENTS data=Format_Money;
RUN;
```

Notice how this code is similar to the code we used in earlier chapters, but what is new is how the commands are combined. After the INPUT command that defines the variable COST, on the next line, we declare a format for this COST variable, and on the next line, we place a label on the COST variable. This code shows how formats and labels can be placed on the fly during a data step where data is being entered using the DATALINES or CARDS commands. The PROC CONTENTS output shows that the label and format were applied properly:

Alphabetic List of Variables and Attributes					
#	Variable	Type	Len	Format	Label
1	COST	Num	8	DOLLAR7.2	Cost of item

Figure 4.6 – PROC CONTENTS showing formats and labels applied to the COST variable

Next, we will run the following PROC DATASETS code to strip the labels and formats from the dataset we created in WORK called Format_Money:

```
PROC DATASETS lib=WORK memtype=data;
    modify Format_Money;
          attrib _all_ label=' ';
          attrib _all_ format=;
contents data=WORK.Format_Money;
RUN;
quit;
```

Let's review how this code works:

- **Declaring** PROC DATASETS: PROC DATASETS is capable of managing more than one dataset in a library in a single PROC. In the first line, we ask PROC DATASETS to look in the WORK library through the command lib=WORK. PROC DATASETS can manage datasets as well as other SAS items, called **members**, that can be stored in SAS libraries. Therefore, memtype=data specifies that this PROC DATASETS command only pertains to the datasets in the WORK library.

- **Using the modify command**: The modify command in PROC DATASETS instructs SAS to modify the member stated in the code, which is the dataset we just generated in WORK called Format_Money.

- **Stripping all labels**: To strip labels from all the variables in the `Format_Money` dataset, the `attrib` command was used to set the label value of all the variables to a consistent character, which is a space. This was achieved through the code `attrib _all_ label =' ',`. The code `_all_` specifically means all the variables in the dataset. Notice that the label was not actually stripped; instead, it was simply set to be a space, which essentially erases it.

- **Stripping all formats**: To strip formats from the dataset, a similar command was used, which was `attrib _all_ format=` followed by a semi-colon to indicate the end of the programming line. This code specifies setting the formats to all the variables in `Format_Money` to having no format, as indicated by `format=` with nothing on the right-hand side of the equal sign.

- **Including a contents command at the end**: Notice that a `contents` command is included at the end of the code requesting that SAS prints the contents of the `WORK.Format_Money` dataset.

Using PROC DATASETS to manage datasets

Most of the editing that is done to datasets using PROC DATASETS could be done to the same dataset using a data step approach. However, PROC DATSETS provides functionality that might be useful to the SAS data warehouse leaders and programmers, because it provides the ability for one PROC to make multiple modifications on multiple datasets in one library at once.

SAS publishes a PROC DATASETS tip sheet that is available under *Further reading*. The tip sheet provides examples of PROC DATASETS being used for the following functions:

- Renaming SAS datasets and moving them to different libraries

- Modifying labels and formats of variables in SAS datasets

- Creating an index in a SAS dataset

- Renaming variables in a SAS dataset

After running the PROC DATASETS code on `Format_Money` and including the `contents` command, we can see from the output that the labels and formats have been stripped from the dataset:

Alphabetic List of Variables and Attributes		
#	Variable	Type	Len
1	COST	Num	8

Figure 4.7 – Contents output from PROC DATASETS showing
labels and formats stripped from the dataset

As is evident from the information in this chapter, managing ETL in a SAS data warehouse involves setting policies that directly relate to data storage and management. These include establishing naming conventions for data, code, and variables, and establishing rules relating to user-level permissions and data access. Although the largest set of policies will relate directly to data storage and transformation, perhaps the most important set of policies to establish in a SAS data warehouse has to do with data transfer into and out of the data system.

Setting data transfer policies

Two types of data transfers will inevitably happen in the data warehouse. First, there is the type of data transfer where data from an external data provider is imported into the warehouse. Second, there is the type of data transfer where data from within the warehouse is transferred out of the warehouse system. In addition to these two types of data transfers, there are likely to be transfers of datasets internally, such as from one internal server to another.

Data transfers that take place internally are usually already covered by policy. For example, in the section talking about user permissions earlier, staff in the data warehouse who perform ETL functions, but who do not read in raw data from the data providers and do not load data into the final warehouse structure, would be given limited access to data storage areas only involved with transformation. By being limited in this way, the staff member would not have the opportunity to bring new data into the system or to export data out of the system because they would be limited by their permissions.

As described earlier, those involved in interacting with data from the data provider, and those involved in loading transformed data into a final data mart, warehouse, or lake structure tend to be a small group of senior programmers and other warehouse leaders who are granted access to these highly secure areas. This group should also be tasked with controlling the transfer of data into and out of the data warehouse system. That way, these types of data transfers will be minimized, as only a few individuals will have the authority to execute them.

With respect to data stewardship policies governing transfers of data into the warehouse, each dataset incorporated into the warehouse will be from a different source, and therefore will have different *policies and agreements controlling the data*. Individuals in the senior programming group may develop subject matter expertise in certain datasets and can be appointed as an official **subject matter expert (SME)** for these datasets. Ideally, the SME would be in charge of data transfers of raw data involving these datasets, as well as developing and updating ETL code and documentation files associated with these datasets. Data receiving policies could instruct the following to occur:

1. For each dataset, an SME is designated, and that person establishes a contact person for the dataset as the data provider.

2. When it is time to transfer a new dataset, the data provider communicates metadata to the SME regarding the dataset being transferred.

3. The dataset is then transferred from the data provider to the SME according to a pre-established mechanism (for example, through a specific **file transfer protocol (FTP)** server)

4. The SME then uses a standardized method to verify that the data received is complete and accurate. If it is, the data is named according to naming conventions, the transfer is logged, and the data is stored in the appropriate storage area. If the data is not complete and accurate, the SME works with the contact person at the data provider to re-transfer the data.

5. In the case of the re-transfer of historical data, the SME and the data provider work out an ad hoc process to complete a one-time refresh of historical data.

There are several reasons why SAS data warehouse leaders should want to control the data being transferred into the warehouse. First, the warehouse does not want to take on liability for storing the *wrong* data, or accepting data to which it is not entitled. Second, the data being transferred often includes sensitive information (such as identifiers or company secrets), and so it needs to be guarded carefully during transfer. Third, any problems associated with the actual data files being transferred need to be contained and immediately identified. Data files can get corrupted during transfer, or other errors can be made, and these need to be identified and resolved as part of the transfer procedure.

SAS data warehouse leaders also have reasons to want to set policies to control the datasets being transferred out of the data warehouse. In some systems, data exported from the warehouse is entirely controlled through an application. **SAP Crystal Reports** is an application that has been used historically to provide users with the ability to build reports from a data system and download tables of data from the reports. Data warehouse developers therefore control data being removed from the warehouse by controlling what users are able to do in the application interface.

In cases where datasets need to be transferred from the warehouse to another environment after ETL without the use of an application, staff in the data warehouse will need to create and develop a **data extract** and transfer it to a contact person for the other environment.

Regardless of whether taking data from the warehouse is controlled by an application or done manually by data warehouse staff, policies should be developed surrounding such a data transfer. Policies that guide the dissemination of data through the application should be established, as well as policies and procedures for establishing, developing, and documenting an extract for transfer out of the warehouse. The following should be considered in the development of policy for transferring data out of the warehouse environment:

- How should data transfers be approved prior to transfer?

- How should data requests be documented and communicated?

- Data transfer procedures

- How should the data transfer be logged?

- How will the warehouse be compensated for the time, effort, and cost involved with each transfer?

Issues to consider with data transfers in and out of the SAS environment:

Before data providers are comfortable providing data to a SAS data warehouse, they need to discuss the prospect and set up a data-sharing agreement. This agreement sets forth many terms, including what the data provider's expectations are for the data warehouse in stewarding copies of the provider's data.

Often, these agreements contain language that limit the data warehouse's ability to transfer any copies of the data from the data provider to others outside of the data system. This language needs to be acknowledged in any policies and procedures for transferring data out of the system.

Finally, it is important to consider the impact of local laws on data transfer. In the US, the **Health Insurance Portability and Accountability Act (HIPAA)** of 1996 is a Federal law that governs data sharing by healthcare providers and other **covered entities (CEs)**. In addition, local state laws that are stricter than HIPAA can also apply. The US has laws governing the transfer of financial and education data as well. It can be helpful to acknowledge these laws and how they will be honored in the data-sharing agreement developed between the data provider and the SAS data warehouse.

Setting other policies

This chapter focused on setting policies relating to data storage, user permissions, naming conventions, and coding approaches, among other topics. All these policies fall under the umbrella of **data stewardship policies**. While the policies on which we focused in this chapter up to now have technical components, other policies that are less technical and more administrative need to exist to ensure the proper functioning of the data system.

Historically, some warehouses experienced issues with involving the correct SMEs as leaders in developing the data warehouse in order to ensure that workable and appropriate data stewardship policies were developed and maintained. If the person appointed to lead the data warehouse has experience in management and business, but no practical experience managing a data system, they will not be able to develop data stewardship policies that are feasible to implement or easy to communicate to the programming team. On the other hand, if the person appointed to lead a data warehouse only has programming experience, but no knowledge of data stewardship, data regulations, and how to manage programming teams, then they will not have the management and leadership skills to be successful in that role.

The data warehouse leader needs to have experience and skills in many areas. On the one hand, they need to perform as a **data ambassador** when out in the community, negotiating the transfer of new datasets into the warehouse, and promoting the use of the data warehouse to end users. On the other hand, they need to perform as a **programming team leader**, setting both administrative policies, such as those that govern data storage and transfer, as well as technical policies, such as those that govern naming conventions. The perfect candidate for the leader of a data warehouse will have both data ambassador and programming team leader experience and skills, as well as subject matter expertise in the topic of the data stored in the warehouse.

Summary

This chapter reviewed the different policies and procedures that need to be set up in a SAS data warehouse to ensure proper management of ETL. First, we reviewed how to arrange storage in the SAS data warehouse, and how to link different storage levels with tasks performed by staff in different user groups. Next, we talked about the status of different datasets in the data warehouse, and how to set naming conventions for these datasets.

Third, we tackled the topic of SAS arrays, and reviewed how using arrays in ETL code can impact variable naming conventions, which we also discussed. Fourth, we covered naming conventions and style for code. Finally, we considered guidance for setting policies for SAS formats and labels. We also reviewed recommendations for implementing data transfers and other policies.

Although these topics are largely administrative, they do have technical components, especially with respect to storage allocation and naming conventions for various elements in the SAS data warehouse. Because data warehouses contain so much data and involve so many team members, it is necessary to set warehouse-wide policies to ensure smooth management of the warehouse. Therefore, it is worthwhile to proactively set these policies to work with the programming team and ensure that those who have to follow the policies and procedures can do so comfortably and efficiently.

In the next chapter, we shift gears away from managing ETL and toward managing reporting in a SAS data warehouse. We cover SAS's **output delivery system (ODS)**, and different ways to develop reports and visualizations from warehouse data using SAS.

Questions

1. Why must raw datasets direct from the data provider be stored in a highly secure area in the warehouse environment?

2. Why is it better to maintain many modular code files instead of one big long code file?

3. In a dataset that contains 10 different cost variables, what is the main advantage and disadvantage of naming the variables COST1, COST2, and so on up to COST10?

4. How is declaring formats in SAS different to declaring arrays?

5. Why is it logical to assign programmers who already have access to raw data from the data provider used in ETL to also maintain SAS label and format code?

6. Why is it helpful to have consistent naming conventions throughout the data warehouse?

7. How is it helpful when senior programmers serve as SMEs for particular datasets?

Further reading

*   Information about SAS names, and rules for naming elements in SAS – available here: https://v8doc.sas.com/sashtml/lgref/z1031056.htm.

*   Charlson Comorbidity Index – more information is available here: https://healthcaredelivery.cancer.gov/seermedicare/considerations/comorbidity.html.

*   PROC DATASETS tip sheet from SAS – available here: https://support.sas.com/rnd/base/Tipsheet_DATASETS.pdf.

5
Managing Data Reporting in SAS

This chapter will introduce you to the different ways in which SAS handles data reporting. First, we will talk about the invention of the **output delivery system (ODS)**. When PROCs run, they create internal tables and files that are not accessible outside of the PROC. First, you will learn how to use the ODS to output internal tables as data tables. Next, you will learn how to use the ODS to save the textual and graphical output from SAS PROCs in `*.pdf`, `*.rtf`, and `*.htm` format.

After that, the chapter will focus on PROCs in SAS designed specifically for reporting that leverage the ODS as part of how the PROC runs. First, you will be shown how to build a tabular report, first in PROC REPORT and then in PROC TABULATE. Next, the family of graphing PROCs, including PROC SGPLOT, PROC SGPANEL, PROC SGSCATTER, and PROC SGRENDER, will be demonstrated. PROCs SGPLOT, PROC SGPANEL, and PROC SGSCATTER provide more options for customization than using the ODS with graphics from PROCs not specifically designed for reporting. Finally, you will learn how to use PROC TEMPLATE with PROC SGRENDER to have even more control over graphical output from SAS.

In summary, this chapter will cover the following main topics:

- What is the ODS and why was it implemented in SAS?

- How to identify internal tables produced by PROCs that can be retrieved through the ODS

- How to use these tables in data processing

- How to create and retrieve SAS output in `*.pdf`, `*.rtf`, and `*.htm` format using the ODS

- What is the format and process for developing `PROC TABULATE` code?

- How to use graphing PROCs with options to control appearance, including `PROC TEMPLATE` with `PROC SGRENDER`

Technical requirements

The dataset in `*.sas7bdat` format used as a demonstration in this chapter is available online on GitHub: `https://github.com/PacktPublishing/Mastering-SAS-Programming-for-Data-Warehousing/tree/master/Chapter%205/Data`.

The code bundle for this chapter is available on GitHub here: `https://github.com/PacktPublishing/Mastering-SAS-Programming-for-Data-Warehousing/tree/master/Chapter%205`.

Using the ODS for data files

SAS software development originally focused on data management and analysis. Exporting files such as graphics and internal datasets from the SAS environment was not originally envisioned. When it was established that SAS users needed some way of doing this, SAS developed the **ODS**.

This chapter will cover several ways in which the ODS can be used. One way in which the SAS ODS can be used is to retrieve internal data files created when PROCs run. Each `PROC` produces a different set of internal tables that can be saved outside of the `PROC` using the ODS so they can be queried later. Here, we will use `PROC UNIVARIATE` to demonstrate identifying which internal tables are available to us through the ODS. Afterward, we will use the ODS to save one of the internal tables from `PROC UNIVARIATE` as a `*.sas7bdat` dataset so we can use it in future data processing. We will practice by using some PROCs in **Base SAS**, which is the base component of SAS.

> **Note**
>
> **Base SAS** is the base component of the SAS program, and includes many commonly used PROCs such as PROC UNIVARIATE. **SAS/STAT** is the component of SAS that includes many of the analytic PROCs, including ones that do regressions. PROCs from these components often have options that can be set to produce a graphic. This will be demonstrated in this chapter, with PROC UNIVARIATE having an option set to produce a histogram. Producing graphics this way is limiting, in that the PROC generally does not provide the programmer with many options for formatting the graphic. Graphics generated this way are generally intended for data diagnostics, and not for publication and reporting.

Identifying available tables in the ODS

SAS's so-called ODS refers to a function added to SAS that provides enhanced power to the programmer. When SAS runs procedures, it creates intermediary calculations, datasets, and other files. Before the ODS, unless datasets created this way were specified to be automatically outputted by the procedure as a dataset, they were not available to the programmer in this format. For example, the list of variables that outputs in PROC CONTENTS comes out as a report, and not a data file. With the implementation of the ODS, the programmer now has the power to use the ODS to identify these data files and facilitate their output into datasets when SAS is running a procedure.

> **Note**
>
> One way to develop metadata about your SAS dataset is to use the ODS combined with PROC CONTENTS. In her SAS white paper, Louise S. Hadden describes a way to—as she puts it—*Build Your Metadata with PROC CONTENTS and ODS Output* (see the *Further reading* section for the link).

Let's look at an example of the ODS in action by using PROC UNIVARIATE, the PROC in SAS used to generate **summary statistics** about **continuous variables**. PROC UNIVARIATE is part of Base SAS.

We will work with variables from the **Behavioral Risk Factor Surveillance Survey (BRFSS)** dataset, an annual phone health survey done in the United States (see *Chapter 1, Using SAS in a Data Mart, Data Lake, or Data Warehouse*, for more details). We are working with a subset of the 2018 dataset comprised of records from only three states:

- Florida (FL, code=12)
- Massachusetts (MA, code=25)
- Minnesota (MN, code=27)

The dataset is named `Chap5_1.sas7bdat` and is in SAS format.

Let's start by running `PROC CONTENTS` on `Chap5_1.sas7bdat`. First, we will map `LIBNAME` to `X`, with `X` being the folder where we put the dataset. Next, we will run `PROC CONTENTS` and look at the variables in the dataset:

```
LIBNAME X "/folders/myfolders/X";
PROC CONTENTS data=X.Chap5_1;
RUN;
```

Here is the list of variables from `PROC CONTENTS`:

#	Variable	Type	Len	Format	Informat
	Alphabetic List of Variables and Attributes				
5	MARIJAN1	Num	8	2.	2.
7	RSNMRJN1	Num	8	2.	2.
2	SEX1	Num	8	2.	2.
6	USEMRJN2	Num	8	2.	2.
3	_AGE80	Num	8	2.	2.
4	_BMI5	Num	8	4.1	BEST32.
1	_STATE	Num	8	2.	2.

Figure 5.1 – List of variables in the dataset from PROC CONTENTS

Of the seven variables listed, only two are continuous variables: `_AGE80`, which is the age in years truncated at 80, and `_BMI5`, which is **body mass index (BMI)**, a continuous measure of a person's size that uses an equation to combine height and weight. Since we are going to demonstrate the ODS using `PROC UNIVARIATE`, which provides summary statistics for continuous variables, we will have to use one of these numeric variables for demonstration. Let's use `_AGE80`, or age:

```
PROC UNIVARIATE data=X.chap5_1;
    var _AGE80;
RUN;
```

Let's take a quick look at the log file from `PROC UNIVARIATE`:

```
1           OPTIONS NONOTES NOSTIMER NOSOURCE NOSYNTAXCHECK;
72
73          PROC UNIVARIATE data=X.chap5_1;
74          var _AGE80;
75          run;

NOTE: PROCEDURE UNIVARIATE used (Total process time):
      real time          1.27 seconds
      cpu time           0.31 seconds


76
77
78          OPTIONS NONOTES NOSTIMER NOSOURCE NOSYNTAXCHECK;
90
```

Figure 5.2 – Log file from PROC UNIVARIATE

Later, when we use the ODS with `PROC UNIVARIATE`, you will notice that the log file will look a lot different. This code produces several tables that print out the output. Instead of looking at the actual numbers in these tables, let's instead step back and consider the titles and contents of the five tables that are present in the output of `PROC UNIVARIATE`:

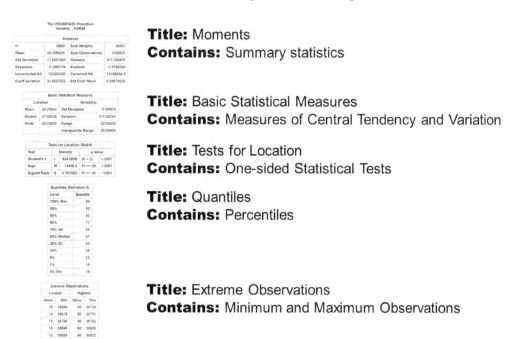

Title: Moments
Contains: Summary statistics

Title: Basic Statistical Measures
Contains: Measures of Central Tendency and Variation

Title: Tests for Location
Contains: One-sided Statistical Tests

Title: Quantiles
Contains: Percentiles

Title: Extreme Observations
Contains: Minimum and Maximum Observations

Figure 5.3 – Titles and descriptions of tables output from PROC UNIVARIATE

Each of the tables' output has `Title`. This title is reproduced in the diagram, along with a description of what the table contains. To better understand how we could use the ODS with `PROC UNIVARIATE`, let's now focus on just the first table of the output, entitled `Moments`:

Moments			
N	38901	Sum Weights	38901
Mean	54.2094291	Sum Observations	2108801
Std Deviation	17.6997863	Variance	313.282437
Skewness	-0.3266154	Kurtosis	-0.9746548
Uncorrected SS	126503585	Corrected SS	12186686.8
Coeff Variation	32.6507522	Std Error Mean	0.08974033

Figure 5.4 – Moments table from PROC UNIVARIATE

The summary statistics in the `Moments` table – such as the number of observations, mean, and standard deviation – are not only useful to know, but might be needed in subsequent programming. Imagine you intended to classify people in the dataset as "more than one standard deviation older than the mean." You would need both the mean and the standard deviation to exist as data points or variables in programming, not printed numbers on output.

Identifying internal tables in the log

When you use the ODS with a `PROC` that produces internal data tables like `PROC UNIVARIATE` does, you are having SAS output the tables as datasets and store them somewhere so you can use the values inside them later by accessing the datasets in subsequent programming. Let's look at how we would modify the code to use the ODS to help us identify the information about the data tables available to us by way of the ODS through `PROC UNIVARIATE`. These are the tables we can output outside of `PROC` using the ODS:

```
ODS TRACE ON / label;
PROC UNIVARIATE data=X.chap5_1;
    var _AGE80;
RUN;
ODS TRACE OFF;
```

The code looks identical to the PROC UNIVARIATE code we ran earlier, except that it is sandwiched between two ODS calls—one at the beginning that turns the ODS trace on, and one at the end that turns it off. ODS TRACE ON is how we tell SAS to turn the ODS on before the PROC starts so it can trace the tables being made as the PROC is running. The label option is used to manipulate the output in the log to include information about the path of the traced tables.

When we ran PROC UNIVARIATE earlier, we saw that the output had five tables. We would expect to see these tables saved in the ODS trace. Let's look at the log file and see whether we can identify the five tables that we saw earlier in the output:

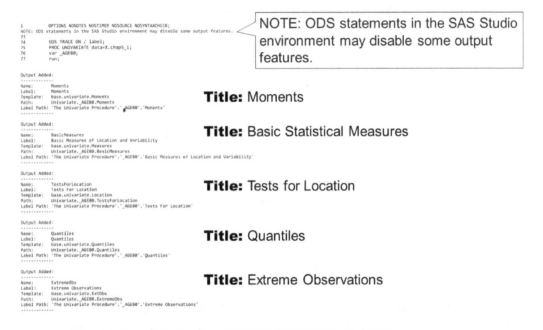

Figure 5.5 – Log from PROC UNIVARIATE with ODS trace on

You will see that turning the ODS trace on and off changes how the log file looks. Notice how the log reports each of the tables being output in order. Also, please be aware that if you are using *SAS University Edition*, ODS statements may disable some output features that normally work in SAS if you are not using the ODS.

> **Note**
>
> The table's output from PROC UNIVARIATE listed in the log file is the default tables that are output. There are even more objects created as part of running PROC UNIVARIATE, which is accessible through the ODS, that are not listed on the output.
>
> A link is provided under the *Further reading* section to the SAS support page that lists all the ODS tables produced with the PROC UNIVARIATE statement. Also, another link is provided under *Further reading* to a SAS support page that provides information on all the tables produced by commands in **Base SAS**, the base SAS software component. You will observe that most PROCs in SAS produce internal tables that can be output by the ODS. Therefore, when using the ODS with PROCs, it is helpful first to read support information about the PROC as well as to use ODS trace to learn about all the objects available to the ODS in the PROC.

As reported in the log file, the first table output through the ODS in our PROC UNIVARIATE code is the Moments table. Let's zoom in on this part of the log file:

```
Output Added:
-------------
Name:       Moments
Label:      Moments
Template:   base.univariate.Moments
Path:       Univariate._AGE80.Moments
Label Path: 'The Univariate Procedure'.'_AGE80'.'Moments'
-------------
```

Figure 5.6 – Close-up of the mention of the Moments table in the log file

Using the ODS TRACE command, we identified the Moments table from PROC UNIVARIATE. In the log, we can see that the name of the table is Moments. That is how we can refer to the table in ODS programming – we will use it to save the table as a dataset during PROC.

Outputting internal tables using the ODS

The next step is to use the ODS to output the Moments table into a dataset we can access and use from SAS after PROC is done running. Please note that this dataset will be in *.sas7bdat format. We will do that with this code, which runs a line of code before PROC UNIVARIATE, allowing us to specify the output of the Moments table to a known location:

```
ODS OUTPUT Moments = X.Age_Moments;
PROC UNIVARIATE data=X.chap5_1;
```

```
      var _AGE80;
RUN;
```

The first line of this code is the one that tells SAS to output the `Moments` table from `PROC UNIVARIATE`. It starts with `ODS OUTPUT`, indicating that the ODS is being given instructions to output an object. `Moments = X.Age_Moments` instructs the ODS to write the data from the `Moments` table being produced in `PROC UNIVARIATE` to an SAS dataset named `Age_Moments`, and it is to be placed in the library mapped to `LIBNAME X`.

After this line is the usual `PROC UNIVARIATE` code. The following table will be available in `*.sas7bdat` format in the directory mapped to `LIBNAME X` after running this code:

VarName	Label1	cValue1	nValue1	Label2	cValue2	nValue2
_AGE80	N	38901	38901	Sum weights	38901	38901
_AGE80	Mean	54.209429	54.20942906	Sum observations	2108801	2108801
_AGE80	Standard deviation	17.699786	17.69978634	Variance	313.28244	313.2824365
_AGE80	Skewness	-0.3266154	-0.32661544	Kurtosis	-0.9746548	-0.97465483
_AGE80	Uncorrected sums of squares	126503585	126503585	Corrected SS	12186687	12186686.78
_AGE80	Coefficient of variation	32.650752	32.65075218	Std error mean	0.0897403	0.089740326

Table 5.1 – Moments table output by PROC UNIVARIATE

This table may look a little odd at first, so it is important to remember that the purpose of this table is to support the output of `PROC UNIVARIATE`. Consider how the `Moments` table looks when it prints on `PROC UNIVARIATE` output. There are four columns of printing:

- In the first column, the labels for the N, mean, standard deviation, skewness, uncorrected sums of squares, and coefficient of variation are printed.
- In the second column, the values for these are printed.
- The third column is again a list of labels.
- The fourth column is the values that belong to those labels.

Let's run PROC CONTENTS on this Moments dataset, and use the VARNUM option to make the variables print in their native order:

```
PROC CONTENTS data=X.Age_Moments VARNUM;
RUN;
```

Let's look at the variable list from the output:

Variables in Creation Order				
#	Variable	Type	Len	Format
1	VarName	Char	6	
2	Label1	Char	15	
3	cValue1	Char	10	
4	nValue1	Num	8	D12.3
5	Label2	Char	16	
6	cValue2	Char	10	
7	nValue2	Num	8	D12.3

Figure 5.7 – Variable list from PROC CONTENTS of the Moments table

The VarName variable is in character format, and the value is set to the name of the variable that was the subject of the Moments table, which was _AGE80. Label1 is also in character format, and refers to the first list of labels, while Label2 refers to the second list of labels. The cValue1 and nValue1 variables appear to have the same values, but it is important to recognize that cValue1 is in character format, and nValue1 is in numeric format. The same can be said for cValue2 and nValue2.

This section demonstrated how the ODS is used to create SAS datasets of results from PROCs. The ODS is also used extensively for outputting graphics files, which will be covered in the next section

Using the ODS for graphics files

So far, we've talked about using the ODS to output internal data files from PROCs. However, we can also use the ODS to save SAS output, including graphical output as files in *.pdf, *.rtf, and *.htm format, which will be described in this section.

Outputting graphics from analytic PROCs

The default output for `PROC UNIVARIATE` does not include any graphics. However, we can add a histogram to the output with the `histogram` option:

```
PROC UNIVARIATE data=X.chap5_1;
    var _AGE80;
    histogram _AGE80;
RUN;
```

This code produces the same output as before, along with this additional histogram:

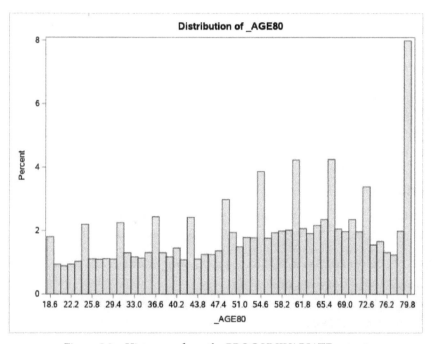

Figure 5.8 – Histogram from the PROC UNIVARIATE output

Outputting graphics in different formats

Imagine we needed to use this histogram graphic outside of SAS. We could use the ODS to output this graphic (along with the rest of the `PROC UNIVARIATE` output) into `*.pdf`, `*.html`, or `*.rtf` format.

PDF format

Let's start by using the ODS to output the results of the code in `*.pdf` format. We will tell SAS to name the `age_histogram.pdf` file and to output it into the directory mapped to `LIBNAME X`:

```
ODS PDF file="/folders/myfolders/X/age_histogram.pdf";
PROC UNIVARIATE data=X.chap5_1;
    var _AGE80;
    histogram _AGE80;
RUN;
```

Please observe the following note from the log file:

```
NOTE: Writing ODS PDF output to DISK destination "/folders/
myfolders/X/age_histogram.pdf", printer "PDF".
```

You will need to replace the path in the code with the path associated with the directory assigned `LIBNAME X`. After you run this code, a PDF named `age_histogram.pdf` should be present in the directory mapped to `X`. If you are using SAS University Edition, the PDF can be downloaded from the SAS environment to your local environment.

HTM format

You can also use the ODS to output graphics into `*.htm` format. Let's practice by outputting the same output from `PROC UNIVARIATE` into `*.htm` format:

```
ODS HTML file="/folders/myfolders/X/age_histogram.htm";
PROC UNIVARIATE data=X.chap5_1;
    var _AGE80;
    histogram _AGE80;
RUN;
```

Notice that the line of code that produces the `*.htm` file looks very similar to the code that produced the PDF file earlier. After the ODS is declared, an `htm` file is created instead of a `pdf` file. Also, in the name of the output file, the extension is `*.htm`, and not `*.pdf`.

When running the preceding code in SAS University Edition, you may see the following errors:

```
1          OPTIONS NONOTES NOSTIMER NOSOURCE NOSYNTAXCHECK;
NOTE: ODS statements in the SAS Studio environment may disable some output features.
73
74         ODS HTML file="/folders/myfolders/X/age_histogram.htm";
NOTE: Writing HTML Body file: /folders/myfolders/X/age_histogram.htm
75         PROC UNIVARIATE data=X.chap5_1;
76         var _AGE80;
77         histogram _AGE80;
78         run;

WARNING: GPATH or PATH is not a writable directory. It will be ignored.
ERROR: Cannot write image to SlantFactor. Please ensure that proper disk permissions are set.
ERROR: Cannot write image to SlantFactor. Please ensure that proper disk permissions are set.
NOTE: The SAS System stopped processing this step because of errors.
NOTE: PROCEDURE UNIVARIATE used (Total process time):
      real time          1.41 seconds
      cpu time           0.67 seconds

79
80         OPTIONS NONOTES NOSTIMER NOSOURCE NOSYNTAXCHECK;
92
```

Figure 5.9 – Error from log file when using the ODS to output *.html files in SAS University Edition

Although the log seems to be giving mixed messages, if you look in the directory mapped to LIBNAME X after running the code, you should find an *.htm file named age_histogram.htm. If you double-click on the file, it will open up in your default internet browser. Like *.pdf, this *.htm file can be downloaded from the SAS University Edition environment.

RTF format

Finally, we can demonstrate using the ODS for output in *.rtf format:

```
ODS RTF file="/folders/myfolders/X/age_histogram.rtf";
PROC UNIVARIATE data=X.chap5_1;
    var _AGE80;
    histogram _AGE80;
RUN;
```

You will observe this additional message in the log file:

```
NOTE: Writing RTF Body file: /folders/myfolders/X/age_
histogram.rtf
```

This code will output a **Rich Text Format (RTF)**, or *.rtf, file named age_histogram.rtf to the directory mapped to LIBNAME X. A *.rtf file is a word processing file that opens in Microsoft Word. Similar to *.pdf and *.htm files, you can download the *.rtf file from the SAS University Edition environment.

When you open the *.rtf file in Microsoft Word, you will find that the tabular results (such as the Moments table) are formatted as Microsoft Word tables, while the histogram graphic is a graphic image object placed in the Word document. It is possible to copy this graphic object and paste it into a graphic editing program such as the Microsoft Windows application Paint, or Adobe's Photoshop (which runs on Windows and Mac operating systems) and continue to edit the graphic. Also, if you are preparing a presentation, you may want to copy the graphic object from *.rtf and paste it into a Microsoft PowerPoint presentation.

The aim of this chapter is to cover topics related to managing data reporting in SAS. **Reporting** actually refers to a general function. How the reporting gets done depends on specifics. If the purpose of your reporting is to produce a professionally published file that is not intended to be edited, you may choose to do reporting by outputting to *.pdf. However, if you are creating a report that users expect to edit in a word processing program, then reporting by outputting to *.rtf would be preferable. Of course, if the goal is to report to the web, you would choose to report to *.htm.

Setting system options

So far, we have discussed setting options as part of running PROCs (such as using the VARNUM option in PROC CONTENTS). However, it is also possible to set system options as global options for the entire SAS session. This is done by running code during the session to change the system options.

We can illustrate setting system options with an example. By default, SAS always prints the current date on the output. Imagine we did not want the date printed on the output. Also, the default size of lines in SAS is 78. Let's say we wanted to set the line size instead to 72. We could run the following code:

```
options nodate linesize=72;
RUN;
```

Running this code would set default behavior for the session. Code running subsequently would produce output with no date on it, and with line sizes of 72, until we run other code to change these system options (or the session ends).

Let's try this experiment:

1. First, view the *.pdf file that was generated by running PROC UNIVARIATE with the ODS. You will see that in the header of each page, the date and time are printed.

2. Next, run the preceding code, setting the system options to nodate.

3. After that, rerun the `PROC UNIVARIATE` code with the ODS to regenerate the `*.pdf` file. Open the `*.pdf` file, and you will see that the date and time are no longer printed on the header of each page.

Running the following code will set the system options back to what they were:

```
options date linesize=78;
RUN;
```

The reason why the topic of setting system options is important to reporting in SAS is that depending upon the desired format of your output, these options can impact the appearance of the output. We saw this in the experiment we did with the `nodate` option.

Systems options can be set and reset during the session as code runs to improve the appearance of reports being generated. A link has been included under *Further reading* for more information about setting SAS system options. Setting systems options in SAS sessions offers even more flexibility in controlling the appearance of SAS output when reporting.

When the ODS was first invented, SAS users had to include ODS code in their programs to obtain files from the SAS environment. Later, SAS designed PROCs specifically for reporting functions that automatically use the ODS. These reporting PROCs will be covered in the next section.

SAS PROCs designed for reporting

Up to now, we have looked at graphics produced by setting options on analytic PROCs. Following the invention of the ODS, SAS invented new PROCs that were designed specifically to produce reporting output that automatically leverages the ODS as part of how they run. For tabular reporting, `PROC REPORT` and `PROC TABULATE` were developed, and for graphical reporting, `PROC SGPLOT`, `PROC SGPANEL`, `PROC SGSCATTER`, and `PROC SGRENDER` were created. These PROCs provide much more ability for the programmer to customize output. `PROC SGRENDER` is intended to be used with `PROC TEMPLATE`, and this provides even more opportunities for customization of reporting. This section will provide examples of how to use these PROCs.

Using PROC REPORT

PROC REPORT is easy to use if your report consists of mostly raw data, comparable to a **select query** in SQL. Imagine we wanted to view the age, sex, and state of the top 10 records in our dataset, Chap5_1. We could first make a reporting dataset using a data step:

```
data X.proc_rpt_data;
    set X.chap5_1 (OBS=10);
    if SEX1 = 1 or SEX1 = 2;
    label _AGE80 = "Age (years)";
    label SEX1 = "Sex";
    label _STATE = "State";
RUN;
```

Let's look at the details of this data step:

1. Observe that we are working with the directory mapped to LIBNAME X, and, from that directory, we are reading in dataset Chap5_1. We are outputting the report dataset proc_rpt_data into the same directory.

2. On the line with the set command, notice that we added (OBS=10) to tell SAS to only take the top 10 observations, or rows, from the source dataset chap5_1 to include in proc_rpt_data.

3. The subsequent if statement restricts the output dataset to those with codes 1 or 2 in SEX1, so records with other codes are filtered out.

4. Finally, we place labels on the three variables that we plan to include in our report using PROC REPORT, which are _AGE80, SEX1, and _STATE.

When we run this code, we create the output reporting datset proc_rpt_data. In our data step, we applied labels, which we will see printed in our output later from PROC REPORT. But we are also reporting on categorical variables that should be decoded on the report output – namely, SEX1 and _STATE.

The *BRFSS Codebook* indicates that the values in SEX1 are as follows:

- 1 = **Male**
- 2 = **Female**
- 7 = **Don't know/Not sure**
- 9 = **Refused**

We also described earlier that we only have three codes for _STATE in the dataset: 12 = **Florida (FL)**, 25 = **Massachusetts (MA)**, and 27 = **Minnesota (MN)**. So, prior to running PROC REPORT, let's first create formats for these variables:

```
PROC FORMAT;
    value sex_f
    1 = "Male"
    2 = "Female"
    ;
    value state_f
    12 = "Florida"
    25 = "Massachusetts"
    27 = "Minnesota"
    ;
RUN;
```

If we run the PROC FORMAT code, we will create formats that will decode the two levels of SEX1 (called sex_f) and the three levels of _STATE (called state_f). But in order for PROC REPORT to leverage these, we will need to attach these loaded formats to the variables as part of the PROC REPORT code. Here is the code:

```
TITLE BRFSS Dataset;
PROC REPORT DATA=X.proc_rpt_data;
COLUMNS _STATE SEX1 _AGE80;
DEFINE _STATE / DISPLAY format = state_f. 'State' CENTER;
DEFINE SEX1 / DISPLAY format = sex_f. 'Sex' CENTER;
DEFINE _AGE80 / DISPLAY 'Age' CENTER;
RUN;
```

Let's review the code:

1. Before the PROC REPORT command, we state a TITLE command saying we want to title the output BRFSS Dataset.

2. On the next line, we declare PROC REPORT and tell SAS to use our reporting dataset, X.proc_rpt_data.

3. The next line is the COLUMNS statement, where we define the arrangement of columns. Since we are reporting data in mostly raw form from three data columns, we state those three columns here: _STATE, SEX1, and _AGE80.

4. Each of the next lines uses the DEFINE command to format the column for each variable. For _STATE, we ask SAS to display the variable by applying format state_f. We tell SAS to label the column State and center the label.

5. For SEX1, we also ask SAS to display the variable, this time using format sex_f. We ask SAS to label the column Sex and center the label.

6. For _AGE80, we do not apply a format, but we tell SAS to label the column Age and center the label.

Once the code runs, we see the following output:

BRFSS Dataset

State	Sex	Age
Florida	Female	76
Florida	Male	72
Florida	Female	46
Florida	Female	51
Florida	Male	53
Florida	Female	80
Florida	Female	68
Florida	Female	80
Florida	Male	80
Florida	Female	62

Figure 5.10 – PROC REPORT output

At the top of the output, we see our title BRFSS Dataset, and we see the column headings we set, which are State, Sex, and Age. We also see that formats are applied, so we see Florida instead of 12, and Male and Female instead of 1 and 2.

PROC REPORT versus PROC TABULATE

Prior to the invention of PROC TABULATE, PROC REPORT was used to achieve similar output to PROC TABULATE. However, the output from PROC REPORT was designed to be displayed on a screen or printed on paper using a fixed-width font. When the ODS was invented, PROC REPORT tables could now leverage the ODS and provide a more finished final product (refer to the SAS white paper, *A Gentle Introduction to the Powerful REPORT Procedure*, by Ben Cochran; the link can be found under the *Further reading* section). Because of this, many users of PROC REPORT transitioned to PROC TABULATE when it was invented because it automatically uses the ODS and also provides more customizable options for display.

We demonstrated using PROC REPORT to display mostly raw data, as is often done with a SQL SELECT query. PROC REPORT is also capable of grouping records and making summary calculations, but, in practice, reports with those features in SAS are usually done using PROC TABULATE.

Understanding the basics of PROC TABULATE

Earlier, we used the ODS to output a dataset from a PROC – specifically, the Moments dataset from PROC UNIVARIATE. The purpose of doing that was to store the values in the Moments table as data so that we could use SAS to query those values later. In that example, we were saving the Moment as data; we were not planning on reporting those values.

Let's say we instead wanted to report summary statistics—such as the mean and standard deviation that are present in the Moments table—in a nicely formatted table. In that case, we could use PROC TABULATE. Many programmers refer to this PROC by its nickname, PROC TAB. PROC TABULATE affords the programmer the ability to use SAS code to modify tabular output so that it is presentable as a report.

Like PROC REPORT, PROC TABULATE leverages the functions of SAS labels and SAS formats. The syntax of PROC TABULATE affords the programmer the ability to design the columns, rows, and summary statistics present in the table output. The code needed to produce well-formatted PROC TABULATE output is structured in three parts:

1. Data step code that transforms the dataset to be reported in PROC TABULATE and adds SAS labels to variables.

2. PROC FORMAT code that creates formats that will be attached to the values in categorical variables reported in PROC TABULATE. These are temporary formats for display purposes.

3. The actual PROC TABULATE code.

To practice using PROC TABULATE, let's build a simple table together using our example dataset, chap5_1.sas7bdat. We will use _AGE80 as our continuous variable. Let's build a table that reports age-related summary statistics on the entire dataset, as well as stratified by men versus women (using the SEX1 variable). This means we will have three columns of statistics about _AGE80—all, male, and female. For our rows, let's add the _STATE variable, and do a bivariate analysis of _AGE80 by _STATE and SEX1. In our statistics, let's include the mean and standard deviation of _AGE80, as well as the frequency and percentage of people in each category subject to the statistics.

Preparing data for PROC TABULATE

Since we want three columns of statistics—one for all, one for male, and one for female—we need to only include records where SEX1 = 1 or SEX1 = 2 in the table we are creating. Also, since we have chosen to use the _AGE80, SEX1, and _STATE variables, we will need to apply labels to them.

As with PROC REPORT, to continue with the first step of PROC TABULATE, the following code uses a data step to create the output dataset named proc_tab_data, which will be stored as a *.sas7bdat dataset in LIBNAME mapped to X:

```
data X.proc_tab_data;
    set X.chap5_1;
    if SEX1 = 1 or SEX1 = 2;
    label _AGE80 = "Age (years)";
    label SEX1 = "Sex";
    label _STATE = "State";
RUN;
```

Note that we ensure that the output dataset, X.proc_tab_data, only has records with the values of 1 or 2 in the SEX1 variable due to the if statement in the data step code. Also notice the label statements attaching labels to the variables we will use in our PROC TABULATE example.

For the second step, we will set up formats we intend to use on the categorical variables SEX1 and _STATE:

```
PROC FORMAT;
    value sex_f
    1 = "Male"
    2 = "Female"
    ;
    value state_f
    12 = "Florida"
    25 = "Massachusetts"
    27 = "Minnesota"
    ;
RUN;
```

Notice in the code how we named the format that we will use for `SEX1` `sex_f` and that the format intended for `_STATE` was named `state_f`. These naming conventions will help us avoid confusion between the variable name and the name of the variable's format.

Formulating PROC TABULATE code

Finally, we arrive at the third step, which is formulating the `PROC TABULATE` code. Wendi L. Wright provides an excellent beginner's guide to using `PROC TABULATE` in her white paper, *PROC TABULATE and the Neat Things You Can Do with It* (link available under the *Further reading* section). Here is a simplification of Wright's characterization of the syntax behind `PROC TABULATE` code:

```
PROC TABULATE <options>;
      FORMAT variables </options>;
      CLASS variables </options>;
      VAR variables </options>;
      TABLE <programming with commas>;
RUN;
```

Figure 5.11 – Syntax behind PROC TABULATE

`PROC TABULATE` code is quite complicated, and therefore, it is usually built in steps. When the programmer is done building the code, the code usually starts with a `FORMAT` statement to attach SAS `formats` to categorical variables, and then includes a `CLASS` statement to specify the categorical variables involved in the table. This is followed by the `VAR` statement. The variables specified in the `VAR` statement must be numeric as they will be the subject of summary statistics. Finally, the complicated `TABLE` statement specifies the structure of the table and what summary statistics to include in it. This statement uses commas as part of its syntax.

> **Details about the TABLE statement in PROC TABULATE**
>
> These are paraphrased from Wendi L. Wright's SAS white paper mentioned earlier.
>
> **TABLE Statement Structure:**
>
> The TABLE statement begins with dimension expressions. There can be up to three dimension expressions. These are followed by table options that appear at the end of the line of code after /.
>
> Commas are used to separate the dimensions specified in the TABLE statement.
>
> The order of the dimensions is page, row, and column. If you only specify one dimension, it will default to column. If you specify two, it will default to row, then column.
>
> **TABLE Statement Features:**
>
> You can have multiple table statements in one PROC TABULATE statement. If you do this, one table will be generated for each statement.
>
> All variables listed in the TABLE statement must be listed earlier in the PROC TABULATE code, either under the CLASS statement if they are *categorical*, or under the VAR statement if they are *numerical*.
>
> There are many statistics that can be specified in the TABLE expression, including row and column percents, counts, means, and percentiles. See the *Further reading* section for a link to an SAS support page listing all the statistics available in PROC TABULATE.

Let's now begin to build PROC TABULATE code to report on our dataset, X.proc_tab_data. Let's start our code simply, where we are reporting on _AGE80 only:

```
PROC TABULATE data=X.proc_tab_data;
    var _AGE80;
    table       ALL,
                _AGE80;
RUN;
```

Before we consider the code, let's see what the output looks like:

	Age (years)
	Sum
All	2105113.00

Figure 5.12 – Output from simple PROC TABULATE code

Now let's look back at our code. We do not include FORMAT or CLASS statements for now, because we have not included any categorical variables yet. Since we are only looking at the numeric variable, _AGE80, in this initial code, we specify this variable under VAR. Finally, for the TABLE statement, we specify two dimensions:

- ALL in the row dimension
- _AGE80 in the column dimension

This code results in a sum of all the ages in the dataset. Let's build upon that code. In our next step, let's change the reporting of the statistics about age. We are not interested in the sum of ages. Instead, let's tell SAS to output the number of records, the column percentage, the mean of _AGE80, and the standard deviation of _AGE80. We will modify the TABLE statement from our earlier code:

```
PROC TABULATE data=X.proc_tab_data;
    var        _AGE80;
    table      ALL,
               _AGE80 *(ALL)*(n colpctn*f=4.1 mean std);
RUN;
```

Let's consider the output:

	Age (years)			
	All			
	N	ColPctN	Mean	Std
All	38826	100	54.22	17.70

Figure 5.13 – Output from simple PROC TABULATE code reporting summary statistics

Notice how the code that was added to the TABLE statement resulted in adding the statistics we desired to the output. _AGE80 is nested with ALL and SEX1. N, COLPCTN, MEAN, and STD are displayed in the same order as specified in the TABLE statement. Now, in the next step, let's add columns for SEX1:

```
PROC TABULATE data=X.proc_tab_data;
    format     SEX1 sex_f.;
    class      SEX1;
    var        _AGE80;
    table      ALL,
               _AGE80 *(ALL SEX1)*(n colpctn*f=4.1 mean std);
RUN;
```

Notice that the code now includes a format statement attaching the format we developed earlier named sex_f to the categorical variable SEX1. Also, notice that the class statement has been added to allow us to specify categorical variable SEX1. Finally, in the TABLE statement, observe that the SEX1 variable has been added in the parentheses after ALL. After adding this format statement and class statement, and modifying the TABLE statement, you will see that the two values of SEX1, 1 and 2, are added as columns in the output:

	Age (years)											
	All				Sex							
					Male				Female			
	N	ColPctN	Mean	Std	N	ColPctN	Mean	Std	N	ColPctN	Mean	Std
All	38826	100	54.22	17.70	18166	100	52.84	17.78	20660	100	55.43	17.54

Figure 5.14 – PROC TABULATE for bivariate output

Let's now add a row for _STATE to the output:

```
PROC TABULATE data=X.proc_tab_data;
    format      SEX1 sex_f.
                _STATE state_f.;
    class       SEX1
                _STATE;
    var         _AGE80;
    table       ALL _STATE,
                _AGE80 *(ALL SEX1)*(n colpctn*f=4.1 mean std);
RUN;
```

This changes our output by adding a set of rows for each level of _STATE:

	Age (years)											
	All				Sex							
					Male				Female			
	N	ColPctN	Mean	Std	N	ColPctN	Mean	Std	N	ColPctN	Mean	Std
All	38826	100	54.22	17.70	18166	100	52.84	17.78	20660	100	55.43	17.54
State												
Florida	15238	39.2	56.14	17.83	6811	37.5	54.89	17.91	8427	40.8	57.15	17.69
Massachusetts	6646	17.1	53.47	17.83	3082	17.0	52.20	18.03	3564	17.3	54.56	17.58
Minnesota	16942	43.6	52.79	17.37	8273	45.5	51.39	17.40	8669	42.0	54.12	17.23

Figure 5.15 – Final PROC TABULATE table

Notice that because we used the `colpctn*f=4.1` command in the `TABLE` statement, the percentages are formatted to one place after the decimal. Now, we have a bivariate analysis of age by state.

The importance of the function of `PROC TABULATE` may not be immediately apparent when considering how difficult it is to program compared with how simple it is to format a Microsoft Excel spreadsheet with the same summary statistics. `PROC TABULATE` is certainly one of the best PROCs available for formatting a table's output to the internet.

For example, the US government put up a BRFSS website that allows the user to query the data over the internet (refer to the link under *Further reading*). Here is a screenshot of a part of the output from an example query looking at the prevalence of smokers in Florida in the 2018 BRFSS:

Florida - 2018

Adults who are current smokers (variable calculated from one or more BRFSS questions)

(Crude Prevalence)

View by: Overall

Response: (All)

	Yes	No
Percent (%)	14.5	85.5
95% CI	13.3 - 15.6	84.4 - 86.7
n	2480	12109

Figure 5.16 – Screenshot of output from the BRFSS website

As illustrated by the screenshot, `PROC TABULATE` basically provides the template for formatting tabular reports that are output in any format, including the web (via the ability to output as `*.htm` via the ODS). You will see later in this chapter that `PROC TEMPLATE` is another `PROC` that can be used for outputting reports to the web, but for tabular output, it requires even more programming than `PROC TABULATE`. It is likely that this output displayed on the web is the result of either a `PROC TABULATE` or a `PROC TEMPLATE` command used in conjunction with the ODS to output an `*.htm` file displayed on the web.

This approach to reporting SAS data on the web poses two problems:

- First, there is the challenge and limitations associated with being able to achieve the desired formatting of output with PROC TABULATE (or even PROC TEMPLATE) code. Notice that the BRFSS output shows large numbers formatted without commas, and percentages formatted without percent signs.

- The second problem specifically with PROC TABULATE is that it is a very inefficient PROC, and therefore, **input/output (I/O)** suffers. This is apparent when you compare the experience of using the BRFSS website for queries to using another commonly queried dataset with a web interface, such as a travel site that sells airplane tickets and hotel rooms.

As described in *Chapter 1*, *Using SAS in a Data Mart, Data Lake, or Data Warehouse*, many PROCs in SAS suffer from challenges to I/O. Combining this problem with the challenges of displaying data on the web is perhaps the biggest obstacle to effectively using web services provided by SAS.

Note

In an SAS white paper, *With a Trace: Making Procedural Output and ODS Output Objects Work for You* (link available under *Further reading*), Louise Hadden describes the following general procedure for using the ODS to access objects from SAS PROCs:

Step #1: Use ODS TRACE with your PROCs to identify and locate output objects or datasets.

Step #2: Using SAS PROCs, study the object or dataset you want to output using the ODS.

Step #3: Using SAS data steps or PROCs, transform or otherwise manipulate the datasets or objects as needed for your purposes.

Step #4: Report on the final graph or dataset.

In the big picture, these four steps illustrate that when it comes to SAS outputting datasets and graphics using the ODS, there are some caveats. First, since the ability of the programmer to access objects and files made during PROCs was a functionality built into SAS in the form of the ODS long after SAS PROCs were invented, using the ODS to access these files requires extra effort. Second, since these files were originally developed to serve the processing needs of the PROC, the programmer using these files for reporting will likely want to modify these files, such as adding labels to graphs, or modifying the formatting of tables. Since those efforts are also challenging, in the big picture, while the ODS supports the functionality of reporting in SAS, reporting in SAS is much more challenging than it is in newer software that was designed to make reporting easy, such as Tableau.

PROC TABULATE is a reporting PROC that automatically uses the SAS ODS to produce nicely formatted tables. Let's now look at some PROCs that automatically use the SAS ODS to produce plots.

Using PROC SGPLOT

As detailed in *Chapter 1*, *Using SAS in a Data Mart, Data Lake, or Data Warehouse*, SAS has historically been a *big data* program. The downside to this is that SAS has also always been challenged in the way it deals with graphical output. Earlier, we looked at the histogram output from the histogram option in PROC UNIVARIATE. Using the HISTOGRAM option with PROC UNIVARIATE is helpful to the analyst who simply wants to quickly view the distribution of a continuous variable in a dataset.

However, actual reporting requires some extra effort, because you are creating an asset intended for an audience to read so as to gain knowledge. Imagine you wanted to make a very detailed and customized set of graphics from some data in an SAS warehouse. Maybe you would want to display a bar plot where you manipulated the resulting format and text in the plot. Perhaps you might want to display a map with rates of disease (or sales) by region.

Being able to do these kinds of tasks in SAS has always been a complex affair. SAS first tried to simplify the process of making camera-ready graphics by developing PROC SGPLOT. PROC SGPLOT is an extremely extensible PROC aimed at providing the programmer with more power when making graphics. Before PROC SGPLOT, SAS had an **SAS/GRAPH** capability that saved graphs in SAS graphics catalogs that could be viewed in a graph window. The SAS/GRAPH component is separate from the base SAS component. GOPTIONS statements were available to control the appearance of the graph in SAS/GRAPH.

By contrast, PROC SGPLOT leverages the ODS as a part of how the PROC runs, so it does not need to be specified with ODS code. The graphs produced by PROC SGPLOT come out in image formats such as *.png and *.jpg, and can be viewed in browsers used for viewing *.htm output. PROC SGPLOT has its own set of code and syntax, and GOPTIONS statements from SAS/GRAPH have no effect.

PROC SGPLOT is capable of making histograms, bar charts, line graphs, box plots, density plots, scattergrams, dot plots, and other visualizations. The SAS white paper, *Using PROC SGPLOT for Quick High-Quality Graphs*, by Susan J. Slaughter and Lora D. Delwiche, provides an excellent resource for designing PROC SGPLOT code to produce SAS visualizations (link available under the *Further reading* section).

Let's practice with PROC SGPLOT by making a histogram of _AGE80 as we did before, using PROC UNIVARIATE with the ODS. First, let's run some very basic PROC SGPLOT code using our SAS dataset, Chap5_1.sas7bdat, saved in the directory mapped to LIBNAME X and making a histogram of the age variable, _AGE80:

```
PROC SGPLOT DATA = X.Chap5_1;
    HISTOGRAM _AGE80;
RUN;
```

The resulting histogram will open in the output window. If you are using SAS University Edition, the **RESULTS** tab will display the graphic. There are buttons on the menu in the upper left that can be used to download the graphic in *.htm, *.pdf, and *.rtf format:

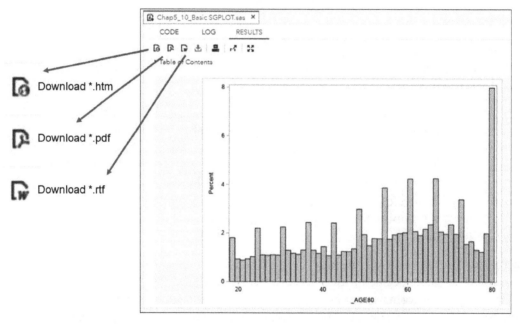

Figure 5.17 – Screenshot of the RESULTS tab from SAS University Edition with the download buttons annotated

Each visualization available to the programmer in PROC SGPLOT comes with its own set of options that can be set. For a histogram, the options available include SHOWBINS, which will add a tick mark to the middle of each bar along the X axis, and SCALE, which specifies the scale used on the Y axis (choices include PERCENT, COUNT, and PROPORTION). For a histogram in PROC SGPLOT, you can also add a density curve, which offers the same SCALE options, and a choice of the type of density curve, NORMAL or KERNAL.

Let's revise our PROC SGPLOT code to add SHOWBINS to our histogram, and set the scale at PERCENT. Let's also add a density curve of the NORMAL type, with the scale also set at PERCENT. We will also add a TITLE to the plot:

```
PROC SGPLOT DATA = X.Chap5_1;
    HISTOGRAM _AGE80 / SHOWBINS SCALE=PERCENT;
    DENSITY _AGE80 / SCALE=PERCENT TYPE=NORMAL;
    TITLE "Age in BRFSS 2018";
RUN;
```

To fetch the graphic, we will use the **Download as *.rtf** button. Next, let's open the *.rtf file and copy the graph into *.jpg (using a graphic editing program). The result will look like this:

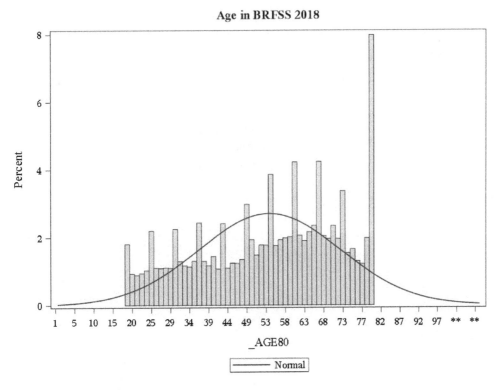

Figure 5.18 – Histogram graphic from the *.rtf file

Notice that the graphic in the *.rtf file looks slightly different to what was printed to screen. Although this PROC SGPLOT output is more attractive that what could be achieved through the HISTOGRAM option in PROC UNIVARIATE, even with a few extra options available, it is not ultimately customizable.

PROC SGPLOT automatically uses the SAS ODS to produce a variety of graphics. Let's now look at two other SAS PROCs that make plots by using the ODS, namely, PROC SGPANEL and PROC SGSCATTER.

Using PROC SGPANEL and PROC SGSCATTER

PROC SGPLOT is actually part of a family of PROCs that includes PROC SGSCATTER, PROC SGPANEL, and PROC SGRENDER, which were the first PROCs designed to produce standalone graphs as their main goal (as opposed to PROCs that simply produced graphs as part of the PROC, such as PROC UNIVARIATE). They are considered part of the SAS ODS **Graphics Template Language (GTL)**. PROC SGPANEL produces plots in a matrix of panels. Here is an example of using PROC SGPANEL with the SAS dataset we used to practice with PROC TABULATE called proc_tab_data.sas7bdat:

```
PROC SGPANEL DATA = X.proc_tab_data;
    PANELBY SEX1;
    HISTOGRAM _AGE80;
    Title "Age Distribution by Sex";
RUN;
```

In the code, the PANELBY command is set to the categorical variable separating the panels (in our case, SEX1), and the HISTOGRAM command is set to _AGE80. A title is added, and the graph taken from the *.rtf output looks like this:

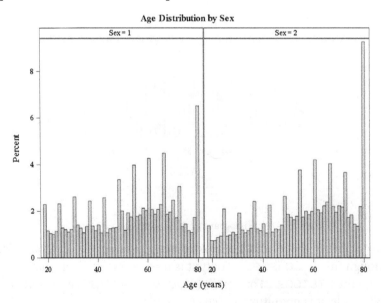

Figure 5.19 – Histogram graphic from PROC SGPANEL

Notice how the two histograms for _AGE80, one for each level of SEX1, are placed side by side.

The third PROC in the family, PROC SGSCATTER, makes a scatter plot. We can demonstrate by using our two continuous variables, _AGE80 and _BMI5. First, we'll use the previously created proc_tab_data dataset to create an example dataset for PROC SGSCATTER called proc_sgscatter_data, and place it in the directory mapped to LIBNAME X. Since the BMI calculation is usually expressed as a two-digit integer, and the _BMI5 variable has four positions—two whole numbers and two decimals—we'll create and graph a different variable, _BMI2, which reflects how BMI is usually expressed. For _BMI2, we'll divide _BMI5 by 100 and round to the nearest whole number. We will also label the new _BMI2 variable, as well as apply the sex_f format we created earlier to SEX1:

```
data X.proc_sgscatter_data;
    set X.proc_tab_data;
        _BMI2 = round(_BMI5/100, 2);
    label _BMI2 ="2-digit Body Mass Index";
    format SEX1 sex_f.;
RUN;
```

Now we are ready to formulate PROC SGSCATTER code to create a scatter plot using the X.proc_sgscatter_data dataset. Let's say we hypothesize that the older we get, the more likely we are to gain weight. This would suggest that our scatter plot should have _AGE80 (age) on the x-axis, and _BMI2 (BMI, a measure of weight status) on the y-axis. This is achieved by means of the following code:

```
PROC SGSCATTER data=X.proc_sgscatter_data;
plot _BMI2*_AGE80 / group=SEX1;
RUN;
```

Interestingly, in order to have _AGE80 appear on the x-axis in the plot, it must be declared second in the statement plot, _BMI2*_AGE80. Please also note the option after the plot statement of / group=SEX1. This tells SAS to plot each level of SEX1 differently. On the screen (and if you download the *.pdf file), the scatter plot appears with different colored open circles, blue for female and red for male:

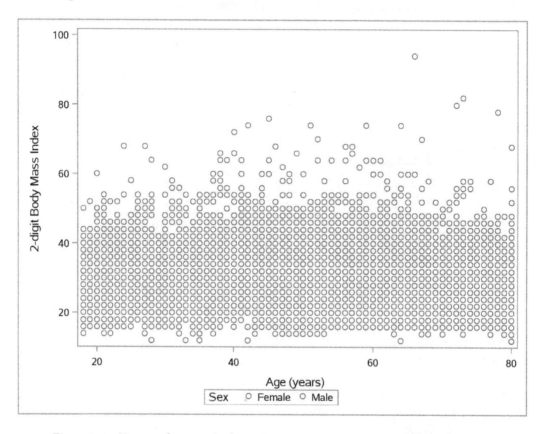

Figure 5.20 – Version of scatter plot from PROC SGSCATTER printed to screen and *.pdf

It is notable that the *.rtf version of the plot looks remarkably different, with the red open circle replaced with a plus, and further layering of points:

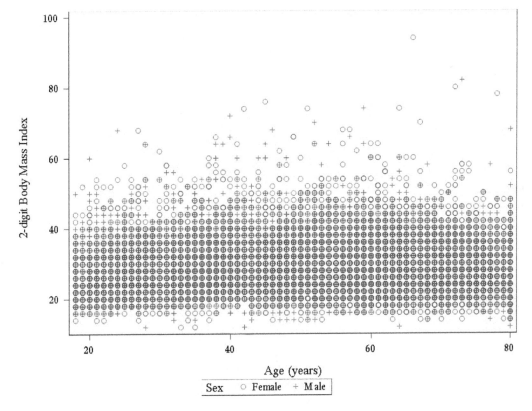

Figure 5.21 – Version of scatter plot from PROC SGSCATTER from *.rtf

It is worth reviewing a note from the log file:

```
NOTE: Marker and line antialiasing has been disabled for at least one plot because the threshold has been reached. You can set
      ANTIALIASMAX=38900 in the ODS GRAPHICS statement to enable antialiasing for all plots.
```

Figure 5.22 – Note from the log file for PROC SGSCATTER

Anti-aliasing has to do with a graphical technique used in visualizing curves, where some pixels along a curve are set to an intermediate color to make the curve look smoother (refer to the blog post, *Ahh, that's smooth! Anti-aliasing in SAS statistical graphics*, by Rick Wicklin – link under *Further reading*). Options can be set on the ODS for anti-aliasing, but a discussion of these is beyond the scope of this book.

PROC's SGPLOT, SGPANEL, and SGSCATTER use the ODS to create plots. Another way to use the ODS for plots in SAS is to use PROC TEMPLATE with PROC SGRENDER, as will be described in the next section.

Using PROC TEMPLATE with PROC SGRENDER

The last PROC from the family of PROCs that leverage the ODS for visualization is PROC SGRENDER, which works intimately with another PROC, PROC TEMPLATE. To use these PROCs, first, PROC TEMPLATE is used to generate a template for a plot. The template is named and saved in a location, and declares specific variables to be rendered, although it is not attached to a specific dataset. After this, PROC SGRENDER is used to specify the dataset to plot, and to apply the template developed in PROC TEMPLATE to generate the graph.

Let's continue with the example of the scatter plot between _AGE80 and _BMI2 using the X.proc_sgscatter_data dataset. First, we will create a template using PROC TEMPLATE named mygraphs.scatter. We will tell SAS to make a template for a scatter plot of _AGE80 and _BMI2 grouped by SEX1. However, no graph will appear because PROC TEMPLATE only makes and saves the template – it does not render the graphic:

```
PROC TEMPLATE;
define statgraph mygraphs.scatter;
begingraph;
     layout overlay;
     scatterplot x=_AGE80 y=_BMI2 / GROUP = SEX1;
     endlayout;
endgraph;
end;
RUN;
```

The log file will report the following: **NOTE: STATGRAPH 'Mygraphs.Scatter' has been saved to: WORK.TEMPLAT**

In other words, the template we named mygraphs.scatter was saved to SAS's WORK directory under WORK.TEMPLAT. We can see that reflected in our first line of code, which says define statgraph mygraphs.scatter;. Notice that in PROC TEMPLATE, there are the begingraph and endgraph commands, while the code in between specifies details pertaining to the template. Notice that within that code, there are the layout and endlayout commands. The code in between those commands specifies the layout of the graph.

Now that we have created our template, the next step will involve using `PROC SGRENDER` to apply the `mygraphs.scatter` template to the `X.proc_sgscatter_data` dataset and observing the resulting plot:

```
PROC SGRENDER data=X.proc_sgscatter_data
                template="mygraphs.scatter";
RUN;
```

Running this code in SAS University Edition will output the following graphic to the screen:

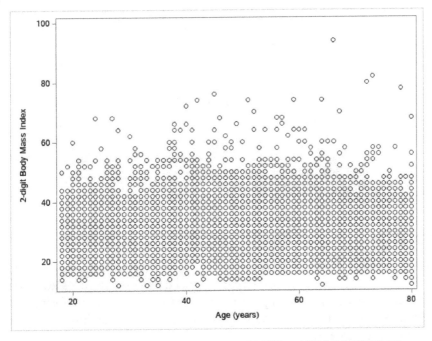

Figure 5.23 – Scatter plot from PROC TEMPLATE and PROC SGRENDER

The result is not superior to the scatter plot we saw before from `PROC SGSCATTER`, mainly because this is just a simple demonstration without any added options. `PROC TEMPLATE` is an extremely extensible `PROC`. For example, in his SAS white paper, *A Programmer's Introduction to the Graphics Template Language*, Jeff Cartier describes ways to make scatter plot templates with *ellipses* and *color-coded density* (link available under *Further reading*). SAS also provides a graphic template tip sheet that lists potential options that can be used with `PROC TEMPLATE`, such as plot statements to modify the plot appearance, options to manipulate the axes, and overall guidance in terms of using the *GTL* (link available under *Further reading*). Together, `PROC TEMPLATE` and `PROC SGRENDER` provide a lot of options for customization in developing SAS graphics.

The amazing flexibility of PROC TEMPLATE is illustrated in the SAS white paper, *Using Styles and Templates to Customize SAS ODS Output*, by Sunil Gupta. In it, the author reviews the different facets of template design and development (refer to the link under *Further reading*). He discusses ways to control style in PROC TEMPLATE, such as being able to change fonts, headers, footers, and margins. He also describes ways to control table formatting in templates, such as being able to change attributes of row headers and columns. He reviews combining these selections with different output formats (such as *.pdf, *.rtf, and *.htm). He discusses template types and defining style elements. At the end of the paper, he provides sets of annotated code for different templates.

Unfortunately, this white paper also illustrates the challenge SAS faces when it comes to delivering visualizations within its native components. The complex code behind PROC TEMPLATE can be attributed to the fact that visualizations in SAS were an afterthought.

Given how long ago SAS was originally developed, even the ODS was an afterthought. On the one hand, it is impressive that such software can stand the test of time, and how PROCs can be developed in the modern day that display data on the web while still utilizing SAS's original engine for processing. On the other hand, compared to data visualization applications developed specifically for the modern era, such as Tableau and dashboard packages deployed in RStudio, SAS visualizations appear clumsy and suffer from low responsiveness due to I/O issues. Workarounds to using SAS for visualization will be described in *Chapter 12, Using the ODS for Visualization in SAS*.

Summary

This chapter introduced you to the approaches SAS takes to reporting. First, we reviewed the function of the ODS, and used it to output internal data tables from a PROC. Next, we used the ODS to save graphical output to *.pdf, *.rtf, and *.htm formats. These examples reflect SAS's first upgrades and modifications aimed at enabling the programmer to output graphics and data from SAS PROCs.

SAS subsequently created PROCs for the express purpose of leveraging the ODS for attractive output. We went over PROC REPORT and PROC TABULATE, which is a PROC used specifically for developing reports in tabular format. Next, we practiced making plots using PROC SGPLOT, PROC SGPANEL, and PROC SGSCATTER. Like PROC TABULATE, these graphical PROCs provide more options for customization than PROCs that produce graphics as an option.

However, the greatest options for customization lie in using PROC TEMPLATE to design templates coupled with PROC SGRENDER to connect datasets to templates and render a graphic. While PROC TEMPLATE provides many opportunities for customization, like PROC TABULATE, it is a complex PROC involving many options and a lot of programming. These issues, coupled with SAS's historically inefficient I/O, pose serious challenges to presenting modern data visualizations from an SAS data warehouse, especially responsive ones on the web.

In the next chapter, we turn our attention away from reporting and back to data management. We revisit using SAS arrays in programming, which we originally discussed in *Chapter 4, Managing ETL in SAS*. Arrays can be helpful to automate processing, especially for data transformation. However, complex array programming can be difficult to maintain and troubleshoot. We will consider these issues carefully as we practice using arrays in SAS programming in the next chapter.

Questions

1. What is PROC TABULATE for?

2. How does the ODS allow the programmer to access tables generated during the running of a PROC?

3. What code needs to be included in order for PROC SGPLOT to use the ODS?

4. The leader of a department wants the SAS report developer to create a quarterly sales report that includes nicely formatted tables. The department leader wants to write paragraphs below the tables in order to interpret them, and then regard this as the department's official quarterly sales report. What is the most useful format in which to provide these tables to the department leader?

5. An SAS programmer receives an SAS dataset with about 200 variables. The programmer wants to output the list of variable names to an SAS dataset so that the variable names can be queried. How might the programmer approach this problem using the ODS?

6. What are the differences between making a histogram in PROC UNIVARIATE, making one in PROC SGPLOT, and making one using PROC TEMPLATE with PROC SGRENDER?

7. You are an analyst on an interdisciplinary team. You run a histogram of a continuous variable and decide, because of the distribution you see, to perform a log transformation on the variable. You want to share your rationale with your team. What do you think would be the most appropriate approach to developing this histogram graphic for communicating with the team?

Further reading

- SAS white paper, *Build Your Metadata with PROC CONTENTS and ODS Output*, by Louise Hadden, available here: `https://support.sas.com/resources/papers/proceedings14/1549-2014.pdf`

- Link to SAS support page listing all the ODS tables produced with the PROC UNIVARIATE statement: `https://support.sas.com/documentation/cdl/en/procstat/63963/HTML/default/viewer.htm#procstat_univariate_sect051.htm`

- Link to SAS support page listing ODS table names produced by base SAS procedures: `https://documentation.sas.com/?docsetId=odsproc&docsetTarget=p037wkiv6e4hqln1snmfk9b7c9it.htm&docsetVersion=9.4&locale=en`

- SAS documentation on setting system options: `https://documentation.sas.com/?docsetId=hosto390&docsetTarget=p0ek9qjzqu2li5n1seq9z0s1gayi.htm&docsetVersion=9.4&locale=en`

- SAS white paper, *A Gentle Introduction to the Powerful REPORT Procedure*, by Ben Cochran, available here: `https://support.sas.com/resources/papers/proceedings/proceedings/sugi30/259-30.pdf`

- SAS white paper, *PROC TABULATE and the Neat Things You Can Do with It*, by Wendi L. Wright, available here: `http://www2.sas.com/proceedings/forum2008/264-2008.pdf`

- SAS support page listing statistics available in PROC TABULATE: `https://documentation.sas.com/?docsetId=proc&docsetTarget=p0n4welprckk8yn1ro9swaef6x0n.htm&docsetVersion=9.4&locale=en`

- Website that allows the user to explore BRFSS data: `https://www.cdc.gov/brfss/brfssprevalence/index.html`

- SAS white paper, *With a Trace: Making Procedural Output and ODS Output Objects Work for You*, by Louise Hadden, available here: `https://www.lexjansen.com/pharmasug/2019/DV/PharmaSUG-2019-DV-003.pdf`

- SAS white paper, *Using PROC SGPLOT for Quick High-Quality Graphs*, by Susan J. Slaughter and Lora D. Delwiche, available here: `https://support.sas.com/resources/papers/proceedings10/154-2010.pdf`

- Blog post, *Ahh, that's smooth! Anti-aliasing in SAS statistical graphics*, by Rick Wicklin, available here: `https://blogs.sas.com/content/iml/2016/10/24/antialiasing-in-sas-graphics.html`

- SAS white paper, *A Programmer's Introduction to the Graphics Template Language*, by Jeff Cartier, available here: `https://support.sas.com/resources/papers/proceedings/proceedings/sugi31/262-31.pdf`

- Graph template language tip sheet from SAS: `https://support.sas.com/rnd/app/ODSGraphics/TipSheet_GTL.pdf`

- SAS white paper, *Using Styles and Templates to Customize SAS ODS Output*, by Sunil K. Gupta, available here: `https://support.sas.com/resources/papers/proceedings/proceedings/sugi29/246-29.pd`

Section 2: Using SAS for Extract-Transform-Load (ETL) Protocols in a Data Warehouse

This section explains how to develop and optimize **extract, transform, and load** (ETL) protocols in a SAS data warehouse. First, we cover using **SAS arrays**, and why **array processing** is a necessary component of a SAS ETL protocol. Next, we go through a process of first designing an ETL protocol, and then developing SAS code to execute the protocol. Then, we describe automating ETL code using the **SAS macro language**, and show how to convert data step code into macros. And finally, we discuss various approaches used in SAS for debugging and troubleshooting, with a specific focus on debugging do loops and macros.

This section comprises the following chapters:

6
Standardizing Coding Using SAS Arrays

In this chapter, we will revisit **SAS arrays** and consider why they are usually necessary in **extract, transform, and load** (**ETL**) when processing code for a SAS data warehouse. First, we will go over examples of where arrays can be useful and demonstrate creating a set of variables as an **output array**. We'll do this by using a set of native variables as an **input array**.

Next, we'll go over when we should add **conditions** to an array, as well as how array processing changes when conditions are added. Arrays are often used when creating **index variables**, so a discussion will be included about factors for SAS data warehouse managers to consider when choosing to serve up index variables or components for their formulas in the SAS data warehouse.

We will then discuss how array processing is documented in a SAS data warehouse, before ending this chapter with a summary of limitations imposed by using SAS arrays, including limitations connected to variable naming and troubleshooting code.

By the end of this chapter, you will have learned how to create a SAS array, as well as use it in a data step to recode a series of similarly coded variables. You will be able to use naming conventions with arrays and array variables that improve processing. You will also know how to document and standardize arrays, as well as what the limitations are in using arrays.

To summarize, this chapter will cover the following main topics:

- Understanding characteristics of SAS arrays

- Recoding a variable based on a native variable using a SAS array

- Considering both advantages and limitations of using arrays in SAS programming, such as naming limitations and the limitation of arrays needing to be called in a data step

Technical requirements

The following are the technical requirements for this chapter:

The dataset in `.sas7bdat` format, which will be used in a demonstration in this chapter, can be found online on GitHub: `https://github.com/PacktPublishing/ Mastering-SAS-Programming-for-Data-Warehousing/tree/master/ Chapter%206/Data`.

The code bundle for this chapter is available on GitHub here: `https://github. com/PacktPublishing/Mastering-SAS-Programming-for-Data- Warehousing/tree/master/Chapter%206`.

Understanding examples of arrays used to create variables

So far, when we have talked about **arrays**, we have focused on the technical aspects of programming them, such as their syntax and how they operate inside a **data step**. In this chapter, we will shift gears and focus on the practical side of using arrays to create variables *en masse*, which is often done when performing ETL protocols as part of managing a SAS data warehouse.

First, this section will focus on the most common use of arrays in SAS data warehouses, which is to create new warehouse variables based on variables that exist in the native dataset. Next, we will focus on how adding **conditions** to array processing can improve the quality of output, but can also add complexity to operations.

Array processing in SAS is fast and provides the warehouse manager with an opportunity to standardize ETL protocols. Standardizing can be done through array documentation and by using standard arrays during ETL. However, using arrays in a SAS data warehouse also presents limitations, all of which will be discussed at the end of this chapter.

Scenarios where arrays are useful

A typical scenario in a data warehouse where array processing is useful is when a data provider regularly gives the warehouse datasets that are structured in the same way. An example of a dataset that fits this description is the **Behavioral Risk Factor Surveillance Survey** (**BRFSS**) dataset, an annual phone health survey that's done in the United States (see *Chapter 1*, *Using SAS in a Data Mart, Data Lake, or Data Warehouse*, for more details).

For this example, we are going to focus on the part of the survey where respondents are asked if they have ever been told by a health professional that they have various health conditions or diseases called **co-morbidities**. We are going to use arrays to create variables that can help us create a **disease index**, or otherwise reduce the number of variables in the analysis about the disease that will ultimately be performed on warehouse data. Since we are doing this activity as warehouse providers rather than analysts, we may simply choose to make the variables that could go into an **index available**. This can help analysts create the index variable on their own.

If we want to go farther as data warehouse designers, we can actually create an index variable ourselves that can be added to the warehouse variables available for request (and documented in warehouse data curation). The pros and cons of creating and serving up an index ourselves as warehouse data providers will be discussed later in this chapter.

Index variables and parsimonious statistical models

SAS data warehouse users include many types of users; some will be statisticians, who perform statistics on a subset of the data in the warehouse. They have a research aim or hypothesis, and will likely use a *regression model*, including warehouse data, as a way of conducting their statistical tests and answering the research aim or hypothesis.

When using big data in statistical models to answer questions, many analysts strive for a **parsimonious model**, or one with the fewest variables in it (for an example, see *The potential of parsimonious models for understanding large-scale transportation systems and answering big picture questions*, the link to which is available in the *Further reading* section). One strategy that's used for building parsimonious models is to include index variables instead of all of the components of the index variables. This helps reduce the number of variables in the model, but still carries the information behind the variables into the model.

Whether we choose to serve up index variables or not, the reason why array processing is particularly applicable to this problem is that the BRFSS is an annual survey, and the questions change very little from year to year (especially questions about co-morbidities, as these are used to establish trends from year to year). This means that a SAS data warehouse hosting the BRFSS survey data would need to process these co-morbidity variables every year, and ensure that they are processed the same way each time. This is a scenario where array processing can be very useful for improving SAS programming efficiency, as well as standardizing ETL rules.

For the examples in this chapter, we are working with a subset of the 2018 BRFSS dataset that's comprised of records from only three states:

- Florida (FL, code = 12)
- Massachusetts (MA, code = 25)
- Minnesota (MN, code = 27)

The dataset is named Chap6_1.sas7bdat and is in SAS format. We will use SAS programming to take a look at what is in the dataset:

1. Let's start by setting a LIBNAME in SAS to X that corresponds to the folder that the dataset resides in.

2. Next, we will use a data step to bring the dataset into the SAS WORK directory. Notice that we will initially name the dataset brfss_a and that as we add variables through processing, we will increment the appended letter at the end of the dataset name as we iterate the dataset, as described when we talked about naming conventions presented in *Chapter 4, Managing ETL in SAS*.

3. Then, we will run a PROC CONTENTS procedure on the brfss_a dataset using the VARNUM option to look at the contents of the dataset:

```
LIBNAME X "/folders/myfolders/X";
data brfss_a;
    set X.Chap6_1;
RUN;
PROC CONTENTS data=brfss_a VARNUM;
RUN;
```

PROC CONTENTS will output a list of 12 variables in the order they appear in the dataset since we used the VARNUM option. Here is some information about the variables listed in the output:

#	Field Name	Definition or Question	Coding	Rename for Array
1	_STATE	Code for state.	12 = Florida (FL), 25 = Massachusetts (MA), 27 = Minnesota (MN)	Not applicable
2	ASTHMA3	(Ever told) you that you have asthma?	1 = Yes, 2 = No, 7 = Don't know/Not sure, 9 = Refused, blank = missing	CM1
3	CVDINFR4	(Ever told) you that you've had a heart attack, also called a myocardial infarction?	1 = Yes, 2 = No, 7 = Don't know/Not sure, 9 = Refused, blank = missing	CM2
4	CVDCRHD4	(Ever told) you that you have angina or coronary heart disease?	1 = Yes, 2 = No, 7 = Don't know/Not sure, 9 = Refused, blank = missing	CM3
5	CVDSTRK3	(Ever told) you that you've had a stroke?	1 = Yes, 2 = No, 7 = Don't know/Not sure, 9 = Refused, blank = missing	CM4
6	CHCSCNCR	(Ever told) you that you have skin cancer?	1 = Yes, 2 = No, 7 = Don't know/Not sure, 9 = Refused, blank = missing	CM5
7	CHCOCNCR	(Ever told) you that you've had any other types of cancer?	1 = Yes, 2 = No, 7 = Don't know/Not sure, 9 = Refused, blank = missing	CM6
8	CHCCOPD1	(Ever told) you that you have chronic obstructive pulmonary disease, C.O.P.D., emphysema, or chronic bronchitis?	1 = Yes, 2 = No, 7 = Don't know/Not sure, 9 = Refused, blank = missing	CM7

#	Field Name	Definition or Question	Coding	Rename for Array
9	HAVARTH3	(Ever told) you that you have some form of arthritis, rheumatoid arthritis, gout, lupus, or fibromyalgia?	1 = Yes, 2 = No, 7 = Don't know/Not sure, 9 = Refused, blank = missing	CM8
10	ADDEPEV2	(Ever told) you that you have a depressive disorder (including depression, major depression, dysthymia, or minor depression)?	1 = Yes, 2 = No, 7 = Don't know/Not sure, 9 = Refused, blank = missing	CM9
11	CHCKDNY1	Not including kidney stones, a bladder infection, or incontinence, were you ever told you have kidney disease?	1 = Yes, 2 = No, 7 = Don't know/Not sure, 9 = Refused, blank = missing	CM10
12	DIABETE3	(Ever told) you that you have diabetes?	1 = Yes, 2 = Yes, but female told only during pregnancy, 3 = No, 4 = No, prediabetes or borderline diabetes, 7 = Not sure, 9 = Refused, blank = missing	CM11

Table 6.1 – Information about variables in the Chap6_1 dataset

This table lists the number and field name of each variable in the dataset in the same order they're shown in the output of PROC CONTENTS. The definition or question from the survey associated with each variable is listed, along with the coding for the answer.

This table shows that all of the variables have the same coding except for _STATE and DIABETE3. The rest of the variables are coded as follows:

- 1 = **Yes**
- 2 = **No**

- 7 = **Don't know/Not sure**

- 9 = **Refused**

- blank = **Missing**

This is logical because in each case, the question that's asked is phrased, *Were you ever told you have _____ disease?*, so answers of *yes*, *no*, and *don't know* seem appropriate. The last column, titled *Rename for Array*, will be discussed at the end of this section.

According to the table, a shorthand way of listing the co-morbidities in the variables could be by stating asthma, stroke, heart attack, heart disease, stroke, skin cancer, other cancer, pulmonary disease, arthritis, depression, kidney disease, and diabetes.

It is important to notice that while DIABETE3, which is the variable associated with the diabetes question, seems to be like the others in terms of the question asked, the answers are coded differently. For DIABETE3, there is a greater range of answers, meaning there's different things available: 1 = **Yes**, 2 = **Yes**, but female told only during pregnancy, 3 = **No**, 4 = **No**, prediabetes or borderline diabetes, 7 = **Not sure**, 9 = **Refused**, and blank (which displays as a period or .) = **missing**. This means that if we use arrays to process these co-morbidity variables, we will need to take into account that the DIABETE3 field is coded slightly differently than the others.

Arrays as temporary objects

In *Chapter 4, Managing ETL in SAS*, SAS arrays were introduced as temporary objects that can be made during a data step. These serve as lists of variables that can be accessed during a data step. The example given was where a list of three similarly coded variables were declared in an array, and then all three variables were recorded in one sweeping action in the data step by calling the array.

The purpose of the demonstration was to show how array processing can speed up ETL because when analysts call arrays in SAS, it allows the SAS engine to run more efficiently, and this reduces processing time. The earlier example focused on the mechanics of how the array actually works, as well as how to properly program arrays into data steps. In this chapter, we will revisit this use of arrays from the more operational standpoint of running a SAS data warehouse.

> **The efficiency of using SAS arrays in data steps**
>
> In his white paper, *Arrays – Data Step Efficiency*, Harry Droogendyk provides explicit examples of using SAS arrays to speed up processing time when using data steps to generate variables *en masse* (a link to this paper is available in the *Further reading* section). He also includes some tips for coding arrays that can improve processing efficiency. In one tip, he shows how to add conditions in order to reduce lines of code SAS has to process when executing the data step (we will talk about this in the *Conditions and index variables in array processing* section).
>
> This white paper is helpful to analysts who are starting out with arrays because it also includes a description of several **special array functions** that can be used during array processing, including VNAME(), VLABEL(), VFORMAT(), and VLENGTH().

One of the features you may remember about an array is that the array is not a standalone function, but is called as part of a data step. Let's refresh our memories about the format of the data step code that includes arrays. The first line declares the dataset to be output with the `data` command, while the second line declares the dataset to be input with the `set` command. If arrays are to be used in the data step, they are typically declared after the `set` command and before any other processing takes place.

Then, the last part of the data step's code includes code that tells SAS how to process the data using the arrays that were called earlier. As a reminder, when the analyst calls the array, the order of the syntax is as follows: first, the `array` command is typed, and then we type in the name that was chosen for the array (we will use the naming convention of ending all array names with the `_list` suffix). Finally, we type the list of variables to be included in the array (with a space between each one), followed by a semi-colon, ending the line of code.

In the next section, we will present a use case for using a SAS array to recode the co-morbidity variables from the BRFSS in our example dataset *en masse*. Although this may initially seem straightforward, implementing arrays in a SAS warehouse for ETL brings up a lot of questions that need to be answered, as well as decisions that need to be made.

Using arrays to create variables

As we described earlier, a popular activity that analysts do with warehouse data is combine multiple similar variables into an index. This is often achieved by manipulating **two-state flags** or variables that are coded as either 0 or 1, depending on whether the row possesses that attribute. In the case of the BRFSS, where the rows represent respondents, we can be assured that the respondents who answered 1 = **Yes** to any of the co-morbidity questions probably had the co-morbidity in question, so it would be possible to make a set of two-state flags indicating whether the respondent had the disease, with 1 = **Yes** and 0 = **all other answers**. Let's do this using arrays.

As per the preceding table, there are 11 co-morbidity variables that we'll want to create flags for; that is, ASTHMA3, CVDINFR4, CVDCRHD4, CVDSTRK3, CHCSCNCR, CHCOCNCR, CHCCOPD1, HAVARTH3, ADDEPEV2, CHCKDNY1, and DIABETE3. This means that we will have to make an array containing a list of those 11 variables that will serve as the list of co-morbidities from the native variables. When we make this array, we'll call it comorbid_native_list.

We also want to create 11 flags, one for each co-morbidity. We can create a naming convention so that each of these flags is named the same as the native variable, except with a suffix appended with _f. We can also decide to name this array comorbid_flag_list.

Now that we have made these decisions, let's formulate some code that makes these flags. We will use the dataset we inputted and named brfss_a, but we will output a dataset named example:

```
1  data example;
2      set brfss_a;
3      array comorbid_native_list ASTHMA3 CVDINFR4 CVDCRHD4 CVDSTRK3 CHCSCNCR CHCOCNCR
4                          CHCCOPD1 HAVARTH3 ADDEPEV2 CHCKDNY1 DIABETE3;
5      array comorbid_flag_list ASTHMA3_f CVDINFR4_f CVDCRHD4_f CVDSTRK3_f CHCSCNCR_f CHCOCNCR_f
6                          CHCCOPD1_f HAVARTH3_f ADDEPEV2_f CHCKDNY1_f DIABETE3_f;
7      do i=1 to dim(comorbid_native_list);
8          if comorbid_native_list{i} = 1 then comorbid_flag_list{i} = 1;
9              else comorbid_flag_list{i} = 0;
10     end;
11 run;
```

Figure 6.1 – Array coding example

> **Tip**
>
> Note that a screenshot of the code has been provided to highlight how the array lists are formatted using tabs.

Let's review the code:

1. Looking through the code, we can see the `data` command, the `set` command, and the two `array` commands being called with the lists we discussed.

2. After that, we can see four lines of code executing a do loop. In the first line, the do loop is declared, and the i variable is set to equal from `1 to dim(comorbid_native_list)`, which tells SAS to run the loop from 1 to as many times as there are members in `comorbid_native_list` (which would work out to be 11 in this example). The next two lines of the loop have the data step code for editing the data.

3. When the code runs, each time it loops, it replaces i with whatever number round the loop is. For example, the fourth co-morbidity in `comorbid_native_list` is stroke, so when the loop is executing for the fourth time, i = 4, and when SAS encounters the code if `comorbid_native_list{i}` = 1, SAS replaces `comorbid_native_list{i}` with the native variable for stroke, which is the fourth member of the array (CVDSTRK3).

4. The full code line says if `comorbid_native_list{i}` = 1 then `comorbid_flag_list{i}` = 1. As is logical, then, the fourth time around, SAS will see if CVDSTRK3 = 1, and if it is, it will set whatever variable is indicated by `comorbid_flag_list{i}` to 1. Since i equals 4 (since we are imagining it is the fourth round of the loop), SAS will replace `comorbid_flag_list{i}` with the fourth member of the array of names we designated for the flags, which is CVDSTRK3_f.

5. The final line of the loop processing says else `comorbid_flag_list{i}` = 0, which sets all of the remaining values in the new flag variable that were not set to 1 in the last line to 0. Again, in the fourth round of the loop, i would be set to 4, and this would cause SAS to replace `comorbid_flag_list{i}` with CVDSTRK3_f when processing the loop. The last line of the loop ends the loop, while the last line of the data step code is a `run` command.

6. As can be seen by this code, the arrays we just declared are very complex and require a lot of typing. Also, they are somewhat harder to process in SAS than they would be if they had simpler names. Imagine that we select a prefix such as CM for co-morbidity and rename the disease variables as CM1 to CM11 (as shown in the *Rename for Array* column in the preceding table). The following data step uses the RENAME option to do this, inputting the `brfss_a` dataset and outputting the `brfss_b` dataset:

```
1  data brfss_b;
2      rename  ASTHMA3 = CM1    CVDINFR4 = CM2      CVDCRHD4 = CM3
3              CVDSTRK3 = CM4   CHCSCNCR = CM5      CHCOCNCR = CM6
4              CHCCOPD1 = CM7   HAVARTH3 = CM8      ADDEPEV2 = CM9
5              CHCKDNY1 = CM10  DIABETE3 = CM11;
6      set brfss_a;
7  run;
8
```

Figure 6.2 – Renaming the variables

Notice that the RENAME command is sandwiched between the data command and the set command in the data step. The dataset to be output is brfss_b because the dataset is iterating. By incrementing the appended letter, it is possible to roll back to brfss_a if desired. In the RENAME command, each variable to be renamed is listed, followed by an equals sign and its new name, and the line ends with a semi-colon. The preceding screenshot highlights how tabs can be used to organize the code visually, making it easier to troubleshoot and edit.

7. Finally, the set command indicates the data step to read in the brfss_a dataset.

8. Now that we have renamed the native variables we will use in the array, let's repeat our array code from before, but this time using the new variable names:

```
data brfss_c;
     set brfss_b;
     array comorbid_native_list CM1-CM11;
     array comorbid_flag_list FL1-FL11;

     do i=1 to dim(comorbid_native_list);
          if comorbid_native_list{i} = 1 then comorbid_
          flag_list{i} = 1;
               else comorbid_flag_list{i} = 0;
          end;
     RUN;
```

Notice that the dataset we created in this code is named brfss_c, and that the only other changes between this code and the previous code are that the array coding is much shorter and easier to work with. Now that the co-morbidity variables are named with numbers at the end, instead of typing them all out in a list, they can be easily called up as a range; that is, CM1-CM11.

By keeping track of which co-morbidity is assigned to which number, such as stroke being 4, this numbering scheme can be carried forward with the second array, which is for the new flag variables being made. This time, we'll choose `FL1-FL11` as the range of names of variables for the new flag variables.

Renaming variables this way can greatly speed up the execution of a data step in a data warehouse when conducting a lot of ETL. It is never surprising to see that SAS code runs much faster if the variables are named the way we renamed the co-morbidity variables, and ranges of variables similarly named ending in incrementing numbers are used to specify all the arrays. However, there can be a significant cost when it comes to renaming the variables in the first place. The `RENAME` command is processing-intensive, so the extra time that needs to be spent on the `RENAME` command needs to be counterposed against the reduction in processing time gained by having the variables in the new naming format.

Efficiency of data step processing

One recurring theme in SAS programming is the necessity of using well-formed code in order to improve processing efficiency, especially data step processing. As SAS processes each line of code sequentially, the **order of code in the data step** is very important, because it can change the order of processing of the lines in the data step code. This can add or subtract seconds, minutes, hours, or even more time to/from processing.

Stephen Philp explains the details of how data step code is processed in SAS in his white paper, *Programming with the KEEP, RENAME, and DROP dataset options* (see the *Further reading* section for a link to this). He describes that there are two types of data step statements: **compile time statements** and **execution time statements**. These refer to the processing phases in SAS that the statements are executed in. The paper includes diagrams indicating how changing the order of code formation changes the order of execution, and how that can speed up or slow down processing. These are minute details about how the SAS engine runs that are not important to know when working with small datasets or independent projects. However, these details are critical to know when running a SAS data warehouse. This is because decisions about data step processing in SAS can greatly impact the cost, time, and effort of operating the warehouse.

After running code that creates variables, it is important to always follow up with checks to ensure that the variables were created properly. After running the code using our new renamed variables, conducting checks is somewhat unintuitive. Let's check the recording of the fourth variable, which was about strokes:

```
PROC FREQ data=brfss_c;
    tables CM4*FL4 /list missing;
RUN;
```

In practice, all variables should be checked during an ETL process, but for brevity, we will just spot check the results of the array processing in this chapter with one or two variables. In the preceding code, we used the PROC FREQ command on our most recent iteration of the dataset, brfss_c, to get the two-way frequency of CM4 (the native stroke variable) and FL4 (the companion flag variable we just created using the array in the data step). We then added the list option to ask SAS to return the result to us in a list (rather than the default table format), and then added the missing option to ensure we did not create any missing variables in our operation. The following table shows a combination of CM4 and FL4 values from the dataset, along with the frequency, percent, cumulative frequency, and cumulative percent:

The FREQ Procedure

CM4	FL4	Frequency	Percent	Cumulative Frequency	Cumulative Percent
1	1	1575	4.05	1575	4.05
2	0	37230	95.70	38805	99.75
7	0	85	0.22	38890	99.97
9	0	11	0.03	38901	100.00

Figure 6.3 – Output from PROC FREQ

The output ensures that the recode is correct in that the FL4 flag is coded as 1 where the native variable CM4 was coded as 1, and all other values of CM4 were coded into FL4 as 0. However, when we notice that there are indeed 1,575 people encoded with a 1, we cannot tell from just looking at the name of the variable what co-morbidity is represented. This dataset includes almost 40,000 people – is it reasonable that only 1,575 should have a yes answer to this question? CM3 represents heart disease, which is a very common problem in the US – much more common than strokes. Therefore, these results might be appropriate for CM4, but not CM3.

So, while renaming variables comes at a processing cost, it also comes at a cost for human brain processing with respect to understanding the variables. Admittedly, the native variable names for stroke (CVDSTRK3) and heart disease (CVDCRHD4) are not very intuitive either. However, you can imagine being able to memorize what they mean from hints in the characters included in the variable names.

Renaming variables with numbers removes a certain level of intuitiveness from the variable naming convention, and while this may seem minor, it can add what can be unnecessary confusion to an already complex process of data warehouse management. Therefore, it is important to weigh up the pros and cons of renaming variables – both from the purposes of improving the efficiency of data processing, but also to maintain a data warehouse and manage its operations.

Conditions and index variables in array processing

This section will continue to demonstrate how array processing is used in SAS warehousing when performing data steps as part of ETL. First, a **conditional array** will be demonstrated, and the reasoning behind adding conditions to array processing will be explained. This will be followed by a discussion on creating and serving up **index variables** from variables that are output by the array. Finally, this section will touch on the topic of documenting and standardizing array processing in a SAS data warehouse.

Adding a condition to array processing

In the previous section, we took all 11 co-morbidity variables as an **input array** and generated a set of 11 flags as part of an **output array**. Even though the DIABETE3 variable was coded slightly differently than the other input variables, the data step that created the flags was based on common coding among all 11 variables, which was where 1 = **Yes**. Let's imagine that this time, we want to code a new grouping variable that combines co-morbidities coded with 7, 9, and missing under one code indicating *unknown*, which would be 9. Therefore, the new grouping variable would be coded as 1 = **Yes**, 2 = **No**, and **all others** = 9.

Unfortunately, this new system of coding is not possible with DIABETE3. DIABETE3 has two different answers for **Yes** that mean two different things, and they are coded 1 and 2. They cannot be combined. Also, there are two different answers for **No** that mean two different things, and these are coded as 3 and 4 and cannot be combined. On the other hand, like with the other co-morbidity variables, in a grouping variable for DIABETE3, we would want to combine 7, 9, and missing under one code; that is, 9.

This is exactly the type of situation that calls for adding a condition to array processing. The following data step code inputs the brfss_c dataset and outputs a dataset named conditional. It calls up two arrays: our comorbid_native_list array as our input array (CM1-CM11) and our comorbid_grp_list array as our output array for the grouping variables we will make using the GR prefix (GR1-GR11). Notice that the condition is added by way of adding a **nested do loop** to the loop of the data step:

```
data conditional;
    set brfss_c;
    array comorbid_native_list CM1-CM11;
    array comorbid_grp_list GR1-GR11;

    do i=1 to dim(comorbid_native_list);
        if i <11 then do;
            comorbid_grp_list{i} = comorbid_native_
            list{i};
        if comorbid_native_list{i} gt 2 then comorbid_grp_
        list{i} = 9;
        end;
        else do;
        comorbid_grp_list{i} = comorbid_native_list{i};
        if comorbid_native_list{i} gt 4 then comorbid_grp_
        list{i} = 9;
        end;
    end;
RUN;
```

First, let's look at `if i <11 then do;`, which executes immediately after the `do` loop starts to recode all the variables in the array. This means that in rounds 1 through 10 of the loop (when `i` is equal to or less than 11), we are telling SAS to execute this code. You may remember that `DIABETE3` was the last in our list of co-morbidities, so it is the one that decodes to `CM11`. This will be processed when `i = 11`. This do loop code is a way of telling SAS that for all the variables in the array before diabetes, this code should be executed.

The code that is executed on the first 10 comorbidities makes the grouping variables coded in the way we described earlier, where 1 = **Yes**, 2 = **No**, and 9 = **all other values**. This is accomplished by copying the native variable into the output variable using code `comorbid_grp_list{i} = comorbid_native_list{i}`. In the next step, if the native variable is greater than two (indicating all other values of the native variable), the new grouping variable is recoded to nine; that is, `if comorbid_native_list{i} gt 2 then comorbid_grp_list{i} = 9`. Finally, this part of the loop ends with an `end` command. Now, it is time to execute the code on the other part of the conditional do loop, where `i = 11`. This is indicated by the next line; that is, `else do`.

The code under the `else do` command indicates what SAS is asked to do when `i = 11`. This means that `CM11` is being processed, which is the `DIABETE3` variable. Notice how the first line of the processing is the same as it is for the other variables, where the native variable is copied into the grouping variable identically. But with the next line, notice that the coding is changed to `if comorbid_native_list{i} gt 4 then comorbid_grp_list{i} = 9`. The change is in the value in the `gt 4` line. This is because, as we discussed earlier, 1, 2, 3, and 4 are all valid answers to `DIABETE3`, and only the ones greater than 4 should be coded as 9. Then, in the next line, we have to place an `end` to end that part of the nested loop. Finally, in the line after this one, we place an `end` to end the overall loop.

Now, we need to check if our recode is correct. We will use the `PROC FREQ` command, just like we did last time, but this time, we will check both the stroke variables, `CM4` and `GR4`, and the diabetes variables, `CM11` and `GR11`:

```
PROC FREQ data=conditional;
    tables CM4*GR4/list missing;
    tables CM11*GR11/list missing;
RUN;
```

Here is the output of the preceding code:

The FREQ Procedure

CM4	GR4	Frequency	Percent	Cumulative Frequency	Cumulative Percent
1	1	1575	4.05	1575	4.05
2	2	37230	95.70	38805	99.75
7	9	85	0.22	38890	99.97
9	9	11	0.03	38901	100.00

CM11	GR11	Frequency	Percent	Cumulative Frequency	Cumulative Percent
1	1	4789	12.31	4789	12.31
2	2	401	1.03	5190	13.34
3	3	32931	84.65	38121	97.99
4	4	712	1.83	38833	99.83
7	9	47	0.12	38880	99.95
9	9	21	0.05	38901	100.00

Figure 6.4 – Output from PROC FREQ

We should be pleased to see that both new grouping variables were coded accurately. GR4, the grouping variable for stroke, has levels of both 7 and 9 from the native variable coded as 9 in the grouping variable, which means this is accurate. For CM11, which is diabetes, we can again see that values 1, 2, 3, and 4 are present separately in the GR11 grouping variable, while values 7 and 9 from the native variable are combined under 9 in the grouping variable.

Even carefully planned array and data step ETL code in a SAS data warehouse often seems very confusing, and it reads like a very long processing plan that needs to be understood sequentially, like a story in a movie or book. It is tempting to try to do things another way with shorter, more modular code snippets. It would be nice to save these code snippets separately, then selectively call them up during processing. Unfortunately, in a SAS data warehouse, this approach loses too much efficiency and creates too much downtime. With SAS, it's usually best to stick with the long compendium of code tucked inside a large data step.

This is why data managers often do small, standalone experiments with SAS data step programming code to determine the most efficient code before rebuilding array code to improve efficiency in SAS data warehouse ETL. Therefore, it is very important to include inline comments and offline documentation, as this ensures that array processing continues to work correctly each time a new dataset is received from the provider and processed. This also ensures that any tweaks that are made to the code to keep it updated continue to be made according to a change management protocol. These also need to be documented properly.

Creating index variables from array outputs

Note that when we use the term **index** in this chapter, we're not using it in the same way we used the it in *Chapter 1, Using SAS In a Data Mart, Data Lake, or Data Warehouse*. Previously, we were talking about how SAS uses indexing as part of its data processing engine. In this chapter, when we say index, we are talking about creating a single composite variable formed to summarize a set of other variables by putting the variables through a formula or algorithm. A common way this word is used is when talking about a **financial index**, or a financial number that represents a summary of other financial numbers that can be used for decision-making. It is not unusual to include **monetary indexes** in SAS data warehouses that are focused on finance and hold many raw financial variables.

As we described earlier, a common operation in a SAS data warehouse is using arrays in a data step to create sets of two-state flags that can later be used to easily calculate various indexes. The reason why this need exists is that, in statistics, often, large groups of variables need to be summarized into one or a smaller group of variables by statisticians for the purpose of **parsimony**, as described earlier. However, combining these variables is not straightforward. It is not unusual to use flags in some way to combine the variables by using a formula on the flags, but which formula to use and which input variables to include are often subjects of debate. Here, we will provide context to the debate by showing a simple example of combining two variables about the same co-morbidity into an index.

Calculating indexes in SAS data warehousing

When managing a SAS data warehouse, it is important to be aware of standard indexes being used in the field where the data warehouse resides, and to be attuned to trying to serve these up in the data warehouse in a way that users will find useful. As an example, dietary data is notoriously difficult to use in its raw form and often needs to be summarized into **diet indexes** in order to be useful in analysis. The **United States Department of Agriculture (USDA)** Food and Nutrition Service provides online SAS files to help analysts calculate a **Healthy Eating Index** (link available in the *Further reading* section).

If a data warehouse included dietary data, it might be worthwhile to include a calculated version of the Healthy Eating Index to save analysts the trouble of calculating it themselves. By doing that, the warehouse leaders can be assured that users of the SAS data warehouse will have at least one high-quality, properly calculated index to use when requesting data, which can improve the reputation of the warehouse.

However, serving up indexes in a SAS data warehouse can also introduce issues, depending on the native data used for calculating the index. In the case of the USDA, the index was developed using the **National Health and Nutrition Examination Survey (NHANES)** dataset, which is a survey similar to the BRFSS, which asks about diet. If the SAS data warehouse is using different input data for calculating the Healthy Eating Index, it will need to document how it operationalized this index, and then publish this documentation so that researchers using the calculated index can both trust it and explain it in reports they write that use the index.

In actuality, the situation tends to be more complex. This is because there are often different versions of the official index formula, and analysts develop their own personal preferences. This is reflected on a web page for **The University of Manitoba Center for Health Policy (MCHP)**, which describes the **Charlson Co-morbidity Index**. This is a way of combining co-morbidities into a disease index (a link to this is available in the *Further reading* section). For that reason, SAS data warehouse managers often find themselves serving up multiple versions of the same index, as well as the inputs for each index (usually as flags or other recoded variables).

In the case of the MCHP, they post official SAS code their organization uses to calculate the Charlson Co-morbidity Index, and they post SAS code that will run algorithms on inputs of co-morbidity variables that have been coded according to two common international formats. This is how the MCHP standardizes their use of indexes, thereby providing evidence on their web page for why they made their choices, as well as links to related information.

Imagine that we wanted to reduce the number of co-morbidity variables we have. To do this, we could do the following:

- We could combine variables about the same condition into a disease index.

- You may have already noticed that the CM2 variable is for heart attacks, while the CM3 variable is for heart disease. These could theoretically be combined into a cardiovascular co-morbidity index.

- You also may have noticed that CM5 is about skin cancer and that CM6 is about other cancer, so these could be combined into a cancer co-morbidity index.

When we ran the array code that made flags, our output dataset was named brfss_c. Now, let's make the disease_index dataset by inputting the brfss_c dataset and creating a cardiovascular index called CVD_index from summing the flags for CM2 and CM3. We will also create a cancer index called CA_index from summing the flags for CM5 and CM6:

```
data disease_index;
    set brfss_c;
    CVD_index = sum(FL2, FL3);
    CA_index = sum(FL5, FL6);
RUN;
```

Let's remind ourselves that when we used an array to create the flags, we named them with the prefix FL, so for our CVD_index, we will be adding together FL2 and FL3 to combine CM2 and CM3, while for CA_index, we will be adding together FL5 and FL6 to represent combining CM5 and CM6. Observe how we use the sum function to summarize the data in the data step. The sum function can also be used in other PROCs, such as PROC PRINT and PROC TABULATE (see the SAS white paper in the *Further reading* section).

Since we chose to sum the index components, each new index variable can only take on the values of 0 (if both input flags were 0), 1 (if one of the input flags were 1 and the other was 0), or 2 (if both input flags were 1). We should run a one-way PROC FREQ on our new disease indexes to verify this:

```
PROC FREQ data=disease_index;
    tables CVD_index;
    tables CA_index;
RUN;
```

Here is the output:

The FREQ Procedure

CVD_index	Frequency	Percent	Cumulative Frequency	Cumulative Percent
0	35477	91.20	35477	91.20
1	2368	6.09	37845	97.29
2	1056	2.71	38901	100.00

CA_index	Frequency	Percent	Cumulative Frequency	Cumulative Percent
0	32061	82.42	32061	82.42
1	5892	15.15	37953	97.56
2	948	2.44	38901	100.00

Figure 6.5 – Output from PROC FREQ

The PROC FREQ output confirms that we coded the indexes properly. However, it also emphasizes the fact that each of the input co-morbidities is weighted equally in the index. Should having heart disease be weighted equally to having had and survived a heart attack as it is in our CVD_index? Should skin cancer and all other cancer be weighted equally in CA_Index?

Making indexes brings up philosophical questions about measurement, and for this reason, SAS data warehouse managers often choose to only serve up indexes produced by the warehouses that are standardized and well-documented, or indexes that are simply passed on in the native data from the data provider. If it is known that users will want to create their own indexes, then data managers will want to ensure they serve up flags or other versions of input variables necessary for index algorithms.

Documenting and standardizing array processing

Let's consider the website from the MCHP, which explains their standard policy for calculating the Charlson Co-morbidity Index. This website is an example of one way to document and standardize an index. But within the data warehouse, there is also the need to simply standardize array processing. There is also the need to create more detailed internal documentation to keep track of the rationale for the different processing steps (such as the conditions).

It is easy to imagine that, in a SAS data warehouse hosting the BRFSS data, the array of the 11 co-morbidities in our example might be used in a lot of different warehouse processing steps for different reasons. However, each time that array is used, it should be formatted in a standard way. It should have the same name, and it should contain the same variables in the same order.

Because arrays are not standalone code – they need to reside inside a data step – code snippets are example of items that can help a SAS data warehouse manager standardize array coding. Here is an example of a code snippet that uses the native variable names and associated flag names for the co-morbidity variables:

```
1  data Y;
2      set X;
3      array comorbid_native_list ASTHMA3 CVDINFR4 CVDCRHD4 CVDSTRK3 CHCSCNCR CHCOCNCR
4                         CHCCOPD1 HAVARTH3 ADDEPEV2 CHCKDNY1 DIABETE3;
5      array comorbid_flag_list ASTHMA3_f CVDINFR4_f CVDCRHD4_f CVDSTRK3_f CHCSCNCR_f CHCOCNCR_f
6                         CHCCOPD1_f HAVARTH3_f ADDEPEV2_f CHCKDNY1_f DIABETE3_f;
```

Figure 6.6 – Array coding snippet

Notice how the output dataset is declared as data Y and that the input dataset is declared as data X. However this code could not run as Y and X are expected to be replaced with names of datasets. Saving this code as a code file, using a documented name in a known location on a server where analysts have access to this code, can allow them to access the code and copy the array code snippet into their code to help standardize array use across the organization.

It is also possible to provide code snippets using renamed variables for arrays:

```
data Y;
    set X;
    array comorbid_native_list CM1-CM11;
    array comorbid_flag_list FL1-FL11;
```

It is possible to post a code snippet like this, but the problem is that, depending on how variables are being renamed and processed as part of the overall processing plan, CM1 and FL1 may have different meanings across all processing code. For example, FL may be a standard name for flags used during all flag-related array processing, and after the final flags are created, they are all renamed to unique names (such as DIABETE3_f). Ultimately, because of the necessity of using arrays in data processing in SAS data warehouses for the processing efficiency they confer, the leaders of SAS data warehouses are forced to make a lot of strategic decisions about how to operationalize array standardization.

Limitations of arrays

Although array processing is usually necessary as part of maintaining a SAS data warehouse, using arrays also introduces limitations, mainly concerning issues surrounding variable naming and renaming, which will be discussed here. Issues associated with troubleshooting array programming will also be covered.

Naming limitations in SAS arrays

To speed up processing, we renamed the input array variables from `CM1`-`CM11`, while to handle recoding the grouping variables, we used a condition in our array processing to accommodate the slightly different native coding of `DIABETE3` compared to the other co-morbidity variables. It's easy to imagine how this situation of renaming native variables and creating long data steps with conditions can become even more complex.

Imagine that our dataset contains 100 disease variables and that they were coded according to five different systems (one or two dominant ones, and three rarely used ones). Theoretically, it would be possible to use a conditional array to recode all 100 variables with grouping variables and flags. Operationally, this would require some careful planning and code design. In order to speed up processing, the variables would need to be renamed, but first, it is important to decide which variables would be in which array to ensure the resulting variable names will be logical and can be leveraged by the array.

Imagine that several heart disease variables were renamed `HT1`, `HT2`, `HT3`, and so on, while several cancer variables were renamed `CA1`, `CA2`, `CA3`, and so on. This would only make sense if they were coded similarly, such as all of the heart disease variables being coded 1 = **Yes**, 2 = **No**, and 9 = **Unknown**. If there are some coded this way and some coded another way, such as 1 = **Agree**, 2 = **Neutral**, 3 = **Disagree**, and 9 = **Unknown**, then maybe it makes more sense to rename the variables according to the system of coding rather than what the variable means. If only a few of the 100 variables are coded according to some of the systems, maybe naming those variables using an array according to the coding system should be done just to speed up processing, regardless of what the variables are actually about.

A particularly challenging example is presented by the **National Health and Nutrition Examination Survey (NHANES)** data, which represents the results of an oral health exam, where the oral health examiner takes a periodontal probe and probes six sites on each tooth in the study participant's mouth to see how far the probe sinks into the gum tissue. This produces a measurement called probing or **pocket depth (PD)**. Given that a full adult mouth includes 32 teeth (28 of which are measured in the NHANES), if there are six measurements of PD for each tooth, this means that for each participant, there are *28 x 6 = 168* variables representing oral health PD measurements. Another measurement that's made as part of the examination is called **attachment loss (AL)**, and this also has six measurements per tooth, so participants have another 168 variables representing AL, for a total of 336 native variables representing PD and AL measurements.

Liang Wei et al. designed a method of taking these 336 variables representing oral health measurements into a very complicated SAS algorithm and producing an output variable that classifies participants as having periodontitis (gum disease) or not. If they are classified as having periodontitis, the algorithm further classifies them by severity (mild, moderate, or severe). The authors documented this SAS algorithm – which includes many different array processing steps – in their SAS white paper, *Array Applications in Determining Periodontal Disease Measurement* (you can find a link to this in the *Further reading* section).

In order to apply the algorithm presented in the white paper, there are a few challenges to overcome. The first challenge that's encountered is the fact that the native variables from NHANES are named according to a naming system that does not include an appended incrementing number. When we tried to recode the BRFSS co-morbidity variables using an array, we had the same problem. This is because our native BRFSS co-morbidity variables were named without an incrementing number at the end, prompting us to rename them CM1 to CM11. In the white paper, the authors give instructions of how to rename all the AL and PD variables first, before executing array processing in the algorithm; for example, the native variable name for AL probing site 1 (which is the disto-facial side of the tooth) for the second molar is OHX02LAD, and the variable for PD for the same site on the second molar is OHX02PCD. The native coding system uses a naming system that indicates the type of exam (OHX), the tooth number according to a dental numbering system (02 in this example), and the type of measurement (AL is coded LA, while PD is coded PC). The final character indicates the probing site (D is for disto-facial in this example). While the naming system is instructive, it is not conducive to providing these variables for use in SAS array processing.

So, the first step in applying the algorithm provided in the white paper is to rename all of the AL and PD measurements using a different naming convention that will append an incrementing number to the end of each variable name. For the example tooth, the authors recommend renaming the AL variable to `loadisf1` and the PD variable to `pddisf1`.

Due to the way their naming convention works, all of the disto-facial measurements are recoded so that they can be grouped into one series of variables ending with incrementing numbers that are named `loadisf1-loadisf28` (since only measurements for 28 adult teeth are processed in the algorithm, as described previously). However, this set of variables is never assembled into its own array. Instead, the final array, named `loa`, contains sets of AL variables for each measurement site. This is represented by six series of variables, including `loadisf1-loadisf28`.

Each series of variables that's entered into the array are named very similarly; for example, `loamesf1-loamesf28` and `loamesl1-loamesl28`. The array probe for the PD measurements is set up the same way. Setting up arrays this way is extremely unintuitive. It might be more intuitive to set up series of variables representing all measurements by tooth, in the same way the original dataset was set up. Unintuitive array and naming approaches add to the confusion of the programming challenge due to the need to use arrays.

The white paper contains data step code that was created using the arrays. This includes complicated nested and conditional do loops. However, the white paper does not include explicit RENAME code – only instructions on how to produce the RENAME code. In practice, the exercise of renaming all these variables in order to enter them into an extremely complex algorithm is daunting, and may be the reason why this approach to classifying a respondent's periodontitis status on the basis of the algorithm presented in this white paper is not more popular.

Naming limitations arrays impose on data storage

There are other issues with renaming variables that can cause headaches in a data warehouse when it comes to data storage. As demonstrated earlier in several examples, we renamed a set of variables according to a system, which was CM1 to CM11. We did this so we could execute a particular type of array processing. But what if we need to input some of these variables into another array in subsequent processing? Do we then re-rename them to input them into the next array? How do we control naming conventions on variables as they go through data step processing? Do we repeat conventions (for example, use FL for all flags), or do we avoid this because there may be collisions in processing, in that a set of flags – `FL1-FL11` – stands for the co-morbidity flags in one part of the processing, but another set of flags in another part of the processing?

These are not impossible questions to answer, and usually the answers come after some initial code has been developed that accomplishes the recoding task inefficiently, and is subsequently rebuilt to be more efficient. Building code this way allows studies to be done on how to make the code more efficient. However, efficiency in coding needs to be balanced against human confusion, which can result from creating code with complex processing with many nested loops where variables are frequently renamed.

The trick to developing the optimal answers to these questions involves remembering that the goal of a SAS data warehouse is to serve up data to users in a user-friendly format, which often means that if native variables such as DIABETE3 are published, their native variable name (for example, DIABETE3) is used. This is because BRFSS users are already familiar with the DIABETE3 variable, and renaming it would introduce confusion to this user base.

Because of this, if DIABETE3 is renamed CM11 during array processing but will still be served up to users, it should be served up with the name DIABETE3. This also suggests that names such as FL11 are equally useless to SAS warehouse users, and that renaming these flags after array processing to a more intuitive name such as DIABETE3_f would greatly help warehouse users.

The ETL protocol for processing data into a SAS data warehouse generally includes many steps. What this situation about arrays implies is that in this ETL protocol, there will likely be more than one **mass renaming step**. It also suggests that one of the last steps in the ETL protocol should be mass renaming variables so as to serve up variables in the data warehouse with the most intuitive names so that users can be optimally served.

Difficulty in troubleshooting

As we described earlier, when processing small datasets, it is possible to avoid using complex array processing. However, with respect to a SAS data warehouse that houses big datasets, it is hard to fathom avoiding the use of a complex assortment of array processing in data steps as part of ETL, simply because of the necessity to gain processing efficiency when conducting ETL on big data in SAS. This conundrum creates challenges with initially building the complex array programming, as well as in amending it when new data formats, structures, or rules show up in the native dataset over time.

When initially building array code, it is necessary to anticipate needing to troubleshoot and amend the code over time. This type of design thinking should go into choices that you make, such as naming arrays and renaming variables for arrays, as well as the way documentation should be developed to help analysts maintain the code. However, even in the best of circumstances, editing the array programming later to accommodate changes in the underlying native data can prove confusing. Strategies for developing array code so that it is easier to troubleshoot later will be covered in *Chapter 9, Tips for Debugging SAS Transformation Code.*

Summary

This chapter presented an overview of how array processing is used in a SAS data warehouse for ETL, and provided context for the challenges and decisions that SAS data managers face when making choices about array processing in a SAS data warehouse. First, we went over scenarios where arrays are useful in processing, and we created flags and grouped variables based on native variables using arrays inside a data step. We discussed how these variables can become part of index variables, as well as issues to consider when serving up index variables or their components in the SAS data warehouse.

Because arrays are important to use in SAS processing to improve processing efficiency, we went over ways to build efficient array code through renaming variables, as well as through using conditions. Examples were provided about ways to document array usage for both internal and external stakeholders. Finally, we reviewed the limitations imposed by the constraints around variable naming in arrays, and the difficulty associated with troubleshooting array code. It is important to know how to use arrays efficiently in SAS because otherwise, processing will take an unnecessarily long time. When dealing with big data in a SAS data warehouse, it is important to use arrays whenever possible to reduce processing time.

In the next chapter, we will continue working on the topic of developing ETL code. We will first discuss ways of studying relevant native variables to facilitate planning a well-designed ETL protocol. Then, we will create transformation code using various ETL strategies for the purpose of adding value to the datasets that are ultimately served up to users of the SAS data warehouse.

Questions

1. Why is it good to include `_list` at the end of an array name?

2. What does adding a condition to an array accomplish?

3. Why is it often necessary to rename variables for arrays?

4. What considerations do SAS data warehouse managers need to make about serving up index variables to users?

5. Why would an organization such as the MCHP publish their SAS code and array programming information for certain indexes they maintain in their datasets for the public to see?

6. Imagine you have a dataset with 250 responses to 20 survey questions, and the variables are all coded as 1, 2, 3, 4, or 5, indicating a spectrum of *strongly disagree = 1* to *strongly agree = 5*. You learn that this is a weekly customer satisfaction survey. You will be receiving hundreds of surveys in this structure every week, and you need to load them and make a report every week. Your supervisor wants you to create flags for each question on each survey, where the answers are coded as either *4 = somewhat agree* and *5 = strongly agree*. You need to include these results in the weekly report. Why would this problem be a good candidate for an array?

7. Since array processing code can introduce complexity in developing and supporting the SAS data warehouse, is it necessary to always include SAS arrays in ETL data step processing? State yes or no, and support your answer with reasoning.

Further reading

- Scientific paper, *The potential of parsimonious models for understanding large scale transportation systems and answering big picture questions* by Carlos F. Daganzo, Vikash V. Gayah, and Eric J. Gonzales, available here: `https://link.springer.com/article/10.1007/s13676-012-0003-z`

- SAS white paper, *Arrays – Data Step Efficiency* by Harry Droogendyk, available here: `https://support.sas.com/resources/papers/proceedings13/519-2013.pdf`

- SAS white paper, *Programming with the KEEP, RENAME, and DROP Data Set Options* by Stephen Philp, available here: `https://support.sas.com/resources/papers/proceedings/proceedings/sugi31/248-31.pdf`

- United States Department of Agriculture Food and Nutrition Service Healthy Eating Index: `https://www.fns.usda.gov/healthy-eating-index-support-files-03-04`

- University of Manitoba's web page, which explains the use of the Charlson Comorbidity Index and how it has been standardized using SAS arrays at the Manitoba Centre for Health Policy: `http://mchp-appserv.cpe.umanitoba.ca/viewConcept.php?conceptID=1098`

- SAS white paper, *Ways to Summarize Data Using SUM Function in SAS* by Anjan Matlapudi and J. Daniel Knapp, available here: `https://www.lexjansen.com/nesug/nesug12/cc/cc35.pdf`

- SAS white paper, *Array Applications in Determining Periodontal Disease Measurement* by Liang Wei, Laurie Barker, and Paul Eke, available here: `https://analytics.ncsu.edu/sesug/2013/CC-15.pdf`

7
Designing and Developing ETL Code in SAS

This chapter demonstrates how to design and develop **extract**, **transform**, and **load** (ETL) code in SAS. The SAS data warehouse serves users, and the main services it provides has to do with presenting data in a manner that is easy to understand, analyze, and visualize. Therefore, successful SAS data warehouses will have ETL protocols that are organized, transparent, and evidence-based, because these practices lead to providing valid, accurate, and easy-to-use data to users.

This chapter will cover these main topics:

- How to conduct preliminary research on categorical and continuous variables to inform the development of ETL code

- How to split up the tasks of planning coding from the actual coding, and how this can lead to better planning and oversight of the data warehouse

- How to develop transformation code that matches the coding plans developed, and check transformed variables for accuracy

Technical requirements

This chapter's dataset in `*.sas7bdat` format is available in this directory on GitHub: `https://github.com/PacktPublishing/Mastering-SAS-Programming-for-Data-Warehousing/tree/master/Chapter%207/Data`.

The code bundle for this chapter is available on GitHub here: `https://github.com/PacktPublishing/Mastering-SAS-Programming-for-Data-Warehousing/tree/master/Chapter%207`.

Planning the ETL approach

In this first section of the chapter, we will plan our *ETL* approach. First, we will begin by developing a data dictionary to help us specify variables – both the variables we would like to extract from the data provider, as well as the variables we would like to develop during transformation. Next, we will use `PROC FREQ` and `PROC UNIVARIATE` to conduct research on the native variables we select to inform the design of our derived variables and ETL process.

Finally, we will use the knowledge we gain not only to design derived variables and the ETL process but also to make decisions about serving up variables and maintaining SAS labels and SAS formats for variables in the warehouse. After we complete this planning phase, we will move on to the next section of the chapter, where we will create the transformation code we planned for in this section. Let's begin by specifying the data.

Specifying data with a data dictionary

Data warehouses are typically assembled around a topic, and this enables them to serve up data to analysts, leaders, and policy-makers that can help them answer questions about that topic. Data warehouses and **data lakes** are often constructed specifically to answer questions about particular geographic locations (for example, states in the **United States (US)**), health topics, and specific financial topics, such as inflation. A common topic around which to assemble a data warehouse or data lake is the **military**. Topics used as themes for data warehouses or data lakes should be specialized to a particular subtopic; for the military, it could be military health or a particular military occupation.

Imagine we were assembling a data warehouse on the topic of US veterans. Our subtopic is assessing the **quality of life (QoL)** of US veterans after they leave active service. US veterans are defined as individuals who served in the US active duty military (for example, the Army) and were discharged (regardless of whether or not they ever were deployed in any armed conflicts). US veterans often suffer more health problems than non-veterans, but they also qualify for special government benefits.

Even though our data warehouse would be focused on veterans, it would be important to include non-veteran data whenever possible to allow for comparisons. And since the subtopic is veteran QoL, it would be helpful to include datasets and variables about veteran health, healthy behaviors, and wellbeing. One dataset that could be helpful to include in the warehouse is the **Behavioral Risk Factor Surveillance System (BRFSS)**. This is an anonymous annual national health survey conducted by phone by the US government. The data is available for download online and is in SAS format. We will be using part of the 2018 dataset for demonstration in this chapter.

The BRFSS survey asks about many topics, but if we were choosing to include portions of this dataset in a data warehouse focused on the QoL of US veterans after leaving active service, there are many variables that would not be of interest. For the sake of demonstration, imagine that the data warehouse designers were interested in the following specific topics:

- Looking at veteran QoL over time and identifying time trends

- Understanding the role of governmental support for veteran QoL at the state level

- Examining experiences of cohorts of veterans of different ages (which often corresponds to when they served, and what military activities were happening at the time)

- Understanding behaviors that could impact QoL, such as sleep quality, occupational success, health conditions, healthcare access, alcohol use, and marijuana use

The objectives of the data warehouse could change over time, but it is important to document them and have them remain static for long enough for the team to collect and process datasets about these topics so analysts can answer questions about them.

As described in *Chapter 3, Helpful PROCs for Managing Data*, in order to select variables from a survey dataset like the BRFSS, it is important to first review the **codebook**, which is a document that explains details about each of the variables available in a source dataset. After reviewing the codebook for the 2018 BRFSS survey data, imagine we choose to select the following native variables to host in our data warehouse about veteran QoL:

Variable Name from Codebook	Question/Description from Codebook	Notes about Variable
_STATE	State identified by FIPS code	This is a two-digit numeric code that can be decoded.
_AGE80	Imputed age value collapsed above 80	Only adults are included in the survey. This is a two-digit number from 18 to 80. All ages above age 80 are truncated to age 80.
VETERAN3	Have you ever served in active duty in the United States Armed Forces, either in the regular military or in a National Guard or military reserve unit?	"Yes" to this question means that they are a veteran, and "no" means they are not.
SLEPTIM1	On average, how many hours of sleep do you get in a 24-hour period?	This is a two-digit number indicating average hours of sleep per night.
FMONTH	File month	This is the month that the phone interview took place and the record was placed in the data file.

Table 7.1 – Variables selected from the BRFSS codebook

As can be seen in the table, we are selecting variables in line with the objectives of the data warehouse. Including the _STATE variable will allow us to study which states have more veterans, and which states have veterans with better QoL. Including the _AGE80 variable will allow us to better understand veteran QoL at different ages, and the VETERAN3 variable will allow us to tell who reported being a veteran on the survey, and who did not, so we can have a comparison group in the data warehouse. SLEPTIM1 will allow us to assess veteran sleep patterns.

The FMONTH variable is kept as an administrative variable in anticipation of keeping data over time about veterans. The BRFSS itself is an annual survey; you could imagine including data from the BRFSS in the data warehouse going back many years. The BRFSS deliberately avoids changing its questions from year to year unless it is necessary, as researchers use the yearly datasets to study trends. Therefore, you can imagine comparing the surveys collected in the August 2016 file with surveys from August in the 2017 file and the 2018 file.

These variables we have chosen to retain from the BRFSS 2018 dataset are considered **native variables** in our data warehouse, and we will need to keep documentation on them that comes from the codebook or other files from the data provider. But we will also need to keep documentation on the variables we transform as part of our ETL protocol. As mentioned briefly in *Chapter 3, Helpful PROCs for Managing Data*, to keep this type of documentation, it is necessary to create a **data dictionary**.

Data dictionaries have different styles, and the one presented here mirrors the style of the US **Military Data Repository** (**MDR**) (see the link in the *Further reading* section), which is in Microsoft Excel. Each data table in the data warehouse that is documented has a main tab in the Excel workbook where the variables in the table are listed and described. Here is our version of this table for the purposes of this demonstration:

Order	Variable Name	Source	Question/Description	Values – See Tab	Serve to Users?
1	_STATE	BRFSS	State identified by FIPS code	FIPS xwalk	Yes
2	_AGE80	BRFSS	Imputed age value collapsed above 80	Agrp	Yes
3	VETERAN3	BRFSS	Have you ever served in active duty in the United States Armed Forces, either in the regular military or in a National Guard or military reserve unit?	YNU	No
4	SLEPTIM1	BRFSS	On average, how many hours of sleep do you get in a 24-hour period?	Sleep	No

Order	Variable Name	Source	Question/Description	Values – See Tab	Serve to Users?
5	FMONTH	BRFSS	File month	1 through 12 for the calendar months	Yes
6	Agrp	SAS	Grouping of _AGE80	Agrp	Yes
7	SLEPTIM2	SAS	Recode of SLEPTIM1	Sleep	Yes
8	Vetgrp	SAS	Grouping of VETERAN3	YNU	Yes
9	Vetflag	SAS	Flag for veteran status	YNU	Yes

Table 7.2 – Main table of the data dictionary for demonstration

Let's go through each column and row in this table to understand their functions:

- **Order**: First, the **Order** column is intended to provide the person developing the dictionary with a column that can be renumbered to allow them to sort the variables in a particular order. The order currently in the table reflects the order in which the variables are to be made in the dataset during the ETL process, with the native variables listed in rows one through five, and the transformed variables listed afterward.

- **Variable Name** and **Source**: Notice that after the second column, **Variable Name**, there is a column called **Source**. The native variables in rows one through five have BRFSS filled in as the source to indicate they are native variables from **BRFSS**. The variables listed in rows six through nine represent variables that do not exist yet but that we would like to make during the ETL process based on the native variables we have obtained from **BRFSS**. You will notice these variables are named agrp, SLEPTIM2, vetgrp, and vetflag. The fact that these are warehouse-derived variables is indicated by the source being listed as SAS – which means we will make the variables part of our ETL process.

- **Question/Description**: The next column, **Question/Description**, is filled in with the exact description copied from the codebook for the native variables, and for the derived variables, we – as warehouse designers – add our own description. From these descriptions, we can have some insight into why the warehouse might want to create these variables. According to the table, `agrp` is a grouping variable for `_AGE80` – in other words, it is a way of putting respondents into age groups. `vetgrp` is also a grouping variable and is based on the native variable `VETERAN3` – however, from this table, we cannot see what the difference in coding is between `VETERAN3` and `vetgrp`, and why the warehouse would want to create `vetgrp` instead of just using `VETERAN3`. We also see that `vetflag` is a **flag for veteran status**, but it is not clear from the table how `vetflag` is coded. Finally, we see that `SLEPTIM2` is a recode of the `SLEPTIM1` variable, and again, from this table, it is not clear why `SLEPTIM1` needed to be recoded into `SLEPTIM2`, and what the difference is between the two variables.

- **Values – See Tab**: The next column, titled **Values – See Tab,** assumes that this table is a tab in an Excel workbook, and the other tabs have the names listed in this column. For example, there should be a tab called **Federal Information Processing Standards (FIPS) xwalk** that provides the crosswalk to decode the two-digit FIPS code to the state that it represents. There should also be a tab called **agrp** that explains the values that are in the `_AGE80` and `agrp` variables. And there should also be a tab called **YNU** to document the coding of `VETERAN3`, `vetgrp`, and `vetflag`, and a tab called **sleep** to document the coding of `SLEPTIM1` and `SLEPTIM2`. However, for `FMONTH`, there is not a tab listed. Instead, there is a short description of what is in the variable (integers `1` through `12` to correspond with the number of the month).

- **Serve to Users?**: The last column in the table is titled **Serve to Users?**. As described in *Chapter 1, Using SAS in a Data Mart, Data Lake, or Data Warehouse*, in a data warehouse or data lake, all of the native variables received from the data provider are stored in the warehouse, but not all of them should be available to warehouse users. Excellent examples of types of variables that warehouse users should not have access to include private codes (such as identification numbers), confusing native variables that are prone to being misunderstood, and variables coded in a clumsy fashion that analysts would need to recode before using anyway. Based on this thinking, both `VETERAN3` and `SLEPTIM1` have a *No* in the **Serve to Users?** column, while the recoded versions of them (`vetgrp`, `vetflag`, and `SLEPTIM2`) have a *Yes*. The back story behind how these decisions were made will also be covered later, in the *Choosing variables to serve to users* section.

Now that we understand the meaning of all the columns and rows in the table that represents the main table in our Excel data dictionary, we can expect that this Excel workbook should have information about the values of these variables on tabs with the following names: **FIPS xwalk**, **agrp**, **YNU**, and **sleep**. Let's review what is on some of these Excel tabs. We will postpone talking about **FIPS xwalk** until *Chapter 10, Considering the User Needs of SAS Data Warehouses*, where we will go into detail about working with **crosswalks**.

Let's first turn our attention to the tabs named **agrp**, **YNU**, and **sleep**, and we will start by looking at **agrp**:

_AGE80	agrp	agrp Description
18 through 34	1	Young
35 through 64	2	Middleage
65+	3	Elder
Coded as unknown or missing (represented by a period)	9	Unknown

Table 7.3 – The agrp tab from the data dictionary

This table has three columns: **_AGE80**, **agrp**, and **agrp Description**. It essentially plans the transformation of _AGE80 into agrp:

- In the first row, the first column indicates that for records where _AGE80 equals 18 through 34, agrp will be coded as 1. This level is described as **Young** under the **agrp Description** column.

- For values of _AGE80 that are 35 through 64, agrp will be coded as 2, and the level will be described as **Middleage**.

- Likewise, an _AGE80 value of 65 and greater will be coded as 3 for **Elder**, and any missing or nonsensical values will be coded to 9, which is **Unknown**.

Let's make a few observations about this table. First, this table represents a plan – not documentation of a variable that already exists. _AGE80 exists, but agrp is just being planned. If all transformed variables are planned using spreadsheets like this prior to developing ETL code, then people other than the programmers developing the SAS code have the ability to weigh in on the choices being made prior to code development, leading to a more efficient and higher quality process. For example, a criticism of the planned coding of agrp might be that the age groups are too large, as 35 through 64 is a broad range. On the other hand, the availability of military benefits shifts at certain ages (such as 65), so there may be empirical reasons to choose the cutpoints designated.

The purpose of developing these Excel tabs with values in the data dictionary prior to creating ETL code is to document the coding of the transformed variables in relation to the native variables (which will later guide the development of ETL code). But making these Excel tabs also serves to provide an opportunity to plan these variables prior to doing the work of coding. Not only does this approach lead to a properly curated data warehouse, but it also enables peers and decision-makers to weigh in on the design of an ETL protocol and variables by simply reviewing an Excel spreadsheet at the planning stage.

The next Excel tab to present is **YNU**, which is named because it pertains to a question whose answers roughly decode to **Yes**, **No**, or **Unknown** (hence the initials **YNU**). Often, datasets will have a large group of variables that share a picklist coded this way, so it can be helpful to make one tab to document the coding levels of all of these variables. As it turns out, in this demonstration, we are only selecting VETERAN3, which is coded this way. But, as you may recall from *Chapter 6, Standardizing Coding through SAS Arrays*, many of the variables about conditions or co-morbidities in the BRFSS survey are coded the same way as VETERAN3, and if they were being included, their levels would be documented on the **YNU** tab.

But for this demonstration, we only have two transformed variables documented on the **YNU** tab, which are `vetgrp` and `vetflag`:

Level description	VETERAN3	vetgrp	vetflag
Yes	1	1	1
No	2	2	0
Don't know / Not Sure / Not asked	7	9	0
Refused	9	9	0
Unknown	`Missing (coded as .)`	9	0

Table 7.4 – The YNU tab from the data dictionary

Let's start by looking at the coding of VETERAN3, the native variable:

1. The coding of 1 = **Yes** and 2 = **No** is not problematic because these are intuitive codes.

2. However, for efficient analysis, it would be easier if the other three categories (7, 9, and missing indicated by a period) were collapsed into one *unknown* category. It is hard to imagine wanting to analyze those coded as **VETERAN3** = 7 separately than those coded as **VETERAN3** = 9. Also, as shown with PROC FREQ output, dealing with missing variables in SAS often requires specific coding to handle the missing values.

3.   Therefore, it would be easier for the analyst handling this grouping variable if it were recoded as described in the column titled **vetgrp**, which collapses the unknown values together so that there are only three values – 1, 2, and 9 for missing.

In some cases, analysts will want to know whether the question was asked and answered, which is information that is available in VETERAN3 and also in vetgrp. For example, if 150 people were asked whether they were veterans, and only 100 gave an answer (meaning they said 1 = **Yes** or 2 = **No**) and of those, 5 said they were veterans (and 95 said they were not), the analyst would report a veteran rate of 5/100 = 5%.

But some analysts might feel comfortable with the assumption that the people who are asked who actually are veterans all jumped at the chance to say 1 = **Yes**. In other words, all the others who did not actually deliberately answer 2 = **No** probably are not veterans anyway. If an analyst felt this way, they would want to say that 5/150 = 3% of the respondents were veterans. Some would argue that the 3% is probably more accurate than the 5% number, but others would disagree. If we were the analysts who agreed with the 3% approach, however, we would not have an easy time calculating our number from vetgrp, because the 9s would first have to be removed.

One way to solve this dilemma in a data warehouse is to **serve up variables in multiple formats** so analysts can choose the format they find most useful to their analysis. As long as these variables are properly documented, and the curation is easy to use, analysts can easily understand the differences between different warehouse variables derived from the same native variable. That way, analysts can be *informed shoppers* of data in the warehouse, and only order the data they really need. They do not have to be like restaurant-goers forced to order an entire *entrée dinner* of BRFSS or other blocks of warehouse data. Instead, analysts requesting data from the warehouse should be able to order off of a *sushi menu* of native and derived variables served up by the warehouse. This is a great way to steward data to prevent the unnecessary transfer of data, miscommunication about data, and misinterpretation of data.

For this reason, it is not unusual for data warehouses to contain multiple derived variables based on the same native variable. In our example, we are using a native categorical variable, VETERAN3, and we are offering a regrouped variable, vetgrp. However, we realize that some analysts may want VETERAN3 recoded a different way so they can calculate the response rate in a different way. To solve the dilemma, let's decide to include a variable that simply indicates the veteran status (yes/no) based on VETERAN3 in the warehouse. We will use VETERAN3 to additionally create a **two-state flag** or **indicator variable** (as initially described in *Chapter 4, Managing ETL in SAS*) called vetflag to facilitate the second percentage calculation described.

Ideally, before planning the transformation of any variable, the analyst would read the data into SAS and do some diagnostic queries, and perhaps even do some research and troubleshooting. This work has already been done with VETERAN3 and _AGE80, and we will go through what was found out about those variables first. Next, the native variable SLEPTIM1 still requires this work to be done, so we will do these steps together. Now, let's proceed by looking at ways to use PROC FREQ and PROC UNIVARIATE as part of planning for our transformations.

Understanding default PROC FREQ

As has been shown in *Chapter 3, Helpful PROCs for Managing Data*, PROC FREQ is the SAS PROC used to calculate one-way and two-way frequencies. PROC FREQ is therefore a very useful PROC for examining data. For example, as part of the decision to recode VETERAN3 as vetgrp and vetflag, it would have been helpful to first run frequencies of VETERAN3 using PROC FREQ.

Let's begin our data exploration by backtracking and doing this now. But first, we need to read in the BRFSS dataset we will be using for this demonstration. To make the dataset more manageable, only data from the states of Florida (FL, _STATE = 12), Massachusetts (MA, _ STATE = 25), and Minnesota (MN, _ STATE = 27) are included in this dataset. The dataset is called Chap7_1 and is in *.sas7bdat format.

Once we read in this dataset, we will use SAS to run our frequencies. First, we will place the Chap7_1 dataset into a directory on our computer, and then we will use SAS to map that directory to LIBNAME X. Next, we will use PROC FREQ to run a two-way frequency between _STATE and VETERAN3 without setting any options to control the output:

```
LIBNAME X "/folders/myfolders/X";
RUN;
PROC FREQ data=X.chap7_1;
tables _STATE*VETERAN3;
RUN;
```

Let's look at the steps in the code:

1. After LIBNAME X is mapped, in PROC FREQ, the dataset is indicated by X.chap7_1.

2. This is followed by the tables statement, where _STATE*VETERAN3 is specified, indicating that a two-way frequency between the _STATE and VETERAN3 variables is requested.

Here is the main table from the output:

The FREQ Procedure

Frequency Percent Row Pct Col Pct	Table of _STATE by VETERAN3				
		VETERAN3			
_STATE	1	2	7	9	Total
12	2440 6.28 16.03 48.66	12747 32.80 83.75 37.74	5 0.01 0.03 29.41	28 0.07 0.18 45.16	15220 39.16
25	687 1.77 10.30 13.70	5960 15.34 89.37 17.65	5 0.01 0.07 29.41	17 0.04 0.25 27.42	6669 17.16
27	1887 4.86 11.12 37.63	15065 38.76 88.74 44.61	7 0.02 0.04 41.18	17 0.04 0.10 27.42	16976 43.68
Total	5014 12.90	33772 86.90	17 0.04	62 0.16	38865 100.00
Frequency Missing = 36					

Figure 7.1 – Default format of PROC FREQ two-way frequency output

Let's observe a few features of this output:

1. First, it is interesting to note that even though the statement in the code was _STATE * VETERAN3, in the output table (also called a **contingency table**), levels of _STATE are listed along the *y* axis, and levels of VETERAN3 along the *x* axis.

2. Next, because we did not attach **SAS labels and formats** as described in *Chapter 3, Helpful PROCs for Managing Data*, the actual variable names (_STATE and VETERAN3) and the actual numeric coding of the levels are displayed on the output, not the values associated with attached labels and formats. This means we will have to refer to our data dictionary to decode these values.

3. Third, the default PROC FREQ does not handle *missing* as a level. You can see this at the bottom of the output, where it says **Frequency Missing = 36**, and does not specify which variables were missing – _STATE, VETERAN3, or both.

4. Finally, observe that this output probably has all the numbers that could ever be needed from a frequency analysis. But for the purposes of troubleshooting data as part of preparing for an ETL process, much less information is needed. In fact, the extra information on the printout can add to confusion and misinterpretation. As can be seen from the key in the upper-left corner of the output, **Frequency** is listed in each cell at the top, but each cell also includes **Percent** (meaning the percentage the cell contributes to the sample), **Row Pct** (meaning row percent), and **Col Pct** (meaning column percent). If these percentages could be suppressed on our output, it would be easier to read and use the information for our purposes, which is troubleshooting data for loading into a data warehouse.

With the understanding of default PROC FREQ, let's explore some options to manipulate PROC FREQ output in the next section.

Using options to manipulate PROC FREQ output

The percentages produced in default PROC FREQ output could be suppressed to make the output easier to read for our purpose. Here is one example of adding options to PROC FREQ to improve the readability of the output:

```
PROC FREQ data=X.chap7_1;
tables _STATE*VETERAN3 /nocol norow nopercent;
RUN;
```

This code is identical to the previous code, except that the following appear after the tables command:

- The options of nocol (suppresses column percent).
- norow (suppresses row percent).
- nopercent (suppresses total percent) is now placed after a slash.

This revised code produces this output:

The FREQ Procedure

Frequency	Table of _STATE by VETERAN3					
		VETERAN3				
	_STATE	1	2	7	9	Total
	12	2440	12747	5	28	15220
	25	687	5960	5	17	6669
	27	1887	15065	7	17	16976
	Total	5014	33772	17	62	38865
	Frequency Missing = 36					

Figure 7.2 – Two-way PROC FREQ output with percentages suppressed

As can be seen from this output, suppressing the percentages improves the clarity of the output. However, because handling missing data in ETL in SAS often requires specific programming, it is not helpful that the missings are still combined together, and not stratified onto their own levels (as can be seen at the bottom of the output).

The following code addresses this issue, and also changes the tabular structure of the output:

```
PROC FREQ data=X.chap7_1;
tables _STATE*VETERAN3 /list missing;
RUN;
```

Let's review how using the /list missing options should impact the output:

1. Because the list option changes the tabular structure of the output, it automatically suppresses the column and row percentages, so nocol and norow are no longer needed.

2. Also, the missing option forces missings to assume their own level for each of the variables involved in the two-way frequency.

We can see these features in the output:

The FREQ Procedure

_STATE	VETERAN3	Frequency	Percent	Cumulative Frequency	Cumulative Percent
12	.	22	0.06	22	0.06
12	1	2440	6.27	2462	6.33
12	2	12747	32.77	15209	39.10
12	7	5	0.01	15214	39.11
12	9	28	0.07	15242	39.18
25	1	687	1.77	15929	40.95
25	2	5960	15.32	21889	56.27
25	7	5	0.01	21894	56.28
25	9	17	0.04	21911	56.33
27	.	14	0.04	21925	56.36
27	1	1887	4.85	23812	61.21
27	2	15065	38.73	38877	99.94
27	7	7	0.02	38884	99.96
27	9	17	0.04	38901	100.00

Figure 7.3 – Two-way PROC FREQ output in list format with missings stratified

The format of this output is particularly helpful when examining categorical variables as part of research for the preparation of ETL protocols:

- Each row represents a cell in the output (with the missing levels included), with the first row representing the state of Florida (_STATE = 12) and the missing value for VETERAN3 (which SAS displays as a period).

- As is shown in the output, the frequency of missing VETERAN3 values in Florida (_STATE = 12) is **22**, which represents **0.06** percent of the dataset.

- This presentation is particularly helpful because it is visually easy to compare missingness in VETERAN3 to the other two states and see that while Minnesota (_STATE = 27) has **14** missing, Massachusetts (_STATE = 25) does not have any missing.

- The output also includes a cumulative frequency and cumulative percentage.

> **More about options in PROC FREQ:**
>
> `PROC FREQ` is one of the most basic commands in SAS and can be run very simply with a short line of code. However, it is also very powerful, in that it allows many options to be set. Options can be set on the actual `PROC FREQ` command, and different options can be set on the `table` command (as we did with `/list missing`).
>
> There is also the option of using a BY command within `PROC FREQ`. For example, if we sorted the dataset by FMONTH (the month the survey was completed), then added the `BY FMONTH` line after our `tables` command, we would create separate `PROC FREQ` output for each month represented in the FMONTH variable.
>
> Other options can request statistical tests from `PROC FREQ`, as well as having `PROC FREQ` output a dataset containing the frequency results. *Joseph J. Guido* provides a beginner's guide to using these options on `PROC FREQ` in his SAS white paper, *Guido's Guide to PROC FREQ – A Tutorial for Beginners Using the SAS® System* (listed in the *Further reading* section).

Using `PROC FREQ`, we have examined the VETERAN3 variable, and are confident that our ETL plans to transform it into `vetgrp` and `vetflag` are reasonable. However, we have not yet examined our two continuous native variables, _AGE80 and SLEPTIM1. For that, we will use `PROC UNIVARIATE`.

Using PROC UNIVARIATE for troubleshooting

In *Chapter 5, Managing Data Reporting in SAS*, we examined the use of `PROC UNIVARIATE`, which provides summary statistics about continuous variables. Our demonstration dataset contains the continuous variables _AGE80 (age in years) and SLEPTIM1 (average hours of sleep per night).

Let's start by using `PROC UNIVARIATE` to evaluate _AGE80 and SLEPTIM1 as native variables to store and possibly serve up to users of our data warehouse:

```
PROC UNIVARIATE data=X.chap7_1;
var _AGE80 SLEPTIM1;
RUN;
```

Let's review our code:

1. Although we have run PROC UNIVARIATE before, what is new in this code is that two continuous variables were listed under the VAR statement: _AGE80 and SLEPTIM1. When we ran PROC UNIVARIATE in *Chapter 5*, *Managing Data Reporting in SAS*, we only included one variable here.

2. As we saw in *Chapter 5*, *Managing Data Reporting in SAS*, the default PROC UNIVARIATE output consists of a series of tables. When specifying more than one variable in the VAR statement, this output consisting of a series of tables comes out about each variable in the order requested (in our case, first _AGE80 and then SLEPTIM1).

For the purposes of evaluating the variables for inclusion in the data warehouse, let's focus on one of the many tables from the PROC UNIVARIATE output: the **quantiles table**. In fact, let's compare the quantiles table between _AGE80 and SLEPTIM1:

_AGE80

Quantiles (Definition 5)	
Level	Quantile
100% Max	80
99%	80
95%	80
90%	77
75% Q3	69
50% Median	57
25% Q1	40
10%	28
5%	23
1%	19
0% Min	18

SLEPTIM1

Quantiles (Definition 5)	
Level	Quantile
100% Max	99
99%	77
95%	9
90%	8
75% Q3	8
50% Median	7
25% Q1	6
10%	5
5%	5
1%	3
0% Min	1

Figure 7.4 – Comparison of quantiles between _AGE80 and SLEPTIM1

Let's first examine the quantiles for _AGE80 based on what we know from the codebook information:

1. First, we expect that the minimum age would be 18 because the codebook says that all the respondents are adults. We see that this checks out because **0% Min** (the minimum) is listed as 18.

2. Next, we expect the maximum age to be 80 because the codebook says that ages above 80 are truncated to 80. Again, we look under **100% Max** (the maximum) and see that it says 80.

3. Although this cannot be seen on the quantiles table, the other information from the PROC UNIVARIATE output says that _AGE80 has no missing values. This is consistent with what is reported about _AGE80 in the codebook.

We learned that _AGE80 is complete with no missing values, and is coded consistently with what we expected from the codebook. These features suggest that _AGE80 is a great candidate to serve up to warehouse users. It is clean, intuitive, complete, well-documented, compliant with documentation, and easy to use in an analysis.

Before interpreting the quantiles for SLEPTIM1, it is first important to consider that sleep research suggests that humans all over the world sleep on average 6 to 8 hours per night. Depending on the person, the average duration of healthy sleep can be between 5 and 10 hours. However, it is hard to believe someone could sleep on average less than 5 hours a night without becoming so exhausted that they would need to have days where they sleep a long time to recover (for example, shift workers), and that would inevitably impact their personal average. Also, people reporting very long average sleep durations of greater than 10 hours either have some special situation that causes them to sleep a lot (such as an illness or disability), or they have an average sleep duration in a more normal range, but due to an error, it is recorded as greater than 10 in the data.

When we reflect on this background information about what is known about sleep duration, the quantiles for SLEPTIM1 look very suspicious:

1. The **25% Q1** line on the output (which is the 25th percentile) registers at 6 hours, which seems reasonable.

2. However, it is unclear which levels below **25% Q1** are reasonable. **0% Min** is 1 hour of sleep per night, and **1%** (the first percentile) is 3 hours of sleep per night. Are these measures correct and valid? Or are they mistakes? Is it even possible to sleep 3 or fewer hours per night, regardless of whether the respondent actually reported this? And how should this be handled in our warehouse data?

3. Next, we can look at the other extreme of SLEPTIM1. We see **100% Max** is 99, and **99%** is 77, and we immediately suspect some sort of error. Both of these numbers – 99 and 77 – look like codes. They seem to violate informatics rules, by including information in the SLEPTIM1 field that does not relate to average hours of sleep per night.

4. We also observe that **95%** (the 95th percentile) is 9 hours, which is not unreasonable but leaves open the question what values are between 9 and 77? If there are values that are greater than 10 that are feasible, such as 15, 18, or even 24, the same questions that need to be addressed about very small average sleep durations will also need to be faced about unusually long (but possible) average sleep durations. Again, decisions will need to be made about how to serve up such data in the warehouse.

Figuring out what we should do in the data warehouse with SLEPTIM1 will require research:

1. First, we need to review the codebook to figure out why these unusual codes – 77 and 99 – seem to be popping up in this variable.

2. Next, we will need to look more carefully at distributions of the values that are actual answers to the question, not codes, and decide whether we *trust* this variable or not – meaning whether we think it is *valid* or not. If we do not feel it is valid, we should not include it in the warehouse, nor should we base any derived variables on it.

3. However, if we think there is valid information available for us in SLEPTIM1 (along with perhaps erroneous or unneeded information), we can instead keep SLEPTIM1 as a native variable that users cannot access. From that, we can create valid transformed variables based on SLEPTIM1 to serve up in the warehouse to users that would be helpful for users.

In the next section, I will demonstrate the kind of research that needs to be done on SLEPTIM1 in order to make these decisions.

Using PROC FREQ to troubleshoot continuous variables

We used PROC UNIVARIATE to examine information about the continuous variable SLEPTIM1 and noticed that there appeared to be codes 77 and 99 in the variable along with what appeared to be valid measurements of average hours of sleep per night. To answer this mystery, we need to look at the codebook.

In our data dictionary, we now focus on the tab titled **sleep** that was mentioned in the main data dictionary table as a way to decode native variable SLEPTIM1 and planned transformed variable SLEPTIM2:

SLEPTIM1	Description	SLEPTIM2
1 through 24	Number of hours [1-24]	1 through 24
77	Don't know/not sure	missing (period)
99	Refused	missing (period)
Unknown (including missing represented by a period)	Missing	missing (period)

Table 7.5 – Tab for sleep from the data dictionary

On the **sleep** tab, we see three columns:

1. In the first two columns, SLEPTIM1 and **Description**, we see information transferred from the codebook.

2. We see that when SLEPTIM1 takes on a value of 1 to 24, the response is considered a valid answer to the question of how many hours on average the respondent sleeps per night.

3. However, also stored in this variable is the code 77 indicating the respondent said *Don't know/not sure* to the question, and the code 99, indicating the respondent refused to answer the question.

This information from the codebook solves the mystery as to what codes 77 and 99 mean but also reveals a serious problem with the underlying data in SLEPTIM1. Storing two different types of data in the same variable (such as storing the answer to *average hours of sleep per night* in the same variable with a code indicating whether or not the respondent answered the question) is a very old practice used in historic times when computer storage space was at a premium. However, it has never been an ideal practice, because it violates informatics rules and makes data confusing. For these reasons, modern data warehouses that maintain historic data that is structured in this old format generally do not make these confusing variables available to the warehouse users. Instead, they derive and serve up valid variables in the warehouse based on the native variable that do not violate informatics rules.

In the third column, the table presents the plans for the derived variable we will serve up in the warehouse named SLEPTIM2:

1. Like SLEPTIM1, SLEPTIM2 will be a numeric variable. This means that we will be able to analyze it in PROC UNIVARIATE just like with SLEPTIM1.

2. However, in order to have the summary statistics refer only to the values that report average sleep duration, the codes 77 and 99 cannot be present in SLEPTIM2. PROC UNIVARIATE will include any numeric values present in a numeric variable. Therefore, recoding the records that have 77 and 99 in that variable as missing will mean that those values will no longer be included in the summary statistics.

3. So, our plan is to copy SLEPTIM1 into SLEPTIM2, but then erase the 77 and 99 values from SLEPTIM2. This action will essentially *clean up* the variable and make it so that the PROC UNIVARIATE output about SLEPTIM2 presents summary statistics about average sleep duration without including the 77 and 99 codes as valid numbers.

Before we can plan to recode SLEPTIM1 into SLEPTIM2, we need to have a better idea of all the values in SLEPTIM1 that we will have to deal with recoding in SLEPTIM2. We already are aware of 99 and 77, but it would be helpful to look at the distribution of the other values, especially the suspiciously high or low ones.

One trick that data warehouse analysts use in a case like this is to actually run PROC FREQ on SLEPTIM1, even though SLEPTIM1 is a continuous variable and PROC FREQ makes frequency tables. This is because the results can be very helpful. We will use the list and missing options as we did previously, with VETERAN3, to demonstrate how we can troubleshoot SLEPTIM1 and prepare to recode it into SLEPTIM2:

```
PROC FREQ data=X.chap7_1;
tables SLEPTIM1 /list missing;
RUN;
```

As you may recall, the `list` option creates output with each frequency cell represented by a row in a table. Since we asked for a one-way frequency about SLEPTIM1, the output lists all the values of SLEPTIM1 in ascending order as the rows. The output is long, so here is a shortened version:

The FREQ Procedure

SLEPTIM1	Frequency	Percent	Cumulative Frequency	Cumulative Percent
1	121	0.31	121	0.31
2	144	0.37	265	0.68
3	297	0.76	562	1.44
4	1103	2.84	1665	4.28
5	2333	6.00	3998	10.28
6	8022	20.62	12020	30.90
7	11591	29.80	23611	60.70
8	11688	30.05	35299	90.74
9	1789	4.60	37088	95.34
10	890	2.29	37978	97.63
11	83	0.21	38061	97.84
12	275	0.71	38336	98.55
23	3	0.01	38474	98.90
24	5	0.01	38479	98.92
77	380	0.98	38859	99.89
99	42	0.11	38901	100.00

Figure 7.5 – Tab for sleep from the data dictionary

Immediately from the output, we can see that only **42** people are coded as 99, and **380** people are coded as **77**, and this represents a very small percentage of the whole dataset (0.11% for 99, and 0.98% for 77). This means that when we develop SLEPTIM2 and suppress these values, we are only removing a very small percentage of values from the variable.

But this output also shows us some complexity with which we will have to wrestle in terms of data warehouse design:

1. At the end of the output, we see that three people say they sleep 23 hours per day, and five people say they sleep 24 hours per day. This means that the part of the output that is not included will show that there are quite a few people reporting between 13 and 22 hours of sleep per day. Are these valid measurements, or are they errors? And if they are valid measurements, are they meaningful to include in this dataset?

2. The same considerations need to be made for the 121 respondents who reported only 1 hour of sleep per night on average, and the 144 who reported only 2, and the other respondents who reported suspiciously low numbers for average sleep per night.

For now, assume we make the decision that we will serve up all of the values from SLEPTIM1 that refer to sleep duration in SLEPTIM2, regardless of how suspicious they may seem. We will only suppress the codes 77 and 99 in SLEPTIM2. This decision requires us to point out in our documentation that the native variable on which SLEPTIM2 is based already has these distributional features (which are not further explained in the documentation). We need to be clear that it is up to the analyst using our warehouse data to review the distribution of SLEPTIM2 and make decisions about what values are valid to include when doing their analysis.

So far, we have examined _AGE80 and SLEPTIM1, our two continuous native variables, using numeric output from PROC UNIVARIATE, and have learned a lot about these variables. We even used PROC FREQ on SLEPTIM1 to learn more about its distribution. As the last step before making final ETL design decisions about these two continuous variables, let's use an option with PROC UNIVARIATE to make plots about both of these variables and review those as well.

Making plots for troubleshooting

Before we finish planning ETL for our continuous variables, _AGE80 and SLEPTIM1, let's use PROC UNIVARIATE once more on them, but this time, to plot their distributions. We will use the same PROC UNIVARIATE code as before, only with the plot option added:

```
PROC UNIVARIATE data=X.chap7_1 plot;
var _AGE80 SLEPTIM1;
RUN;
```

Notice that the code used is the same as before, but on the first line of the code and before the semi-colon, the `plot` option is added. The output includes the summary statistics tables normally included in `PROC UNIVARIATE` output for each variable (`_AGE80` followed by `SLEPTIM1`), along with a set of plots for each variable, which are shown here:

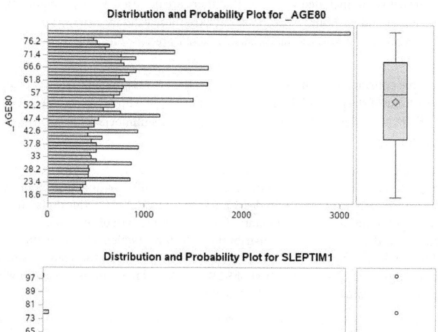

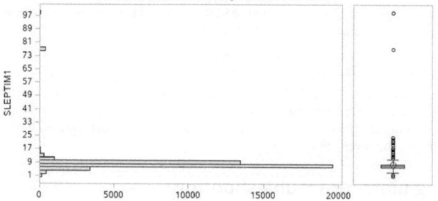

Figure 7.6 – Plots from PROC UNIVARIATE

As can be seen in the figure, the automatic default plots that come out include a horizontally oriented histogram (on the left) and a vertical box plot (on the right). These plots are not easily configurable or appropriate for presentation, but they are extremely helpful for the analyst diagnosing data distributions when preparing for ETL. For example, we observe by the box plot for `_AGE80` that the data is normally distributed with a slight left skew. We can also observe from the histogram that there are spikes of particular frequencies at regular intervals.

Further inquiries into the data will show that respondents have the propensity to round their age to the nearest number ending in zero or five if their age is close to that number (for example, people aged 64 and 66 will say they are age 65). Let's run this PROC FREQ code on _AGE80 to understand this phenomenon:

```
PROC FREQ data=X.chap7_1;
tables _AGE80 /list missing;
RUN;
```

Here is a truncated version of the output:

The FREQ Procedure

_AGE80	Frequency	Percent	Cumulative Frequency	Cumulative Percent
18	360	0.93	360	0.93
19	343	0.88	703	1.81
59	785	2.02	21639	55.63
60	893	2.30	22532	57.92
61	752	1.93	23284	59.85
62	803	2.06	24087	61.92
63	741	1.90	24828	63.82
64	842	2.16	25670	65.99
65	915	2.35	26585	68.34
66	871	2.24	27456	70.58
67	780	2.01	28236	72.58
68	797	2.05	29033	74.63
69	765	1.97	29798	76.60
70	914	2.35	30712	78.95
71	764	1.96	31476	80.91
72	712	1.83	32188	82.74
73	602	1.55	32790	84.29

Figure 7.7 – Truncated output from PROC FREQ with the _AGE80 variable

Let's focus on the frequencies and percentages for ages 59 through 73 from the output, which are formatted using the list option:

1. In that age range, the three ages that end in either 0 or 5 are 60, 65, and 70. As can be seen from the circles placed on the output, those ages make up **2.3%**, **2.35%**, and **2.35%** of the dataset, respectively.

2. All of those percentages are much higher than the other ones on the output for numbers not ending in 0 or 5. The closest one is **2.24%**, representing age 66 (1 year after the US federal retirement age).

Now let's move away from _AGE80 and change our focus to the plots for SLEPTIM1:

1. As could be predicted from our earlier frequency analysis, the box plot for SLEPTIM1 handles the 77 and 99 codes as outliers. Since SLEPTIM1 is stored as a numeric variable, SAS does not know that 77 and 99 are actually codes representing categories.

2. Even with this error in the plot, it is still possible to see something of the distribution among values we will be considering valid (1 through 24). There, we see a skew toward the right side, with longer durations of sleep more prevalent in the dataset.

Admittedly, it will be more instructive to rerun this plot from PROC UNIVARIATE on the variable we will make during ETL called SLEPTIM2, which will not have the codes 77 and 99 in it. Running the same plot on SLEPTIM2 later would provide a clearer picture of the distribution of the sleep duration values.

> **Plots for troubleshooting across time:**
>
> In this chapter, we are examining variables cross-sectionally that are being transferred (presumably annually) from the BRFSS into our hypothetical warehouse. In a real scenario, we additionally would want to conduct consistency checks on the new data we are loading from 2018 against the previous years of BRFSS data we are already maintaining in the warehouse. That way, we can see whether there are any time-related anomalies in our data load.
>
> For example, when we ran PROC FREQ about SLEPTIM1, we found that about 1% of the data was coded as 77 or 99. This rate should not change much from year to year. If it does, it could either be an indication of an error in the data or of a profound shift in the environment over the past year (such as refusal rates increasing due to political or social issues, thus increasing rates of missingness in the data).
>
> Plots to help troubleshoot data across time over time are outside the scope of this chapter, but those of you who are interested can learn about some of these plots from *Robert N. Rodriguez* and *Tonya E. Balan's* SAS white paper, *Creating Statistical Graphics in SAS 9.2: What Every Statistical User Should Know* (see the link in the *Further reading* section).

Up to this point, we have accomplished a lot of design work. We have selected native variables from the BRFSS to store in our warehouse, and we have reviewed the distribution and content of some of the native categorical and continuous variables using PROC FREQ and PROC UNIVARIATE. We have even designed some new variables based on these native variables and documented both the native variables and the newly designed variables in a data dictionary.

We still have some more design decisions to make, however. Just because we are storing a native (or even transformed) variable in our data warehouse does not mean that we should necessarily serve it up to the users of our data warehouse. The next section will cover how to think about which of these variables in the warehouse should be published to be used by warehouse users, and which should be hidden from users, and only kept behind the scenes for the purpose of warehouse maintenance.

Choosing variables to serve to users

Referring to the main table in our example data dictionary, we can consider the native variables we extracted and the transformed variables we are planning. In considering these variables, we can try to make informed decisions about which variables to keep behind the scenes, and which variables to serve up to users.

Serving up variables about veterans

Let's start by thinking about the veteran variables. We already have decided that it would be unlikely anyone would use VETERAN3 in its current form because the unknown values are separated into several categories, which makes the variable clumsy to use in analysis. In fact, if an analyst was sloppy and did not realize this, they may make a mistake in PROC FREQ or other analyses, and miscount the number of records with an unknown value for VETERAN3.

This situation provides a strong argument behind not serving up VETERAN3 to users, and only providing access to vetgrp and vetflag. If we do this, we must publish documentation, so analysts can see how vetgrp and vetflag are related to VETERAN3. This is because they may already be familiar with VETERAN3 from using the *BRFSS* dataset in other projects. In reference to these decisions, this is why when you look at our main data dictionary table, under the column **Serve to Users?**, the answer to VETERAN3 is *No*, and the answers for vetgrp and vetflag are *Yes*.

Serving up age variables

Although we have chosen to avoid serving up native variable VETERAN3 to users, we do not give the same treatment to the age variables. Instead, we choose to serve up both _AGE80 and agrp (our transformed age group variable). The reason we are making the native variable available along with the grouping variable is that _AGE80 is already in good shape in terms of being consistent with the codebook, as we found from our PROC UNIVARIATE and PROC FREQ analysis. _AGE80 does have some interesting spikes in its distribution, but the data in _AGE80 is logical and appears valid.

Unlike SLEPTIM1 and VETERAN3, _AGE80 does not contain any confusing or missing data. Therefore, an analyst might find a use for _AGE80 in its native form. It is reasonable, then, to serve it up in its native form in the warehouse, along with the grouping variable we plan to make in the ETL process, agrp.

Serving up sleep variables

Now, we will apply the same line of reasoning as we did the age and veteran variables to our sleep variables. These include native variable SLEPTIM1 and transformed variable SLEPTIM2. Remember, our PROC UNIVARIATE and PROC FREQ analysis showed that SLEPTIM1 included measurements about two different concepts, violating informatics rules. This provides evidence that SLEPTIM1 is a confusing variable that would need to be *cleaned up* before being used in any analysis anyway.

Given this situation, it is best for both the user and the warehouse to not serve up SLEPTIM1 to users. Instead, it should serve up SLEPTIM2 only, and explain its relationship to SLEPTIM1 in the documentation. That way, users who are familiar with the SLEPTIM1 variable can already understand what they are getting in SLEPTIM2.

Serving up other variables based on continuous variables

So far, thinking of our continuous variables, we have chosen to develop agrp based on _AGE80, and SLEPTIM2 based on SLEPTIM1. At this point in the design stage, it is important to weigh the pros and cons of developing other potential variables derived from native continuous variables such as _AGE80 and SLEPTIM1, especially categorical classifications (such as agrp). We would do this expressly for the purpose of serving the needs of users of the warehouse. An example can be seen in healthcare, where many measurements (including laboratory values) are stored as continuous variables. However, the actual value must be placed in some sort of classification system to facilitate interpretation.

Healthcare laboratory reports provide an easy explanation for this situation. Anyone familiar with laboratory reports in healthcare knows that the actual value of the result from the laboratory test is not nearly as important to know as the interpretation as to whether the value is *too high*, *too low*, or *in the normal range*, which is actually a categorical classification. This is why, as a practical matter, laboratory reports in healthcare tend to print both the actual laboratory value along with the categorical interpretation in the report. That way, both of these pieces of information are conveniently available to the patient and the clinician.

This challenge with needing to store classification variables alongside continuous variables in order to keep track of their interpretations is routinely encountered in the design of the storage of **blood pressure data** in a healthcare data warehouse:

- Blood pressure is expressed with two continuous numbers written as a fraction, and the one on top is called **systolic blood pressure** (**SBP**).

- Values of SBP are typically between 80 and 200, with higher values being associated with a greater health risk. SBP tends to be normally distributed, but the mean of this curve falls in different places, depending upon the underlying demographic group.

- Among younger people, the mean SBP is lower, and SBP falls in a smaller range. Therefore, analysts who want to make comparisons of SBP among young people will typically create a grouping variable that codes **quartile** levels of SBP, and will then compare quartile four with quartile one.

- If it is known that users will approach the continuous data this way, then the warehouse designer should consider providing not only the native SBP variable but also a classification variable for quartiles of SBP.

The reason why researchers often use quartiles (or **quintiles**, **cuts at the median**, or other data-driven ways) to classify continuous data is that it is actually rather difficult to interpret continuous data without first classifying it, as seen with lab values and SBP. The challenge of comparing young people with a small range of low values of SBP to each other could be addressed by creating a quartile classification as described. But there is another scenario where classifications are needed to aid interpretation, and that is when it comes to empirical classifications:

- According to the **American Heart Association** (**AHA**), an SBP of under 120 is considered *normal*, an SBP of 120 to 129 is considered *elevated*, and there are three more categories above this one indicating increasing clinical risk (see the link in the *Further reading* section).

- Some analysts using *SBP* data may want the actual SBP value, but most would likely find a categorical variable with the levels of *normal*, *elevated*, and the other three categories as more useful.

This situation is parallel to the lab values issue. However, it differs in the sense that many patients and clinicians have a working understanding of high and low SBP numbers, while many people involved in healthcare – including both patients and clinicians – may not have a working understanding of the scale on which laboratory values are placed. Hence, while actual SBP values are somewhat meaningful to a large group of people including clinicians and blood pressure patients, many actual lab values are meaningful to only a small group of people (who are more likely to be clinicians than patients). This discussion shows that while offering an SBP categorical variable based on AHA classifications to users could be seen as an effort to accommodate users that they might appreciate, providing interpretation variables for laboratory values should be seen as a critical function of the warehouse as a good faith effort to prevent misuse or misinterpretation of the data.

Serving up identification variables

Now, let's turn our attention away from our continuous variables to the two administrative variables we obtained from the BRFSS dataset, _STATE and FMONTH. In the case of both these variables, they are administrative codes that are widely decodable. _STATE is coded with a two-digit **FIPS** code assigned by the US government, which is decoded in the public domain, and FMONTH simply refers to the number of the month when the survey was administered (1 through 12). Therefore, if we store these numeric codes in our data warehouse, we do not need to worry that bad actors may hack the data environment and steal the decoding scheme for _STATE and FMONTH. On the contrary, we expect that users will find these codes useful, and will use their own methods to decode them (such as creating a variable based on FMONTH that groups records by quarters).

However, imagine we instead were storing a private identification number as a code to uniquely identify each of the respondents in the data, such as a **medical record number (MRN)** assigned uniquely to patients at a healthcare organization. Or, even more controversially, we could store a **social security number (SSN)**, which is a governmentally assigned unique identification number to track income for people in the US for taxation purposes. Because of the ability of the SSN to uniquely identify a specific US resident, the SSN tends to be stored in many datasets unrelated to taxes (for example, in insurance datasets).

As a part of ethics in data warehousing, it is important to consider the following points:

- Many countries have a similar situation, where residents are assigned a unique identification number for administrative purposes, and subsequently, the number then takes on a role outside of the original intention. It becomes useful as a unique identifier for that individual in datasets across both governmental and private organizations.

- Unfortunately, when unique personal identification numbers such as the SSN takes on such an outsized role, this also means that those engaging in cybercrime try to illegally target obtaining these numbers (especially SSNs). This is because the criminals can then easily steal the identities of these individuals by using the SSN as a way of gaining access to other data about the individual (such as their bank accounts).

- Obviously, private identification numbers, in general, should not be served up to a user group in a data warehouse, because if they are, anonymized individuals in the data could have their data breached, stolen, and used for criminal activity. It is highly unethical for us as data stewards to increase the likelihood of this possibility, even if we trust our users. This is because our users might accidentally lose control of the data, or hackers could get past our safeguards and obtain the data illegally.

Our responsibility to protect such private identification numbers might be obvious. However, what is less obvious is whether or not the numbers should be stored at all in the live warehouse environment, where they could be hacked or breached accidentally. After all, any data stored on any network with internet access could get hacked, and any data shared legitimately with trusted users could be unlawfully breached. On the other hand, anyone who has designed a data warehouse knows that these identifiers are very important to have in data to link records across entities (such as different patients in health insurance). This topic is important, so we devote a whole discussion to it in *Chapter 10, Considering the User Needs of SAS Data Warehouses*.

Serving up indexes and their components

In *Chapter 6, Standardizing Coding Using SAS Arrays*, we discussed developing **index variables**, or **composite variables** derived from a larger group of source variables. The purpose of creating index variables is to reduce the number of variables used in an analysis, which is often the goal of analysts. The example given in *Chapter 6, Standardizing Coding Using SAS Arrays*, was a co-morbidity index that calculated a number based on respondent reports of having different diseases or conditions in the BRFSS survey. A two-state flag coded as 1 = **Yes** and 0 = **All other values** was created for the different conditions, then the flags were summed together to create an index variable.

As was mentioned in *Chapter 6, Standardizing Coding Using SAS Arrays*, the data warehouse needs to make several decisions when it comes to index variables:

- First, if an index variable is passed to the data warehouse, does the data warehouse serve up the variable as it is? Or should the warehouse recalculate the variable?

- Normally, index variables are continuous, so they can undergo the scrutiny under which we placed _AGE80 and SLEPTIM1 to be evaluated for their distributions and other logic checks. But the reality is that the data warehouse may not have the component variables to calculate the index, so it only has the choice of serving it up to users, or not publishing it and making it available.

Another set of decisions surrounds the issue of the warehouse calculating its own indexes from data in the warehouse:

- If this occurs, how the index is calculated needs to be well-documented, and the documentation needs to be available to users.

- Further, the source data variables that are used in the index should be well documented.

- The data warehouse may or may not make these source variables available. For example, data warehouse leaders may not want two-state flags about respondents having certain health conditions being available to users. But, they may feel more comfortable about publishing a composite index, which still provides a measure of co-morbidity level, but obscures the underlying details in the data.

- If the source variables are made available for the published index, users should be able to replicate the generation of the index and achieve the same results as the published index. This is a necessary check that should be included in any published index variables where the source variables are also available to users in the warehouse.

This section covered how to make decisions about choosing which variables to serve to users, and documenting them in the data dictionary. At this point, we will have chosen all the native variables we want to serve up, as well as having designed the new variables we want to generate during the ETL process. This is a good point at which to also consider whether to use SAS formats and labels.

Creating and maintaining formats for variables

In *Chapter 3, Helpful PROCs for Managing Data*, we talked about **SAS labels** and **SAS formats**. As a brief reminder, labels are descriptors that can be attached to variables that can be seen on output (such as attaching the label *Veteran Status* to `vetgrp`, our transformed variable derived from `VETERAN3`). **Formats** are descriptors that can be attached to values of variables to decode them on output (such as attaching formats to `vetgrp` so that the values 1, 2, and 9 display as, *Yes*, *No*, and *Unknown*, respectively). *Chapter 3, Helpful PROCs for Managing Data in SAS*, demonstrated that labels are attached to a SAS dataset during a data step. *Chapter 3, Helpful PROCs for Managing Data in SAS*, also demonstrated that, by contrast, formats can be created as standalone files and then attached to a SAS dataset during a data step, or during the calling of certain PROCs.

At this point, we are now in the position as data warehouse designers of contemplating involving labels and formats in an ETL protocol. We already have the variables and values of the levels of variables well-documented in our data dictionary. Would we benefit from including labels and formats in our data warehouse? And if so, how do we operationalize this?

If, after ETL, data will be exported from SAS into a different language (such as **SQL**), there is no point in making SAS labels and formats. This is because only SAS programs can use them. On the other hand, if SAS will be used for the visualization of the data, then it is strongly encouraged to attach SAS labels and formats to data:

- Format files can be developed and run, and also made available to users in curation documentation.

- During ETL, after the last transformation step and before the data is loaded into the warehouse, a data step can be executed that attaches all the formats to the appropriate variables, and labels all the variables.

- This approach allows data warehouse leaders to work to synchronize the data dictionary with the label and format code. It also minimizes the use of labels and formats during transformation, when they are more likely to get in the way and not be helpful.

Please observe that up to this point, we have only done design work. Any coding we have done has been for data diagnosis, exploration, and troubleshooting. We used commands such as `PROC FREQ` and `PROC UNIVARIATE` to study the data and plan our *ETL* approach.

This work was important for a few reasons. First, this work served to gather and document evidence about native variables and datasets. This itself is work that could be split up. For example, we could have assigned one programmer to study the continuous variables and make recommendations for warehouse variables, and have another programmer study the categorical variables. Next, because the results of what they found would be documented, this would allow us to assign the work of actually designing the new variables based on what was found out about the native variables to someone else.

It is possible to imagine being a manager of a team of programmers and being able to split up work like this. At this point, it would be time to make transformation code. In the next section, we refer to our plans and actually create the transformation code. Along the way, we continue to update our documentation, which continues to maintain the ability for the team to cross-train and split up work.

Creating transformation code

Now that we have planned our analytic dataset, we will prepare our transformation code. First, we will deal with creating categorical grouping variables such as `agrp` from `_AGE80`. Next, we will clean up the continuous variable `SLEPTIM1` into a new transformed variable, `SLEPTIM2`. We will go on to make indicator variables where needed, as we have planned for `vetflag`. Finally, we will talk about how dates and numerical values can be used as indexes in SAS data warehouses, and then we will demonstrate exporting the transformed dataset.

Designing categorical grouping variables

Earlier, we presented our plan for coding `agrp` from `_AGE80`. Let's start our transformation code by making `agrp`:

1. We will start by copying our demonstration dataset `Chap7_1` from `LIBNAME X` to the `WORK` directory and naming it `brfss_a`.

2. In the data step, we will add `agrp` using `if/then` statements.

3. We will then output the `brfss_b` dataset.

4. We will follow this operation with a data check using `PROC FREQ` to check our coding.

Here is our code:

```
data brfss_a;
    set X.Chap7_1;
RUN;
data brfss_b;
    set brfss_a;
    agrp = 9;
    if _AGE80 le 34
        then agrp = 1;
    if _AGE80 ge 35 and _AGE80 le 64
        then agrp = 2;
    if _AGE80 ge 65
        then agrp = 3;
PROC FREQ data = brfss_b;
    tables _AGE80 * agrp / list missing;
RUN;
```

Let's review our code:

1. As described earlier, we copied the X.Chap7_1 dataset into the WORK directory and named it brfss_a.

2. Then, we executed a data step to transform brfss_a into brfss_b.

3. In that data step, we were guided by the data dictionary tab we created, and used **arithmetic operations** and **if/then statements** as discussed in *Chapter 3, Helpful PROCs for Managing Data*, and *Chapter 6, Standardizing Coding Using SAS Arrays*, to code the agrp variable.

4. Finally, we called PROC FREQ with the list and missing options to check our recode of the data.

Here is a shortened version of the PROC FREQ output:

The FREQ Procedure

_AGE80	agrp	Frequency	Percent	Cumulative Frequency	Cumulative Percent
18	1	360	0.93	360	0.93
19	1	343	0.88	703	1.81
20	1	363	0.93	1066	2.74
21	1	345	0.89	1411	3.63
22	1	365	0.94	1776	4.57
77	3	477	1.23	35024	90.03
78	3	432	1.11	35456	91.14
79	3	341	0.88	35797	92.02
80	3	3104	7.98	38901	100.00

Figure 7.8 – PROC FREQ output for _AGE80 and agrp

The output verifies that agrp was coded correctly, for the younger ages, agrp = 1, and for the older ages, agrp = 3. However, as mentioned before, it is possible to make an argument that other age groupings might be needed in the data warehouse because after all, the warehouse has many different users with different needs. If requests from users come in for an additional age grouping variable, the warehouse designers could choose to add it and support it, in which case they would need to update their data documentation and ETL code simultaneously.

Naming convention policies for derived variables:

Chapter 4, Managing ETL in SAS, discusses setting naming conventions for variables. It should be mentioned how the naming conventions for variables were applied in this demonstration. First, native variables were planned to keep their native names (such as `_AGE80`). Next, grouping variables were planned to be named with the suffix `grp` (`agrp`, `vetgrp`), and indicator variables were planned to be named ending in a flag (`vetflag`). Finally, `SLEPTIM2` is named to be intuitive as a recode of `SLEPTIM1`. This numbering convention can be extended. For example, if multiple age groups are wanted in the warehouse, `agrp2`, `agrp3`, and `agrp4` could be developed.

One final note should be mentioned with reference to SAS and variable naming. It is that the default sort order of variables in `PROC CONTENTS` follows the sort order as described in *Chapter 3, Helpful PROCs for Managing Data,* which is ASCII for SAS University Edition. This means that variables starting with capital letters, such as `SLEPTIM1` and `VETERAN3`, will sort first, and variables starting with underscores, such as `_AGE80`, will sort next. Variables starting with lowercase letters, such as `agrp` and `vetgrp`, will sort last. Sometimes analysts keep this default sort order in mind when choosing the names of derived variables so that it is easy to identify derived variables on `PROC CONTENTS` output, as the output can become very long as many variables are transformed.

In terms of the categorical grouping variables we planned, we created `agrp`, and we still need to create `vetgrp` (which we will do in the *Designing indicator variables* section when we also create `vetflag`). We did not plan any other categorical variables in this demonstration, so we can move on to transforming our continuous variables. Now that we have transformed `agrp`, let's move on to transforming `SLEPTIM1` into `SLEPTIM2`.

Cleaning up continuous variables

The main point of transforming `SLEPTIM1` into `SLEPTIM2` is to *clean up* the variable. In other words, we want to make `SLEPTIM2` the version of the variable that we can easily use in our SAS programming. We have chosen to keep all values of `SLEPTIM1` in `SLEPTIM2` that BRFSS considered valid, which are 1 through 24. However, if we had made other decisions, we could have chosen to suppress values of `SLEPTIM1` from `SLEPTIM2` that we did not feel were valid to include.

In our code, we will do this in a series of steps:

1. First, we will use a data step to transform brfss_b into brfss_c.

2. During the data step, we will add SLEPTIM2. We will start by setting SLEPTIM2 equal to SLEPTIM1.

3. Then, we will follow up by setting all values of SLEPTIM2 that are greater than 24 to missing.

4. We will follow this with a PROC FREQ using the list and missing options to check the recode of the data.

Here is our code:

```
data brfss_c;
    set brfss_b;
    SLEPTIM2 = SLEPTIM1;
    if SLEPTIM2 gt 24
        then SLEPTIM2 = .;
RUN;
PROC FREQ data = brfss_c;
tables SLEPTIM1 * SLEPTIM2 /list missing;
RUN;
```

Let's review a shortened version of the PROC FREQ output to assure ourselves that the recode was done correctly:

The FREQ Procedure

SLEPTIM1	SLEPTIM2	Frequency	Percent	Cumulative Frequency	Cumulative Percent
1	1	121	0.31	121	0.31
2	2	144	0.37	265	0.68
3	3	297	0.76	562	1.44
23	23	3	0.01	38474	98.90
24	24	5	0.01	38479	98.92
77	.	380	0.98	38859	99.89
99	.	42	0.11	38901	100.00

Figure 7.9 – PROC FREQ output for SLEPTIM1 and SLEPTIM2

As can be seen by the output, the strategy of setting the values of SLEPTIM2 that were greater than 24 to missing worked to suppress the data for the codes 77 and 99. Now, running PROC UNIVARIATE with a plot option will produce a plot that provides us with a distribution of the values of SLEPTIM2 we are planning to serve up as valid:

```
PROC UNIVARIATE data = brfss_c plot;
var SLEPTIM2;
RUN;
```

Let's review the plot output by PROC UNIVARIATE for SLEPTIM2 and look at the distribution now that the 77s and 99s are removed:

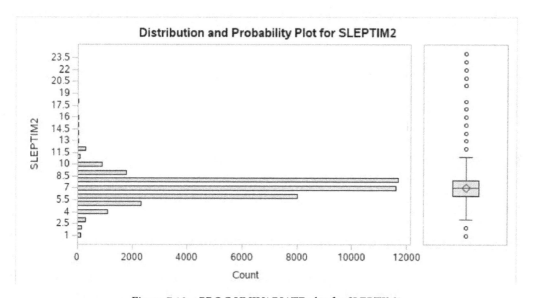

Figure 7.10 – PROC UNIVARIATE plot for SLEPTIM2

Although the distribution of SLEPTIM2 looks relatively normal, we can see that there are a lot of outliers, especially at the higher end of the distribution. While this may pose challenges when using the data in analysis, the warehouse can still serve the data to the user. The warehouse can also help the user by documenting this distribution and guiding the user to the source documentation from BRFSS. That way, the user can be provided with as much information as possible on which to base analytic decisions about SLEPTIM2.

Designing indicator variables

In our planning, we designed grouping variable `vetgrp` as a *cleaned-up* version of VETERAN3, and we also designed `vetflag` as an **indicator variable**, or **two-state flag**, derived from VETERAN3 (and consistent with the coding in `vetgrp`). When designing grouping and indicator variables together, a good strategy is to transform both of them in one set of code:

```
data brfss_d;
    set brfss_c;
    vetgrp = VETERAN3;
    if VETERAN3 in (7, 9, .)
        then vetgrp = 9;
    vetflag = 0;
    if VETERAN3 = 1
        then vetflag = 1;
RUN;
PROC FREQ data = brfss_d;
tables vetgrp * VETERAN3 /list missing;
tables vetflag * VETERAN3 /list missing;
RUN;
```

Let's review the steps in this code:

1. The code begins with a data step copying `brfss_c` into `brfss_d`. In this data step, the new variable `vetgrp` is first created by setting it equal to the coding of VETERAN3. Next, an `if` statement is used to modify that coding, and recode `vetgrp` as 9 (unknown) if VETERAN3 contains the values of 7, 9, or missing (represented by a period), which indicate *unknown*.

2. Next, while still in the data step, the `vetflag` variable is created by setting it equal to 0. In the following lines, `vetflag` is modified and set to the value of 1 if VETERAN3 is set to 1. This means that if VETERAN3 is *yes*, `vetflag` is set to 1 (forcing both *no* and *unknown* from `vetgrp` to be grouped together under `vetflag = 0`).

3. Finally, PROC FREQ is run, which checks both the new variables. First, `vetgrp` is checked against VETERAN3, and then `vetflag` is checked against VETERAN3. In both cases, the `list` and `missing` options are used.

The output produced by PROC FREQ shows that the coding was done properly according to the plans we made in our data dictionary:

The FREQ Procedure

vetgrp	VETERAN3	Frequency	Percent	Cumulative Frequency	Cumulative Percent
1	1	5014	12.89	5014	12.89
2	2	33772	86.82	38786	99.70
9	.	36	0.09	38822	99.80
9	7	17	0.04	38839	99.84
9	9	62	0.16	38901	100.00

vetflag	VETERAN3	Frequency	Percent	Cumulative Frequency	Cumulative Percent
0	.	36	0.09	36	0.09
0	2	33772	86.82	33808	86.91
0	7	17	0.04	33825	86.95
0	9	62	0.16	33887	87.11
1	1	5014	12.89	38901	100.00

Figure 7.11 – PROC FREQ to check the coding of vetgrp and vetflag

As can be seen from the output, both variables were recoded properly. It is important to observe that the output is ordered by ascending value of the first variable specified in the tables command, so for vetflag, the records coded as zero (representing *no* and *unknown*) sort first.

Indicator variables can also serve as *helper* or *warning* variables in a data warehouse:

- A variable like SLEPTIM2 poses a dilemma to data warehouse managers because even though it apparently contains valid data, the data is suspicious in some ways.

- On one hand, it is technically possible for people to sleep 22, 23, or 24 hours per day, and BRFSS is a very high-quality operation, so these values probably reflect what the respondent actually reported.

- On the other hand, some people might have been not being honest with BRFSS, such as at least some of the people saying they sleep 1, 2, or 3 hours per night.

If we want to serve up a variable like SLEPTIM2 that has such distribution anomalies, we can consider including additional indicator variables that could assist analysts who want to use some standardized approach to remove outliers. Imagine we set some data warehouse business rules that determined that SLEPTIM2 values of 4 through 12 are considered *valid*. We could make a companion variable called slp_valid, which is essentially an indicator variable that codes the valid values as 1, and the invalid values as 0.

We will do this in code:

1. We will first use a data step to create the valid_example dataset from the brfss_d/ dataset.

2. Next, based on the values of SLEPTIM2, and using *arithmetic operators*, we create the slp_valid indicator variable.

3. Finally, we will check it with PROC FREQ.

Here is our code:

```
data valid_example;
    set brfss_d;
    slp_valid = 1;
    if SLEPTIM2 > 12 | SLEPTIM2 = . | SLEPTIM2 < 4
        then slp_valid = 0;
RUN;
PROC FREQ data = valid_example;
    tables SLEPTIM2 * slp_valid /list missing;
RUN;
```

Let's view the top of the output from PROC FREQ:

The FREQ Procedure

SLEPTIM2	slp_valid	Frequency	Percent	Cumulative Frequency	Cumulative Percent
.	0	422	1.08	422	1.08
1	0	121	0.31	543	1.40
2	0	144	0.37	687	1.77
3	0	297	0.76	984	2.53
4	1	1103	2.84	2087	5.36
5	1	2333	6.00	4420	11.36
6	1	8022	20.62	12442	31.98

Figure 7.12 – PROC FREQ to check the coding of slp_valid

The output shows that we coded slp_valid correctly:

- As is shown in the output, sleep values of missing (represented by a period), 1, 2, and 3 that were deemed invalid by data warehouse business rules are flagged in slp_valid with a 0.

- By contrast, those coded as 4 or higher in SLEPTIM2 have slp_valid = 1.

- Creating this variable affords analysts choices when using the SLEPTIM2 variable. They can either include all values in their data or just include the ones where slp_valid = 1.

So far, we have talked about dealing with categorical and continuous variables, but we have not focused on dates. Because date variables are important to maintain datasets over time, it is necessary to consider hosting date-related variables in the data warehouse.

Considering dates and numerical variables

As described in *Chapter 3*, *Helpful PROCs for Managing Data*, date variables in SAS are essentially numeric variables with some sort of date format applied. This can be contrasted with other applications, such as Microsoft Excel and **R** (the open source statistical application), where variables converted into a date format behave differently in calculations. This feature of date handling in SAS can have both advantages and disadvantages. One advantage is that if the programmer is not interested in dates to the day, and is only interested in dates to the month, it is fairly easy to make a unique, sortable index variable. This variable can be used in a **longitudinal** data table, which keeps data over many months and years, to uniquely identify monthly records.

The BRFSS files provide an opportunity for an example. Although the demonstration dataset we are using is from the year 2018, BRFSS is an annual survey, so datasets from 2017, 2016, and earlier are available for inclusion in the warehouse. This means that each time the data is loaded, both native variables such as VETERAN3 as well as transformed variables such as vetgrp and vetflag will be appended to a table already existing in the data warehouse. In order to keep track of the source file date and month for each record, an index variable can easily be created by concatenating the year and month together.

Here we have a demonstration of creating such a variable in our 2018 BRFSS dataset. Let's start by running a one-way PROC FREQ on the FMONTH variable using the most recent iteration of our dataset, brfss_d:

```
PROC FREQ data = brfss_d;
    tables FMONTH /list missing;
RUN;
```

The PROC FREQ output is not shown here, because we will see the same information shortly after we make our concatenated variable and run another PROC FREQ. The reason why it is demonstrated here is that it is necessary to first run PROC FREQ on a documented variable such as FMONTH before using it as part of creating another variable to verify it really contains the values reflected in the documentation. As we will see when we run a two-way PROC FREQ later involving FMONTH, FMONTH indeed includes only values of 1 through 12 (with no missing values) as promised in the documentation.

Now let's make that concatenated index variable. We will do that as part of a data step, where we read in the brfss_d dataset and read out a dataset into work named longitudinal. During our data step, we will make two variables:

1. First, we will make a variable for the file year called FILEYR, which equals 2018. The reason for this variable is that we are working with the 2018 file. We can imagine when the 2019 file is available, we will be creating a variable called FILEYR set to 2019. Having FILEYR set to 2019 will uniquely identify all the records from the 2019 dataset once all the processed records are appended together into a long dataset.

2. Second, we will make the concatenated index, which we will call YRMONTH. We will call it that to remind ourselves that it is a concatenation of the year of the file (FILEYR) and the month of the record (FMONTH). The command for concatenation is two pipes together (||).

Let's run this data step, followed by a two-way PROC FREQ between the month variable, FMONTH, and the new variable we are making, YRMONTH:

```
data longitudinal;
    set brfss_d;
    FILEYR = 2018;
    YRMONTH = FILEYR || FMONTH;
RUN;
PROC FREQ data = longitudinal;
    tables FMONTH * YRMONTH /list missing;
RUN;
```

Here is the PROC FREQ output, where you can see the distribution and levels of not only FMONTH, but also YRMONTH:

The FREQ Procedure

FMONTH	YRMONTH	Frequency	Percent	Cumulative Frequency	Cumulative Percent
1	2018 1	3493	8.98	3493	8.98
2	2018 2	4574	11.76	8067	20.74
3	2018 3	3568	9.17	11635	29.91
4	2018 4	3189	8.20	14824	38.11
5	2018 5	2897	7.45	17721	45.55
6	2018 6	3142	8.08	20863	53.63
7	2018 7	2879	7.40	23742	61.03
8	2018 8	2765	7.11	26507	68.14
9	2018 9	2625	6.75	29132	74.89
10	2018 10	3258	8.38	32390	83.26
11	2018 11	3331	8.56	35721	91.83
12	2018 12	3180	8.17	38901	100.00

Figure 7.13 – PROC FREQ to check the coding of FMONTH and YRMONTH

Variables such as YRMONTH become very helpful when looking at longitudinal data over years:

- First, YRMONTH provides a useful sorting variable, in that it puts the records in chronological order if used as a sort composite index as described in *Chapter 1, Using SAS in a Data Mart, Data Lake, or Data Warehouse.*

- Secondly, YMONTH can work as a time-based **key**, or an index variable used to link to other datasets.

Please note that this approach to storing a date in a data warehouse focuses on using *administrative dates* as variables for warehouse maintenance. It does not talk about configuring dates to the day, adding formats to them, and serving them to users as data about a topic (such as veterans' QoL). If you are interested in this, please consider reviewing the information provided in *Chapter 3, Helpful PROCs for Managing Data,* which discusses storing numeric characters with formats, including date formats.

Exporting the transformed dataset

Reflecting back on our data dictionary we designed earlier in the chapter, we successfully obtained our native variables from the BRFSS dataset, designed variables to transform on the basis of those native variables, then created the transformation code and created those variables. As we transformed the variables, we did checks to ensure we coded the new variables properly. Each time we did a data step, we named the new dataset with an incrementing letter after an underscore, and the last one we made happened to be named `brfss_d`.

At this point, `brfss_d` has all the variables in our data dictionary, and now, it is time to export it as an analytic dataset. When exporting it, we want to omit the variables we do not want to publish. We also want to place the exported dataset in a location where it can be loaded into the data warehouse (or accessed in a data lake). Let's plan our code:

1. Let's pretend that the directory mapped to `X` is the location where we want to place the analytic dataset.

2. We will designate the name for this analytic file as `vet_analytic`.

3. Therefore, we will use a data step to copy `brfss_d` from `WORK` into `X.vet_analytic`.

4. According to our data dictionary, we decided not to serve up `VETERAN3` and `SLEPTIM1`, so those variables should be dropped from the analytic dataset when we copy it over.

5. Then we will use `PROC CONTENTS` with the `VARNUM` option on `X.vet_analytic` to assure ourselves that our final analytic dataset follows the plans in our data dictionary.

Here is the code we made:

```
data X.vet_analytic (drop = VETERAN3 SLEPTIM1);
    set brfss_d;
RUN;
PROC CONTENTS data=X.vet_analytic VARNUM;
RUN;
```

Let's review our code:

1. Notice how we added the (drop = VETERAN3 SLEPTIM1) command in the first line of the data step. This tells SAS to omit those two variables from being included in our X.vet_analytic dataset.

2. We can see from the PROC CONTENTS output that our analytic dataset is correct and placed in the correct location.

Here is the list of variables from PROC CONTENTS:

Variables in Creation Order					
#	Variable	Type	Len	Format	Informat
1	_STATE	Num	8	BEST12.	BEST32.
2	_AGE80	Num	8	BEST12.	BEST32.
3	FMONTH	Num	8	BEST12.	BEST32.
4	agrp	Num	8		
5	SLEPTIM2	Num	8		
6	vetgrp	Num	8		
7	vetflag	Num	8		

Figure 7.14 – Contents of the final analytic dataset

We have now completed transforming our analytic dataset and publishing it to the warehouse server. In other words, we have completed our ETL protocol. If we need to make any edits to the published dataset, we can always add modular code that could transform a dataset into brfss_e or brfss_f before naming the final version vet_analytic and then publishing it. This approach to naming allows flexibility for the warehouse programmer, who may need to manipulate variables in the dataset behind the scenes. It also provides consistency for the warehouse user, who may have all their programming reflect the name of a dataset as being the published version, vet_analytic.

Summary

This chapter provided an overview of planning the *ETL* approach and creating transformation code. First, we selected native variables to include in our SAS data warehouse, then designed transformed variables to derive during ETL. We used `PROC FREQ` and `PROC UNIVARIATE` to study our native variables and make good design decisions about our transformed variables. We documented our decisions in a data dictionary, which we then used as a guide when we created our transformation code. We used SAS data steps to create grouping variables and two-state flags and recoded continuous variables. We checked the variables for accuracy in recoding as we created them, and then we exported a final analytic dataset.

These skills are important to know when running a data warehouse. Making a data dictionary is a great skill to have for planning variables for transformation, as well as for keeping documentation about both native and transformed variables. It is helpful to become adept at using `PROC FREQ` and `PROC UNIVARIATE` for studying and checking variables. It is also necessary to know how to create data step code that successfully transforms the variables designed to be derived by the warehouse.

In the next chapter, we will continue discussing ETL, but we will also introduce the **SAS macro language**. SAS macros can help automate ETL, and we will explore how they can do that in the next chapter.

Questions

1. How does an analyst tell what the variable names mean and what the coded levels for categorical variables mean in a dataset?

2. `PROC FREQ` is for creating frequency tables about categorical variables, and `PROC UNIVARIATE` is for producing summary statistics about continuous variables. Therefore, why would a person preparing data for loading into a data warehouse ever use `PROC FREQ` on a continuous variable?

3. Why is it helpful to plan transformed variables in a data dictionary before developing ETL code?

4. How does suppressing values as missing in a continuous variable impact `PROC UNIVARIATE` output?

5. Imagine you were working on a data warehouse with stock market data. In your data warehouse, you had the value of the stock market at the time of closing every day. What are some classification variables you could make that might improve how the users of the data warehouse were served?

6. Imagine you had some weather data from a tropical region. After checking a data field called TEMP, you found there were some values that seemed very cold. When you talked to the scientists who gave you the data, they said that some of their temperature monitors can get wet and register an erroneous low reading from time to time. If you wanted to serve up TEMP to users in a data warehouse, how might you deal with this data anomaly?

7. Imagine after publishing vet_analytic that one of the users found a problem with agrp. How would naming conventions help you troubleshoot the problem?

Further reading

- US **Military Data Repository (MDR)** data dictionary: https://www.health.mil/Military-Health-Topics/Technology/Support-Areas/MDR-M2-ICD-Functional-References-and-Specification-Documents

- SAS white paper, *Guido's Guide to PROC FREQ – A Tutorial for Beginners Using the SAS® System* by *Joseph J. Guido*, available here: https://www.lexjansen.com/nesug/nesug07/ff/ff07.pdf

- Information about blood pressure categories from Harvard Health available here: https://www.health.harvard.edu/heart-health/reading-the-new-blood-pressure-guidelines

- SAS white paper, *Creating Statistical Graphics in SAS® 9.2: What Every Statistical User Should Know* by *Robert N. Rodriguez* and *Tonya E. Balan*, available here: https://support.sas.com/resources/papers/proceedings/proceedings/sugi31/192-31.pdf.

8
Using Macros to Automate ETL in SAS

Up to now, we have learned how to run PROCs and data steps in SAS, especially for **extract, transform, and load (ETL)** protocols, which make up a lot of the work of running a SAS data warehouse. In this chapter, we will learn about SAS **macros** and **macro variables**, and how they can help us with ETL.

First, we will learn how to build base code, then convert it to a macro. We will do this with both PROCs and data steps. Next, we will talk about different ways of storing and calling macros. Finally, we will look at an example of using macros to automate loading transformed data.

This chapter will teach you the following skills:

- How to make transformation macros by developing standalone data step code first, and then rebuilding it into a macro

- How to standardize ETL in a data warehouse using SAS macros, and the advantages and disadvantages of using macros for this

- How to use macros to load transformed datasets into a SAS data warehouse, and different ways this can be done to gain efficiency

Technical requirements

The datasets in `*.sas7bdat` format used as a demonstration in this chapter are available online on GitHub: `https://github.com/PacktPublishing/Mastering-SAS-Programming-for-Data-Warehousing/tree/master/Chapter%208/Data`.

The code bundle for this chapter is available on GitHub here: `https://github.com/PacktPublishing/Mastering-SAS-Programming-for-Data-Warehousing/tree/master/Chapter%208`.

Creating macros out of data step code

When automating ETL, we would like to use macros in data step code. However, to understand how to do this, it is important to learn about other topics first. This section will start by explaining the difference between macros and macro variables, and describing how to set macro variables with the `%LET` command. Next, we will move on to writing code, then rebuilding it into macros. First, we will do that with PROCs, and then we will do it with data steps. Finally, this section will demonstrate how to add conditions to macros to further enhance automated processing.

Choosing to use macros and macro variables

SAS is said by some to not be one language, but a collection of languages. Those who say that point to the fact that data step language is different than PROC syntax, and that `PROC SQL` is a different set of commands altogether. Another language in SAS is called the **macro language**.

In their SAS white paper, *SAS Macro Programming for Beginners*, Susan J. Slaughter and Lora D. Delwiche explain the basics of macro programming in SAS (see *Further reading* for the link to the paper):

- First, they make the point that macros are used for automation, so you should only convert code to the macro language if you want to automate a process.

- Automating a process using macros makes programming very complex and hard to troubleshoot. Therefore, careful consideration should be given before converting code to the macro language.

- Macros take in statements that allow the analyst to set values to variables that can change each time the code is run. These are called **macro variables**. This is facilitated through the `%LET` command.

- Programming an actual **macro**, however, is bigger than just using macro variables. It involves using some special commands in the macro language, including `%IF-%THEN/%ELSE` and `%DO-%END`. We will go over examples of macros that use these commands in the *Adding conditions to macros* section.

- Because SAS macros usually (but not always) use macro variables, the two topics are taught together.

In SAS data warehouses, the two main places that macros are used are in ETL and in reporting. This chapter will focus on the use of macros in ETL, while *Chapter 12, Using the ODS for Visualization in SAS*, will cover using macros for reporting using the SAS ODS. Before we focus on using macro variables in data steps, let's first look at a simpler case, which is using macro variables in PROCs with the `%LET` command.

Using macro variables with the %LET command

Before we start, let's look at our demonstration dataset. It is called `Chap8_1`, and it contains the same variables we used in *Chapter 6, Standardizing Coding Using SAS Arrays*, when practicing arrays. The data comes from the **Behavioral Risk Factor Surveillance System (BRFSS)**, an annual health survey conducted anonymously on the phone in the US. The variables we will be working with are listed in this table:

#	Field Name	Definition or Question	Coding
1	_STATE	Code for state	12 = Florida (FL), 25 = Massachusetts (MA), 27 = Minnesota (MN)
2	ASTHMA3	(Ever told) you had asthma?	1 = Yes, 2 = No, 7 = Don't know/Not sure, 9 = Refused, blank = missing.
3	CVDINFR4	(Ever told) that you had a heart attack also called a myocardial infarction?	1 = Yes, 2 = No, 7 = Don't know/Not sure, 9 = Refused, blank = missing.
4	CVDCRHD4	(Ever told) you had angina or coronary heart disease?	1 = Yes, 2 = No, 7 = Don't know/Not sure, 9 = Refused, blank = missing.
5	CVDSTRK3	(Ever told) you had a stroke?	1 = Yes, 2 = No, 7 = Don't know/Not sure, 9 = Refused, blank = missing.

#	Field Name	Definition or Question	Coding
6	CHCSCNCR	(Ever told) you had skin cancer?	1 = Yes, 2 = No, 7 = Don't know/Not sure, 9 = Refused, blank = missing.
7	CHCOCNCR	(Ever told) you had any other types of cancer?	1 = Yes, 2 = No, 7 = Don't know/Not sure, 9 = Refused, blank = missing.
8	CHCCOPD1	(Ever told) you have chronic obstructive pulmonary disease, C.O.P.D., emphysema, or chronic bronchitis?	1 = Yes, 2 = No, 7 = Don't know/Not sure, 9 = Refused, blank = missing.
9	HAVARTH3	(Ever told) you have some form of arthritis, rheumatoid arthritis, gout, lupus, or fibromyalgia?	1 = Yes, 2 = No, 7 = Don't know/Not sure, 9 = Refused, blank = missing.
10	ADDEPEV2	(Ever told) you have a depressive disorder (including depression major depression, dysthymia, or minor depression)?	1 = Yes, 2 = No, 7 = Don't know/Not sure, 9 = Refused, blank = missing.
11	CHCKDNY1	Not including kidney stones, bladder infection, or incontinence, were you ever told you have kidney disease?	1 = Yes, 2 = No, 7 = Don't know/Not sure, 9 = Refused, blank = missing.
12	DIABETE3	(Ever told) you have diabetes?	1 = Yes, 2 = Yes, but female told only during pregnancy, 3 = No, 4 = No, prediabetes or borderline diabetes, 7 = Not sure, 9 = Refused, blank = missing.

Table 8.1 – Information about variables in the Chap8_1 dataset

Let's review some features of these variables:

- The variable _STATE refers to a two-digit code of the state of the US where the survey respondent lives. The only codes included in this demonstration dataset are **Florida (FL)** = 12, **Massachusetts (MA)** = 25, and **Minnesota (MN)** = 27.

- All of the other variables refer to the respondent answering a question about whether or not they have a certain health condition (for example, asthma, whether they have had a heart attack, and so on).

- All of these variables are coded the same way except for the question about diabetes (DIABETE3). The other variables are coded this way: 1 = **Yes**, 2 = **No**, 7 = **Don't know/Not sure**, 9 = **Refused**, blank (represented by a period) = **missing**. But for the DIABETE3 variable, although the other coding is the same, both 1 and 2 refer to a *yes* answer, and 3 and 4 refer to a *no* answer.

The dataset is called Chap8_1 and is in *.sas7bdat format. To follow along, place this dataset into the directory and we will map this to LIBNAME X. We will use a data step to copy it into work and name it brfss_a. Because we have already reviewed the variables from the table, we will skip running PROC CONTENTS. Instead, we will begin to get to know our dataset by doing a one-way frequency on the _STATE variable using PROC FREQ with the missing option:

```
LIBNAME X "/folders/myfolders/X";
data brfss_a;
    set X.Chap8_1;
RUN;
PROC FREQ data=brfss_a;
tables _STATE / missing;
RUN;
```

Let's review our code:

- We begin by mapping our LIBNAME X to the folder where we put the dataset, Chap8_1.

- In the next two lines, we use a data step to copy X.Chap8_1 into a dataset called brfss_a in the WORK directory.

- Next, we run PROC FREQ on brfss_a. We ask for a one-way frequency on _STATE with the tables command and add the missing option, so if there are any missing values, they will be included in percentages calculated in the output.

Let's look at the distribution of _STATE in the output from PROC FREQ:

The FREQ Procedure

_STATE	Frequency	Percent	Cumulative Frequency	Cumulative Percent
12	15242	39.18	15242	39.18
25	6669	17.14	21911	56.33
27	16990	43.67	38901	100.00

Figure 8.1 – One-way frequency of the _STATE variable in the Chap8_1 dataset

In the dataset, the state with the most records is Minnesota (16,990), followed by Florida (15,242), with Massachusetts having the smallest number at 6,669. The reason to point this out is that our first demonstration using a macro variable will center around filtering by state.

Imagine we wanted to look at the frequency of diabetes (DIABETE3), but we only wanted to see the results for Florida (_STATE = 12). The simplest way to do it would be to add a WHERE clause to PROC FREQ. We will demonstrate that here. We will also demonstrate using the title command:

```
PROC FREQ data=brfss_a;
tables DIABETE3 / missing;
WHERE _STATE = 12;
title "PROC FREQ with Florida only";
RUN;
```

Let's go over this code:

- As with the previous PROC FREQ, this is a one-way frequency using data brfss_a.

- However, this time, we use DIABETE3 with the tables command instead of _STATE. We also use the missing option as we did last time.

- To filter results by Florida = 12, we add the WHERE clause specifying WHERE _STATE = 12;.

- We also add the title command with the manually written title "PROC FREQ with Florida only".

This code produces this output:

PROC FREQ with Florida only

The FREQ Procedure

DIABETE3	Frequency	Percent	Cumulative Frequency	Cumulative Percent
1	2379	15.61	2379	15.61
2	136	0.89	2515	16.50
3	12368	81.14	14883	97.64
4	332	2.18	15215	99.82
7	22	0.14	15237	99.97
9	5	0.03	15242	100.00

Figure 8.2 – One-way frequency of the DIABETE3 variable for Florida only

Let's look at a few features of this output:

- Our manually written title showed up at the very top of the output, above the header saying **The Freq Procedure**.

- The WHERE clause appears to have worked to filter in only records from Florida because the last value in the **Cumulative Frequency** column is 15,242, which was the total number for Florida we got from our previous PROC FREQ.

But what if we wanted to automate this a little by using a **macro variable**? The way the code is now, it is hardcoded to filter by Florida. One way we could make this code more flexible is to create a macro variable for _STATE, so we could change the state each time we ran the code.

%LET is the command that allows you to create an empty named variable, set this named variable to a particular value, and then call up the variable later in processing. Let's plan how we will write our code:

1. We will call the empty named variable selected_state.

2. In the %LET command, we will set this variable to 12 (for Florida).

3. Then, we will run our PROC FREQ again, but this time, instead of referring to the value 12 for Florida, we will refer to our named variable selected_state.

Also, as a bonus, we will reuse the `selected_state` variable in our `title` command, allowing that to be automated as well:

```
%LET selected_state = 12;
PROC FREQ data=brfss_a;
tables DIABETE3 / missing;
WHERE _STATE = &selected_state;
title "PROC FREQ with _STATE=&&selected_state only";
RUN;
```

Let's go over what's happening in this code:

1. Notice how the `%LET` statement sits alone on a line ending in a semi-colon without a `run` statement. It sets the variable name we chose, `selected_state`, to the value we chose, which is `12`.

2. Next, `PROC FREQ`, which we programmed before, is presented. However, where it used to say `WHERE _STATE = 12`, it now says `WHERE _STATE = &selected_state`. The `&` signals to SAS that we are referring to a variable that was set with the `%LET` command. Therefore, it fills in `&selected_state` with `12`.

3. After that, we run the `title` command. However, we changed it a little so we could automate it. Now, we reuse the variable `&selected_state`. Because we are making a statement in quotes for our title, we have to use two `&` before `selected_state` to tell SAS that we are talking about a value from a macro variable.

4. Since `selected_state = 12`, we had to change the title wording so it makes sense as a sentence. Therefore, we changed it to `"PROC FREQ with _STATE=&&selected_state only"` because that way, when `&&selected_state` is filled in with `12`, the title will make sense. For macro variables to resolve correctly in titles, double quotes are required.

Let's look at the new output we have now that we have used our macro variable, `selected_state`:

PROC FREQ with _STATE=12 only

The FREQ Procedure

DIABETE3	Frequency	Percent	Cumulative Frequency	Cumulative Percent
1	2379	15.61	2379	15.61
2	136	0.89	2515	16.50
3	12368	81.14	14883	97.64
4	332	2.18	15215	99.82
7	22	0.14	15237	99.97
9	5	0.03	15242	100.00

Figure 8.3 – One-way frequency of DIABETE3 variable for Florida only using the macro variable

As we can see from the output, the numbers are the same as the previous output, so the rearrangement of the code into using the macro variable worked. Also, we can see that the title changed so that 12 was filled in where we had put &&selected_state in the coding.

This demonstration of using a macro variable provides some insight into how the **macro processor** works in SAS. This diagram provides a visual explanation:

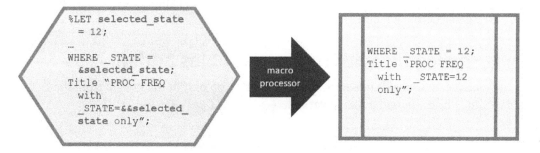

Figure 8.4 – Diagram of how the macro processor works in SAS

As can be seen from the diagram, in the shape on the left, we have some snippets of our code using macro variables. We see that %LET tells SAS to set selected_state to 12, and then the rest of our code tells SAS to use that value in PROC FREQ. When we run this code, SAS first puts the code through the **macro processor**, which is depicted with an arrow in the diagram. The macro processor rewrites the SAS code, filling in the values for the variables. As can be seen in the box on the right, once the code is rewritten, it looks like it did before we added the macro variable. Once SAS rewrites the code, it runs it.

> **Note:**
>
> Notice that when we decided we wanted to add a macro variable to allow us to change which state was being reflected in PROC FREQ, we started by hardcoding PROC FREQ for a particular state, which is Florida. This is the right process to use when adding macro variables. It is first important to find a use case that works with the code before a macro variable is added. In our case, we made sure the code worked with Florida before rebuilding our code with the %LET command. Then, we reran it for Florida using the macro variable to ensure we got the same results.
>
> Imagine we had not started by building code that was hardcoded for Florida and ran without errors. Imagine we had just started by building PROC FREQ with the macro variable and then filling it in with values using %LET. If we hit errors when running the code, they could come from two different sources: either our coding is wrong, or our data doesn't make sense with the code. For example, if to test the macro, we had written %LET selected_state = 13, which is the state of Georgia, there would have been an error, because we only have codes 12, 25, and 27 in our dataset. If we had not built the code using the steps we took, we would not have realized that the reason this would not work is we have an unallowable code in the %LET statement. Approaches to macro coding to prevent errors are discussed more extensively in *Chapter 9, Debugging and Troubleshooting in SAS*.

So far, we have learned how to use a macro variable. Now, let's explore how we can use a macro variable within a macro.

Using the log file with macro variables and macros

Before we move on to making macros, let's review the top part of the log file from the last code we ran, where we used a macro variable:

```
1        OPTIONS NONOTES NOSTIMER NOSOURCE NOSYNTAXCHECK;
72
73          %LET selected_state = 12;
74          PROC FREQ data=brfss_a;
75          tables DIABETE3 / missing;
76          WHERE _STATE = &selected_state;
77          title "PROC FREQ with _STATE=&&selected_state only";
78          run;

NOTE: There were 15242 observations read from the data set WORK.BRFSS_A.
      WHERE _STATE=12;
NOTE: PROCEDURE FREQ used (Total process time):
      real time           0.12 seconds
      cpu time            0.11 seconds


79
80       OPTIONS NONOTES NOSTIMER NOSOURCE NOSYNTAXCHECK;
92
```

Figure 8.5 – Log file from running code with a macro variable in SAS

The boxes were placed on the log file to call attention to some code that was added automatically by SAS that we did not write. We can see our code starting with the statement `%LET selected_state = 12`. The code before that says `OPTIONS NONOTES NOSTIMER NOSOURCE NOSYNTAXCHECK;` and was added automatically by SAS.

The reason it is important to talk about this line of code is that in SAS University Edition, when using the macro language, the programmer sometimes must add this programming themselves manually to their macros, or else they will not run. This line of code sets **SAS system options**. These options were added automatically by SAS when we used the macro variable, but may not be added automatically when we make an actual macro.

The purpose of these options is to suppress log output. If the programmer is using SAS University Edition and forgets to include these system options in the macro, the log output might cause errors, and stop the macro from running. The upside to suppressing the log output is that it results in the log being very short. The downside is that a lot of helpful information must be suppressed from the log in order to get the macro to run. This makes it hard to troubleshoot errors in coding.

SAS system options

Here is what the system options we are setting mean:

NONOTES: This suppresses notes from the log. Notes include copyright information, licensing and site information, and the number of observations and variables in the dataset. SAS will also throw up a note when it halts a processing step due to errors.

NOSTIMER: The system option `STIMER`, which is the default, writes to the log a list of computer resources that were used for each step, and also, for the entire SAS session. `NOSTIMER` turns this off.

NOSOURCE: This suppresses the listing of SAS statements to the log.

NOSYNTAXCHECK: The default setting is `SYNTAXCHECK`, where the syntax of code is checked before the code is run. When using SAS and getting errors, you may notice that some come right away, and some come after SAS does some processing. The ones that come right away have to do with syntax errors in coding (such as missing semi-colons). This fast response comes from the `SYNTAXCHECK` function, which we turn off with `NOSYNTAXCHECK`.

SAS has a notoriously mysterious error and other messages in the log, but because they are consistent and well-documented, they can be used for troubleshooting. You are encouraged to explore ways to use the log for troubleshooting in this SAS white paper, *Logging the Log Magic: Pulling the Rabbit out of the Hat* by Adel Fahmy (available under *Further reading*).

Because so much information is suppressed from the log, it is important to continue to use a programming process that builds code one step at a time to prevent errors. Let's use this process and make our first macro.

Making macros with PROCs

As described before, the macro will be built in steps. In fact, we will continue with the code we were building, where we added the macro variable `selected_state`. Now, we will turn that entire snippet of code into a little program or macro. That way, we can run the macro with one line of code, like it's a command. We will make the following modifications:

- We will start by sandwiching our code between a %MACRO statement at the beginning and a %MEND statement at the end.

- This means that before our code, we will use a %MACRO statement, and give a name to the macro. This is the name we will use later when we want to run it.

- Placing %MEND at the end of the macro code tells SAS the end of the programming intended to be included in the macro.

- After the %MACRO statement, we will include the line setting system options to suppress log output to ensure our code will run in SAS University Edition.

Since the macro is about frequencies by state, and since we will be building the macro in steps, let's call the first one we build `state_freq1`, so we can call the next one `state_freq2`, and so on. Here is the code we formulated:

```
%MACRO state_freq1;
OPTIONS NONOTES NOSTIMER NOSOURCE NOSYNTAXCHECK;
%LET selected_state = 12;
PROC FREQ data=brfss_a;
tables DIABETE3 /list missing;
WHERE _STATE = &selected_state;
title "PROC FREQ with _STATE=&&selected_state only";
RUN;
%MEND state_freq1;
```

Let's run the macro, then observe a few things about our code:

- Only the highlighted code changed. The rest is identical to the previous PROC FREQ code where we used the macro variable.

- At the beginning, we name the macro `state_freq1` in the line `%MACRO state_freq1`, and at the end, we close the macro with `%MEND state_freq1`. It is not necessary to repeat `state_freq1` in the `%MEND` statement, but it can be helpful to repeat it if storing many macros in a file together for easy identification. It is good programming practice to include the macro name at the end of the macro.

- There is a `run` statement inside the macro, but there is no `run` statement outside of the macro.

- When we run the macro, we see a log file entry that looks like the one we saw before. It has very little information in it because most of it has been suppressed. Therefore, it is hard to tell what happened.

The macro has been loaded into SAS's memory. Now, we use this code to call the macro:

```
%state_freq1;
```

In this one line of code, we use the `%` sign to indicate to SAS that we are calling a macro:

- By using the name `state_freq1`, we tell SAS to look for a loaded macro of that name. If it does not find it, it will not run it.

- However, if it finds it and the code executes properly, it should produce the same output as we saw before because we did not change the `PROC FREQ` code; we just turned it into a macro.

Because the results of `PROC FREQ` are identical to *Figure 8.3*, they will not be reprinted here.

We may be pleased to see that our first macro, `state_freq1`, ran and produced the output we expected. However, we might notice that embedding the `%LET` statement in the macro defeated the purpose of creating it in the first place. After all, we wanted to be able to run the macro using different values of `_STATE`.

This is where macro programming can become very powerful. We will rewrite our macro and call it `state_freq2`. We will plan these changes:

- Instead of only naming the macro in the `%MACRO` command, we will add code that makes the macro automatically expect the user of the macro to specify a macro variable value.

- Because we will do this, we will get rid of the `%LET` statement because we will not need it anymore.

- Not only will we make changes in the programming of the macro, but we will also change how we call it up. That way, we can pass it the value of the macro variable when calling it up.

Here is our modified code that makes the macro state_freq2:

```
%MACRO state_freq2(selected_state=);
OPTIONS NONOTES NOSTIMER NOSOURCE NOSYNTAXCHECK;
PROC FREQ data=brfss_a;
tables DIABETE3 /list missing;
WHERE _STATE = &selected_state;
title "PROC FREQ with _STATE=&&selected_state only";
RUN;
%MEND state_freq2;
```

Let's look at the changes we made in this code:

- Notice how the code no longer has a %LET statement that sets the value of the macro variable selected_state. Instead, the first line says %MACRO state_freq2(selected_state=).
- By putting selected_state= in parentheses, we are telling SAS that the macro wants the user to set a macro variable when they run the macro.

By making this change, we make it so we can run the macro for another state other than Florida. However, let's continue to test it by running it with Florida to ensure that our modifications did not create any errors in the macro:

```
%state_freq2(selected_state=12);
```

Notice that this time, when we called up our macro state_freq2, we set the macro variable by adding selected_state = 12 in parentheses. Because running the code produces the same output as the previous code, we will not reproduce it here. This macro statement can be repeated for each state code to run the macro for each state.

So far, we have demonstrated how to program macro variables and macros that automate PROCs. However, most SAS data warehouse maintenance consists of ETL protocols, and those involve data steps. Therefore, we will now turn our attention to making macros with data steps.

Making macros with data steps

Before we make a macro with a data step, let's first consider what might be a useful process to automate. An operation we did in *Chapter 6*, *Standardizing Coding Using SAS Arrays*, and in *Chapter 7*, *Designing and Developing ETL Code in SAS*, was to create **two-state flags** (coded as 1 = **yes** and 0 = **all other answers**) to indicate the presence of a disease condition. Let's do this again with this dataset. As you may remember, our demonstration dataset has many variables about health conditions that are all coded the same way: 1 = **Yes**, 2 = **No**, 7 = **Don't know/Not sure**, 9 = **Refused**, blank (represented by a period) = **missing**. For these, we could create two-state flags.

Let's consider our first variable, ASTHMA3, which asked the respondent about having asthma. Let's say we want to create a variable called asthma_flag that is coded 1 where ASTHMA3 was coded 1, but is coded 0 for all other answers. Although other variables are coded the same way, we will use ASTHMA3 as our example variable so we can make sure the code runs before we try to turn it into a macro. We will make some data step code that creates asthma_flag, and follow it with PROC FREQ to check our recoded variable:

```
data brfss_b;
    set brfss_a;
    asthma_flag = 0;
    if ASTHMA3 = 1
        then asthma_flag = 1;
PROC FREQ data=brfss_b;
    tables asthma_flag * ASTHMA3 /list missing;
RUN;
```

Let's unpack what is happening in this code:

1. First, we observe that we are reading in dataset brfss_a and outputting dataset brfss_b.

2. Next, we see that we create our flag, called asthma_flag, by setting the variable to 0.

3. After that, we use an if statement to modify the value in asthma_flag. If the value of the native variable, ASTHMA3, is 1, then asthma_flag gets updated to 1. Otherwise, it stays at 0.

4. In the next step, PROC FREQ is run on the output dataset, brfss_b. In the tables command, we have a two-way frequency between the new flag variable and the original native variable.

Let's take a quick look at the PROC FREQ output:

The FREQ Procedure

asthma_flag	ASTHMA3	Frequency	Percent	Cumulative Frequency	Cumulative Percent
0	2	33933	87.23	33933	87.23
0	7	90	0.23	34023	87.46
0	9	15	0.04	34038	87.50
1	1	4863	12.50	38901	100.00

Figure 8.6 – PROC FREQ results for asthma_flag and ASTHMA3

As could be expected, most of the dataset is coded with asthma_flag = 0, with only 12.5% falling in the asthma_flag = 1 category. Also, we see our recode was accurate.

This code should seem completely familiar to you at this point in the book. But when it comes to turning this code into a macro, some design decisions need to be made. We know from our experience with making macros that we can allow variables to be set when the macro is called. So the first thing to consider about our code is which parts of it should be macro variables that are set when the macro is called. Let's think about it this way:

- The first thing that happens is that we read in an input dataset (brfss_a), and we state the dataset we want to output (brfss_b). We might want to adjust these each time we run the macro. Therefore, we will choose macro variables for these: data_in and data_out, respectively.

- Another thing we will likely want to change each time we run the macro is which native variable we are processing, and the name of the flag we are creating. Therefore, let's create two more macro variables: native_var and flag_var.

- Let's call our macro %make_flag.

Here, we will present the finished macro, but please note that we skipped using the proper step-by-step process to build it. Because this process works hand in hand with debugging, we will go over the step-by-step process we used to build this macro in *Chapter 9, Debugging and Troubleshooting in SAS*. In the current chapter, we will just present the finished macro that runs properly and can take in all these macro variables we described.

Here is the final %make_flag macro:

```
%MACRO make_flag(native_var=, flag_var=, data_in=, data_out=);
data &data_out;
    set &data_in;
```

```
      &flag_var = 0;
      if &native_var = 1
            then &flag_var = 1;
PROC FREQ data=&data_out;
      tables &flag_var * &native_var /list missing;
run;
%MEND make_flag;
```

Let's review the features of this code:

- This time, we do not set system options for our macro, because this macro does not cause errors in SAS University Edition for whatever reason.

- As we promised, we named our macro make_flag in the first line where we have our %MACRO command. However, notice that in the parentheses that follow, all the macro variables are declared. Their order does not matter, but each must be followed by an equals sign, and they must be separated by commas.

- We carefully went through our code and replaced all of the original code where we wanted to add macro variables. In the first and second lines of the data step, we replaced brfss_a with macro variable &data_in and replaced brfss_b with macro variable &data_out.

- We did something similar with ASTHMA3 and asthma_flag. We replaced ASTHMA3 with macro variable &native_var and asthma_flag with &flag_var.

- In PROC FREQ, we reused the macro variables. We used &data_out to specify the dataset, and &flag_var and &native_var to specify the variables in the two-way frequency.

The macro must be run to be put in memory in SAS before it can be called up. So let's make sure we run the macro code before we call it up.

Now, we will state the code to call up the macro. We will replicate what we did with creating a flag for ASTHMA3 – but this time, we will use the macro we made called %make_flag:

```
%make_flag(native_var=ASTHMA3, flag_var=asthma_flag, data_
in=brfss_a, data_out=brfss_b);
```

Although this is one line of code, it has several components:

- First, the macro is called by `%make_flag`, but there is also the work of defining all the macro variables in parentheses. This is essentially back-engineering how we built the macro in the first place, which we are doing for testing purposes. Defining macro keyword parameters offers better documentation of macro variables than the alternative macro position parameters.

- In the parentheses, we set macro variable `native_var` to `ASTHMA3` and `flag_var` to `asthma_flag` based on our original coding that we turned into a macro.

- We also set `data_in` to `brfss_a` and `data_out` to `brfss_b`.

Running the `%make_flag` code will produce the same output as in *Figure 8.6*, so it will not be repeated here. This reassures us that our conversion of code to macro worked – at least for the asthma variable.

Let's use our `%make_flag` variable to make some more flags and ensure the macro continues to work with the different variables:

- The next two native variables on our list are `CVDINFR4`, which is the answer to the question of the respondent ever having had a heart attack, and `CVDCRHD4`, which is whether or not the respondent has heart disease. When running the macro, we'd set `native_var` to these variables.

- For `CVDINFR4`, let's create a flag variable called `HA_flag`, and for `CVDCRHD4`, let's create `HDis_flag`. These will be what we fill in for the macro variable `flag_var`.

- We will transform `CVDINFR4` first, so when we call that macro, we will set `data_in` to `brfss_b` because that was the last iteration of the dataset. In addition, `data_out` will be set to `brfss_c`.

- Since we will run the macro on `CVDCRHD4` next, for that one, we will set `data_in` to `brfss_c`, and `data_out` to `brfss_d`.

Here is the code we made calling the macro on the basis of this plan:

```
%make_flag(native_var=CVDINFR4, flag_var=HA_flag, data_
in=brfss_b, data_out=brfss_c);
%make_flag(native_var=CVDCRHD4, flag_var=HDis_flag, data_
in=brfss_c, data_out=brfss_d)
```

As we planned, we call the `%make_flag` macro twice, first to transform `CVDINFR4`, and next to transform `CVDCRHD4`. Let's look at the `PROC FREQ` output that is produced:

The FREQ Procedure

HA_flag	CVDINFR4	Frequency	Percent	Cumulative Frequency	Cumulative Percent
0	2	36423	93.63	36423	93.63
0	7	145	0.37	36568	94.00
0	9	12	0.03	36580	94.03
1	1	2321	5.97	38901	100.00

The FREQ Procedure

HDis_flag	CVDCRHD4	Frequency	Percent	Cumulative Frequency	Cumulative Percent
0	2	36460	93.73	36460	93.73
0	7	268	0.69	36728	94.41
0	9	14	0.04	36742	94.45
1	1	2159	5.55	38901	100.00

Figure 8.7 – PROC FREQ results for calling %make_flag on CVDINFR4 and CVDCRHD4

The output suggests that our `%make_flag` macro is working, which is wonderful. However, not all of the variables are coded the same way. Remember how for the diabetes variable, `DIABETE3`, the coding was a little different? In `DIABETE3`, both values 1 and 2 refer to a *yes* answer. If we used `%make_flag` on `DIABETE3`, it would produce erroneous results, because it would only take into account 1 = **yes**, and not 2 = **yes**.

Of course, we could make a special macro just for `DIABETE3`, but that would defeat the purpose of automation, which is why we are making macros. To continue with the automation, a situation like this calls for adding a **condition** to the macro.

Addition conditions to macros

Earlier, we mentioned using the commands `%IF-%THEN/%ELSE` and `%DO-%END` when making and using SAS macros. These commands allow us to add conditions to our macros. Let's continue to build onto our macro, `%make_flag`, but we will change its name to `%make_flag_conditional`. Here is how we will plan to redesign our macro:

- In our redesign, the first line of our code to build the macro will be the same as previously, but we will just change the name to `make_macro_conditional`. It will still need to take in the same macro variables, so we will keep what is in parentheses the same.

- If needed, we will modify the system options to get the macro to run in SAS University Edition.

- Following that, we will have two sets of data step code. The first set of code will tell SAS how we want it to make the flag if the native variable we provide is `DIABETE3`. The second set of code will tell SAS how we want it to make the flag if the native variable is any other one we name.

- Of course, we do not want the macro to run both sets of code. We want it to either run the `DIABETE3` code (if `DIABETE3` is the native variable), or we want it to run the other code if the other disease variables are used. So we will insert an `%IF` statement to add a condition before the first set of code is executed. The `%IF` statement will be based on the macro variable `native_var`. We will tell SAS that `%IF native_var = DIABETE3`, we want it to run this first set of code – in SAS macro language, that translates to `%THEN %DO`.

- Because we are putting a `%DO` statement at the beginning of the macro, we have to close it at some point. Therefore, after we present our first set of code tailored to `DIABETE3`, we have to put `%END` so SAS knows to close the loop, and go on and look for more conditions.

- SAS processes code in a linear fashion. If SAS finds that `native_var = DIABETE3`, it will execute the first set of code, `%END` the loop, then see the `%MEND` statement and end the macro. It will skip over the second set of code.

- But if SAS finds that `native_var` is not `DIABETE3`, it will skip over the first set of code and continue to look to see if it has any other work to do in the macro. If it sees `%ELSE %DO`, then it will continue processing code in the macro.

- Therefore, we will put `%ELSE %DO`, and after that, we will place the code we already wrote for `%make_flag` that works on all the other disease variables.

- Because we are putting another `%DO` statement, we have to end that loop with another `%END` statement.

- We place our `PROC FREQ` between `%END` and the end of the macro so it runs after the new variable is generated.

- Finally, we put `%MEND` to end the macro.

Even though we made very few changes, adding just one condition made our code a lot longer. Let's take a look at it:

```
%MACRO make_flag_conditional(native_var=, flag_var=, data_in=,
data_out=);
OPTIONS NONOTES NOSTIMER NOSOURCE NOSYNTAXCHECK;
%IF &native_var = DIABETE3 %THEN %DO;
data &data_out;
    set &data_in;
    &flag_var = 0;
    if &native_var = 1 | &native_var = 2
        then &flag_var = 1;
RUN;
%END;
%ELSE %DO;
data &data_out;
    set &data_in;
    &flag_var = 0;
    if &native_var = 1
        then &flag_var = 1;
%END;
PROC FREQ data=&data_out;
    tables &flag_var * &native_var /list missing;
run;
%MEND make_flag_conditional;
```

Let's look at some features of this code:

1. Notice the statement %IF &native_var = DIABETE3 %THEN %DO. This statement evaluates the first condition, which is to see if native_var = DIABETE3. If it does, then it will execute the first set of code.

2. Also highlighted in the code is where the coding for DIABETE3 differs from the coding for the other flags. For DIABETE3, both 1 and 2 are accepted as *yes* answers and set the flag to 1. For the rest of the native variables, only the value of 1 qualifies for setting the flag to 1.

3. We can see highlighted in the code where the second set of code starts. It starts with the line %ELSE %DO.

4. We can also see in the code where the two different %DO loops are closed with %END statements.

5. We see our PROC FREQ between the last %END statement and the end of the macro.

6. Finally, we observe the use of %MEND to end the macro.

In order to use the macro we programmed, we must run the macro to put it into SAS memory. Next, we will test the macro by calling it with variables that meet either condition. We will start with the variable CVDSTRK3, which is the native variable for stroke. This variable should only meet the second condition of the macro, and not the first. Therefore, only the second set of code should be executed. Let's try this:

```
%make_flag_conditional(native_var=CVDSTRK3, flag_var=stroke_
flag, data_in=brfss_d, data_out=brfss_e);
```

Let's look at the code more carefully:

- As we call our new macro, %make_flag_conditional, we set the native_var macro variable to CVDSTRK3, and flag_var to stroke_flag. That means we should expect this macro to generate a flag called stroke_flag.

- When we were using our previous macro, %make_flags, we processed data up to dataset brfss_d. Therefore, we set data_in as brfss_d, and data_out as brfss_e.

After running the macro, we can look at the PROC FREQ output to see if stroke_flag looks correct:

The FREQ Procedure

stroke_flag	CVDSTRK3	Frequency	Percent	Cumulative Frequency	Cumulative Percent
0	2	37230	95.70	37230	95.70
0	7	85	0.22	37315	95.92
0	9	11	0.03	37326	95.95
1	1	1575	4.05	38901	100.00

Figure 8.8 – PROC FREQ results for calling the %make_flag_conditional on CVDSTRK3

The results show that the second set of code was executed, and the flag was recoded properly. Now, let's try it again with DIABETE3 and see whether SAS executes the first set of code instead of the second:

```
%make_flag_conditional(native_var=DIABETE3, flag_var=diab_flag,
data_in=brfss_e, data_out=brfss_f);
```

Let's look at how we called the macro:

- We set native_var to DIABETE3 as planned. Then we decided to name the flag being generated diab_flag, so we set flag_var to diab_flag.

- Because we had just output brfss_e from the last time we executed the macro, we set data_in to brfss_e. We set data_out, then, to brfss_f.

Let's run the code and look at the output:

The FREQ Procedure

diab_flag	DIABETE3	Frequency	Percent	Cumulative Frequency	Cumulative Percent
0	3	32931	84.65	32931	84.65
0	4	712	1.83	33643	86.48
0	7	47	0.12	33690	86.60
0	9	21	0.05	33711	86.66
1	1	4789	12.31	38500	98.97
1	2	401	1.03	38901	100.00

Figure 8.9 – PROC FREQ results for calling the %make_flag_conditional on DIABETE3

As we can see, the first set of code must have executed on DIABETE3, because both native values of 1 and 2 are coded as 1 in the flag variable. This demonstrates the power of such automation. As long as BRFSS does not change the name or coding of the native variables over time, this macro could be used to mass-process these native variables into flags, taking into account any anomalies or nuances in coding as we see with DIABETE3.

> **The difference between the macro language and data step language**
>
> In this chapter, we have talked about **macro expressions** such as %IF/%THEN, and %DO loops. Other macro expressions can be used, and you are encouraged to study SAS documentation for the functions and meanings of macro expressions.
>
> At a quick glance, a lot of macro language looks the same as data step language, only the commands are preceded by %. While the commands %IF, %THEN, and %DO look very similar to data step language, as a technical matter, macros are code formulated to be fed into the macro processor, not directly into the SAS engine like data step code. That is why it is important to recognize that these two languages are different.
>
> For example, in data step code, we could have said if DIABETE3 in (1,2) then diab_flag = 1. However, the in command does not work in macros, so we had to state the code another way when it became part of a macro. In fact, this issue was addressed in the SAS 9.2 upgrade, according to Warren Repole Jr., in his white paper, *Don't Be a SAS® Dinosaur: Modernizing Programs with Base SAS 9.2 Enhancements* (link available in the *Further reading* section). By adding the minoperation option on the line of code initiating the macro, the programmer could now use the %if condition like this:
>
> %if &native_var in 1 2 then &flag_var = 1;
>
> SAS programmers who choose to rebuild data steps into macro language should take time to study the differences and make sure to formulate code that runs in the macro environment.

Now that we have an understanding of building macros, let's explore different options for storing and calling macros in SAS.

Storing and calling macros

There are two ways to store and call macros in SAS. Either the macro can be placed within the same code as where it is called, as we have demonstrated, or it can be placed in a separate file. This section will provide guidance about the choice of whether to store the macro in the same code as where it is called, or in a different set of code.

Storing and calling macros in the same code

So far in this chapter, we have been storing and calling macros from the same code:

1. First, we program the macro, then run it. This places the macro in SAS memory.
2. Then, we call the macro in the same code by its name, making SAS pull it from memory.

3. We also add values to macro variables when we call the macro, and that allows the macro processor to fill in the values, rewrite the SAS code for us, and execute it.

While this is a fine way to do it for demonstration purposes, in reality, if we were automating ETL, we would want to store the macros in a separate place as snippets of code. That is because we would not want to have to go hunting around for macros embedded in a long compendium of ETL code if we wanted to read them or revise them. We would want these snippets of code to be stored in a centralized place where they could be easily managed and called from other code. That way, we can have one file that just holds the macro and another file that is filled with calls to the macro. The next section will explain how to do this in SAS.

Storing macros separately and calling them from code

Here are the basic steps for storing macro code separately from the SAS code that calls the macro code:

1. First, create the macro and make sure it runs properly.

2. Next, save just the macro code in a SAS file in a particular location. Keep track of the name of the SAS file and the location where it is stored.

3. Now, start a new SAS file for calling the macro. At the beginning of the file, use an `%INCLUDE` command to map to the SAS file that stores the macro.

4. Next, in the new SAS file, call the macro.

We can do these steps together with the `%make_flag` macro we created. As you may recall, we already did the first step, which is to create the macro `%make_flag` and make sure it runs properly. So, we will proceed from the second step:

1. We need to create a particular location to save our macro code files. If using SAS University Edition, we'll create a folder within `myfolder` called `Macros`.

2. Let's copy the `%make_flag` macro code into a file, and save it in the `Macros` folder. We will call the file `make_flag.sas`.

Now, let's make the new SAS code that calls the macro. First, we will add the %INCLUDE statement. Next, we will call the macro on the next disease variable, which is CHCSCNCR, or a report of ever having skin cancer. Because of this, we will call the flag variable generated, skinCA_flag:

```
%INCLUDE "/folders/myfolders/Macros/make_flag.sas";
run;
%make_flag(native_var=CHCSCNCR, flag_var=skinCA_flag, data_
in=brfss_d, data_out=brfss_e);
```

Let's go through the code:

1. The %INCLUDE line of code is highlighted. This %INCLUDE statement alerts SAS to the fact that the current code will be calling macros from the file that is specified in the %INCLUDE statement.

2. Next, as we have done before, we call %make_flag on the native variable CHCSCNCR to produce the flag variable skinCA_flag. We read in dataset brfss_d and write out dataset brfss_e.

Running the code produces PROC FREQ (not shown here). As can be seen from this approach, theoretically, one large SAS file filled with all the macros for an ETL project could be present in a central, accessible location. Then, analysts could use %INCLUDE to call these macros from this centralized file. This is a way of splitting up the work among teams. One team could be focused on maintaining the integrity of ETL protocols by keeping macros up to date. Another team could be actually doing the ETL by launching the macros when it is time to process and load the data.

> **Alternative to %INCLUDE:**
>
> There is actually another way to handle SAS macros by managing them in an automated way without using the %INCLUDE approach. We could sum it up by calling it the **SASAUTOS** approach. SASAUTOS refers to an environment variable that can be set to manage macro libraries. The SASAUTOS approach provides efficiency in calling macros through being able to reuse them efficiently in SAS's memory. An in-depth discussion on how this works is outside the scope of this book, and the curious reader is encouraged to check out the SAS white paper, *A SASAUTOS Companion: Reusing Macros* (link available in the *Further reading* section).

Loading transformed data

As described in *Chapter 4, Managing ETL in SAS*, in a SAS data warehouse, it's not uncommon to receive monthly or annual files that require regular ETL. Imagine receiving the BRFSS files from 2016, 2017, and 2018, and needing to process them. If the datasets are all named according to a particular naming convention, we can use macro code to automatically load the data files and put them through an ETL protocol.

The easiest way to demonstrate this with a simple exercise is to have us generate the multiple files we will later read in. That way, we can concentrate on writing the load macro, and not whether the data will cooperate. For this exercise, we will use the dataset Chap8_2. This dataset has only two variables: _STATE and FMONTH. As with the previous file, the only states included in _STATE are codes 12, 25, and 27, for Florida, Massachusetts, and Minnesota, respectively. FMONTH contains the month the survey was conducted coded as a number, 1 through 12.

Imagine we worked for the government of Florida, and therefore, we only wanted data from files pertaining to Florida (_STATE = 12). If we had 12 monthly files from 2018 that contained data from all three states, we might want to make a macro to do the following:

1. Read them in one at a time.
2. Retain the Florida records and delete the other ones.
3. Then save the transformed SAS dataset in an output location.

In a real data warehouse, we would likely do more operations (such as calling ETL macros), but for this demonstration, this is all we will do.

Let's start by placing Chap8_2 in the folder mapped to LIBNAME X. Then we will run a big data step that will split the file into 12 monthly files and place each of them in X. Each dataset will be named according to a convention of the three-character month followed by a four-digit year (for example, X.JAN2018). Then, we will create a load macro to read in these files and just keep the Florida records.

Here is the code to split Chap8_2 into monthly files:

```
data X.JAN2018 X.FEB2018 X.MAR2018 X.APR2018 X.MAY2018
     X.JUN2018 X.JUL2018 X.AUG2018 X.SEP2018 X.OCT2018
     X.NOV2018 X.DEC2018;
   set X.Chap8_2;
   if FMONTH = 1 then output X.JAN2018;
   if FMONTH = 2 then output X.FEB2018;
   if FMONTH = 3 then output X.MAR2018;
```

```
     if FMONTH = 4 then output X.APR2018;
     if FMONTH = 5 then output X.MAY2018;
     if FMONTH = 6 then output X.JUN2018;
     if FMONTH = 7 then output X.JUL2018;
     if FMONTH = 8 then output X.AUG2018;
     if FMONTH = 9 then output X.SEP2018;
     if FMONTH = 10 then output X.OCT2018;
     if FMONTH = 11 then output X.NOV2018;
     if FMONTH = 12 then output X.DEC2018;
run;
```

After running this code, all of the monthly files should be in the directory mapped to X and named accordingly. Here is a screenshot of what this would look like in SAS University Edition:

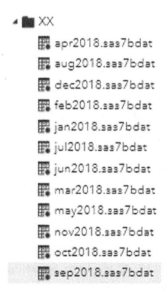

Figure 8.10 – Monthly input files in the X directory

Next, let's start by using the file X.JAN2018 as a test dataset for making our base code for our macro:

1. First, we will map libname to a different directory, Y. This will be the directory where SAS saves the transformed datasets.

2. Next, we will choose to name the transformed datasets with the _FL suffix (for example, Y.JAN2018_FL).

3. Finally, we will use data step code to keep only Florida records in the output dataset.

Here is the base code we wrote for our macro:

```
LIBNAME Y "/folders/myfolders/Y";
data Y.JAN2018_FL;
     set X.JAN2018;
     if _STATE = 12;
RUN;
```

Running this code successfully reads in dataset X.JAN2018, keeps only the records with _STATE = 12, and saves the transformed dataset named JAN2018_FL to the directory mapped to Y. This code runs successfully, so the next challenge is to convert it into a macro that will walk through each of these monthly files – X.JAN2018 through X.DEC2018 – and run the same operation.

For brevity, we will skip the macro-building steps, and present a final macro built from this base code. We call the macro keep_FL:

```
%MACRO keep_FL;
OPTIONS NONOTES NOSTIMER NOSOURCE NOSYNTAXCHECK;
%LET M=JAN FEB MAR APR MAY JUN JUL AUG SEP OCT NOV DEC;
%DO J=1 %TO 12;
     %LET MONTH=%SUBSTR(%STR(&M),(&J*4-3),3);
     %LET outfile = Y.&MONTH.2018_FL;
     %LET infile = X.&MONTH.2018;
     data &outfile;
          set &infile;
          if _STATE = 12;
     RUN;
%END;
%MEND keep_FL;
```

This code is very complicated, so we will cover only the highlights here:

1. The %LET statement sets the macro variable &M to a string of three-character month abbreviations separated by spaces.

2. Next, the %DO loop is declared. This is what will happen each time we read in a monthly file.

3. Immediately after the %DO command, a series of %LET commands are used to set macro variables.

4. &MONTH is set to equal a substring (represented by the command %SUBSTR). What this code actually does is read the three-digit character string for the month of the current iteration of the do loop (stored in J) from the &M macro variable. This code will cycle through the list so that each month is selected to be processed. So, for the month of May, %LET would set the &MONTH variable to the part of the &M macro variable that says MAY.

5. Using the variable &MONTH, the other two %LET statements assemble the names of the output file (under the macro variable outfile) and the input file (under the macro variable infile).

6. Decoding the macro, the input file will be named whatever is in the &MONTH variable followed by 2018, and will be located in the directory mapped to X. The output file will be placed in the Y directory and will be named whatever is in the &MONTH variable in that iteration, followed by 2018_FL.

This macro is named keep_FL. If launched using the command %keep_FL, it will successfully read in each of the monthly files we created in X, keep only the Florida records, and write the resulting datasets according to our naming convention in the output directory mapped to Y. We can verify this with a screenshot from SAS University Edition:

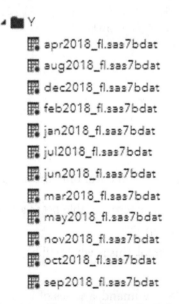

Figure 8.11 – Monthly output files in the Y directory

As mentioned earlier, this code is very simplistic. A much more realistic view of using macros to load data – specifically BRFSS files – can be understood in the SAS white paper *Using SAS Macro Language and SAS Pipe to Process State Health Care Survey Data* by Xiaoting Qin and Cathy M. Bailey (link available in *Further reading*). In this report, the authors develop three separate macros: one that sets up a drive environment, one that reads the files, and one that aggregates them. The drive environment macro actually calls up the other two macros, which is an example of having **nested macros**. The authors also use a PIPE function as part of reading the data. This white paper provides a more appropriate use case for developing SAS macros for loading datasets than is presented here, which is only an introduction to the topic.

As can be seen, using macro programming in SAS has both advantages and disadvantages:

Advantages	Disadvantages
• Coding is automated, so ETL is simplified into running a series of code. • Automating coding with macros this way standardizes it across programmers. • Macros can be stored in separate files and called up from another file. This can modularize coding. • By using conditions, coding can be highly automated in macros, because macros can take into account anomalies in the data and still do automatic processing. • If the original code developed, on which the macro was based, was efficient the macro will be efficient.	• Existing macros used on new datasets may encounter errors, and they may not be easy to identify. • It is extremely hard to edit previously developed macros that are found to not be compliant with the structure of current data warehouse extracts due to evolution in the data structure. • Regular code must be developed before it is converted into macro code. This takes an enormous amount of time. • Macros must be built step by step and can contain a lot of %LET variables. This makes them confusing and hard to document.

Knowing this, it is important for the data warehouse manager to weigh up the pros and cons of developing particular portions of SAS ETL code into macros, supporting them, and incorporating them into standard ETL protocols.

Summary

This chapter guided us on how to create **macros** and **macro variables** to automate ETL code in SAS. We started this chapter by developing base code that ran PROC FREQ. We further developed that base code into a macro that allowed us to set a macro variable for selected_state so we could filter by state in our PROC FREQ. We went on to develop a macro called %make_flags that used data steps to automatically generate two-state flags based on variables from a survey about health conditions. When we found one of the variables, DIABETE3, had slightly different coding, we rebuilt our macro to handle this conditional coding and called it %make_flags_conditional. After this, we covered different alternatives for storing and calling up macros. Finally, we ended by developing a macro that read SAS files automatically.

Although there is a lot to know about macro coding, an important lesson in this chapter was to develop base code first and then build it into macro code step by step. Becoming patient with this process will lead to better-documented and higher-quality macros. Once the decision is made to actually use macros to automate a process, it is important to constantly maintain them to ensure standardization and integrity. Storing the macro code separately from ETL code can facilitate sharing this work across teams. And while many times it will not be necessary to use macros to load data, if you have tens or hundreds of files, it is worth considering developing macros for automated inputting and outputting of data files.

Base PROCs and data steps are easier to program than macros, but troubleshooting in SAS is necessarily challenging no matter what the code. The next chapter will focus on approaches to debugging and troubleshooting in SAS.

Questions

1. What is the difference between a **macro variable** and a **macro** in SAS?

2. What do the system options NONOTES, NOSTIMER, NOSOURCE, and NOSYNTAXCHECK do?

3. Why is it necessary to build SAS macros in a step-by-step manner? Why not try to compose the perfect macro code from scratch?

4. Why are conditions added to macros?

5. What is the advantage of storing macros in separate files than the SAS files that call them?

6. Imagine you were working at a data warehouse and were told that you would be assigned to design and conduct ETL on some historic datasets that were being added to the warehouse. What would you need to know in order to decide whether or not to develop a macro to load the files?

7. Imagine you are an analyst at a SAS data warehouse, and you learn that you are being asked to add a new set of data to the warehouse. The dataset consists of responses of small business owners to a survey done by the local chamber of commerce. The chamber does this survey every 5 years. Some of the questions are the same each time, but many change. Since the survey started 10 years ago, there are only two sets of data so far. The warehouse has decided the first set is too old, so you will be loading the second set. They are planning a new survey coming up this year that you will also add. Should you make macros for this ETL project?

Further reading

- *SAS Macro Programming for Beginners* by Susan J. Slaughter and Lora D. Delwiche, available here: https://support.sas.com/resources/papers/proceedings/proceedings/sugi29/243-29.pdf

- *Logging the Log Magic: Pulling the Rabbit out of the Hat* by Adel Fahmy, available here: https://www.lexjansen.com/pharmasug/2010/TT/TT08.pdf

- *Don't Be a SAS® Dinosaur: Modernizing Programs with Base SAS 9.2 Enhancements* by Warren Repole Jr., available here: https://analytics.ncsu.edu/sesug/2009/FF011.Repole.pdf

- *A SASAUTOS Companion: Reusing Macros* by Ronald Fehd, available here: https://www.lexjansen.com/nesug/nesug05/pm/pm12.pdf

- *Using SAS Macro Language and SAS® Pipe to Process State Health* by Xiaoting Qin and Cathy M. Bailey, available here: https://support.sas.com/resources/papers/proceedings11/179-2011.pdf

9
Debugging and Troubleshooting in SAS

This chapter will provide guidance for debugging and troubleshooting code in SAS. Because much of the code in a data warehouse focuses on **extract, transfer, and load (ETL)** protocols, we will focus on debugging and troubleshooting data step code. First, we will discuss general strategies for developing, maintaining, and managing data step code and practice troubleshooting when something goes wrong. Next, we will examine ways of debugging the do loop code using SAS debug functions. Finally, we will cover how to prevent errors in SAS macros, and ways to troubleshoot if errors are identified.

We will go over the following practical approaches:

- Considering the placement of the RUN command in data step code, and strategically developing long data step code

- Forming code in both data steps and macros that facilitates finding and closing %DO loops

- Implementing a step-by-step approach to developing finalized ETL macros, and storing the research and development code as documentation

Technical requirements

The datasets used for demonstration in this chapter are available online from GitHub: `https://github.com/PacktPublishing/Mastering-SAS-Programming-for-Data-Warehousing/tree/master/Chapter%209/Data`.

The code bundle is available on GitHub here: `https://github.com/PacktPublishing/Mastering-SAS-Programming-for-Data-Warehousing/tree/master/Chapter%209`.

Debugging data step code

As you must realize by now, well-formed data step code is critical to efficiency in SAS. Therefore, once we write data step code for an enduring process (such as ETL), it seems like we are always editing it to improve it. We may even rewrite it to add arrays or macro processing.

However, each time we write and rewrite data step code, we create the opportunity for bugs. For this reason, and because the data step code is so complicated to begin with, there is an extensive array of strategies for troubleshooting and debugging SAS data steps. Many are covered in the SAS white paper *Errors, Warnings, and Notes (Oh My) A Practical Guide to Debugging SAS Programs* by Lora D. Delwiche and Susan J. Slaughter (link available in the *Further reading* section).

Because this topic is so extensive, we cannot give it a full treatment here. Instead, we will just cover the basics of debugging data step code and refer to the white paper and other resources for further information.

Writing well-formed and well-formatted code

Up to now, we have created data step code, but we have not talked about the details of how SAS processes this code. It is helpful to know these details to write well-formed code. These details are covered in SAS documentation as well as the SAS white paper, *How SAS Thinks and Why the Data Step Does What it Does* by Neil Howard (link available in the *Further reading* section). When a programmer writes data step code that ends in RUN and then runs it, the processing is split into two phases – **compile** and **execute**:

1. First, SAS compiles the data step code. If it does not compile, SAS stops the processing and throws up an error in the log.
2. If the data step successfully compiles, then SAS executes the data step code. If there is an error in the execute process, SAS prints an error to the log. Otherwise, it executes, and one or more datasets are output.

To do each phase of the processing, SAS starts at the beginning of the data step, then proceeds forward through the RUN command. Because data step processing is in two phases, it means that different errors are possible in the compile phase compared to the execute phase.

Let's look more carefully at the compile phase. Here are some of the steps in the compile phase:

1. The syntax of the code is scanned.

2. SAS source code is translated to machine language.

3. Input and output files are defined.

4. The following are created: an input buffer (if the data being read is not in SAS format), a **Program Data Vector (PDV)**, and dataset descriptor information.

5. Attributes for the variables in the output datasets are established.

6. Variables that are to be initialized to missing are captured.

It is easier to describe the **PDV** by explaining how it works. Let's reflect on how we have seen data step code behave when we have run it throughout the different chapters. Part of the reason we build data step code using smaller snippets of code, and then paste those snippets together later, is that we want each snippet to run properly before we put it with more code so we can be sure we have no errors. If we build three transformation steps in separate data step code and they all run properly, we feel we can put all three together in one big data step with one RUN command at the end. This is because we know that as SAS processes the large data step, it will complete the first transformation successfully, adding the transformed variable to the PDV, before moving on to the second transformation. After the second transformation, it will add the second transformed variable to the PDV, and move on.

Once the compile phase is complete, the execution phase begins, which includes the following steps:

1. The **input/output (I/O)** engine supervisor works to optimize the execution through controlling looping.

2. The initialize-to-missing instruction is handled.

3. Observations to be read are identified.

4. Variables in the PDV become initialized.

5. Data step programming is called.

6. User-controlled data step machine code statements are executed.

7. The default output of observations is completed.

Knowing the breakdown of the data step compile and execute phases can help the SAS programmer troubleshoot errors in the log, and create well-formed and efficiently-executing code.

In addition to having SAS code be well-formed, it should also be well-formatted. In SAS, especially in a data warehouse environment, code should be written with the idea that it will probably be edited, especially data steps. Here are some tips for making code more easily editable:

- Do not put more than one SAS statement on each line.

- Use comments at regular intervals. It is recommended that a comment policy be developed so there are some guidance and standardization as to the verbosity, placement, and content of comments.

- Use indentation to visually show which statements belong together.

- Make short, modular code files instead of long ones when possible.

- The **SAS Enterprise Guide** offers a `Format Code` function from the `Edit` menu that can help format code. It will separate code into lines with one statement per line, and apply indentation automatically.

In order to write quality code, especially about ETL, the analyst needs to understand the underlying data. We have gone over—especially in *Chapter 7, Designing and Developing ETL Code in SAS*—how to study data in order to plan well-formed ETL code. But admittedly, it is hard to study a big dataset.

Think of our approach in this book. The exercises in this book are on a smaller subset of data from a much larger dataset. The source dataset is the **Behavioral Risk Factor Surveillance System (BRFSS)**, an annual anonymous health survey conducted on the phone by the **United States (US)** government. Each year, this dataset includes surveys from each state in the US, and the source dataset contains about 450,000 records. This number of records could not be used with SAS University Edition, which has a limitation on the raw dataset size. Further, without a highly efficient SAS environment, tuning would be necessary in order to get the examples to run in a reasonable amount of time.

Therefore, for most of the demonstration datasets in this book, only records from three states were taken, reducing the dataset to only 38,901 records. With datasets that size, there will likely be no problems with SAS data processing, regardless of how inefficient the code or environment is.

This highlights a conundrum for the SAS data warehouse manager. It is necessary to study the entire dataset—not a subset—to establish an evidence-based ETL protocol. But it takes a long time to run any big data queries in SAS. With poorly-written code, the processing takes too long, so well-formed code is necessary, which itself takes time. This is why it is typical to take subsets of SAS datasets and study those. Imagine we had a problem where a variable was missing on some records when it was not supposed to be missing. We could make a smaller dataset of a sample of those records to see if we can figure out some patterns as to why that variable had a missing value in all of them.

When debugging data step code, it is very important to pay attention to the contents of the log file, which we will discuss next.

Using log information as guidance

Technically, there are three kinds of messages in the log, although many find these classifications unintuitive:

- **Error** messages, which are preceded with the word **ERROR:** and appear in red. Error messages indicate your code stopped processing.

- **Warning** messages, which are preceded by **WARNING:** and appear in green. These indicate that the code kept processing, but SAS thinks there may be an underlying problem that the programmer should check out.

- **Notes** messages, which are preceded by **NOTE:** and appear in blue. These provide information about the dataset. We most commonly see these when we read in a dataset, and SAS tells us how many observations it has.

As can be seen from their description, the content of these messages is more important than their classification. Here are a few common messages we see when using SAS data steps that may indicate errors to us:

Message	What it means	Common reasons for message	Possible actions
WARNING: The dataset XXX may be incomplete. When this step was stopped there were 0 observations and N variables.	This means that SAS halted a data step.	SAS cannot find the input dataset.	Check to see whether any variables or dataset names are spelled wrong. Check to see whether the input dataset is actually in the location specified by the set statement.
NOTE: The SAS System stopped processing this step because of errors.	This means that SYNTAXCHECK was passed, but when SAS started executing the code, something went wrong, and it stopped.	The underlying data does not make sense with the coding.	Check to see whether any variables or dataset names are spelled wrong. Check to see whether coding is referring to values not in the underlying data.
NOTE: Variable XXX is uninitialized and ERROR: The variable XXX in the DROP, KEEP, or RENAME list has never been referenced.	SAS couldn't find the variable specified in the code in the underlying dataset.	The variable is spelled wrong.	Ensure all variables specified in the code are actually in the underlying dataset and spelled correctly.
NOTE: Missing values were generated as a result of performing an operation on missing values. Each place is given by: (Number of times) at (Line): (Column) N at N:N.	A formula was done on at least one numeric column where the result was a missing value.	Creating a new variable by applying a formula to a continuous variable where there are missing values.	Nothing needs to be done if there is no problem with the resulting variable also being missing.

Message	What it means	Common reasons for message	Possible actions
NOTE: Character values have been converted to numeric values at the places given by: (Line): (Column) N:N N:N	You asked SAS to do a numeric operation on a character variable when a numeric variable is required.	SAS thinks it figured out that you did a numeric operation on a numeric variable stored as a character, and is letting you know it tried to do it by converting the character variable to numeric first.	SAS may have solved the issue, but you have to check. It might be better to convert the underlying variable to numeric first to avoid getting this message and completely control the results. The SAS data step commands INPUT() and PUT() can be used for converting variables (see link for reference in the *Further reading* section).

Table 9.1 – Common log messages seen when doing data steps in SAS

In addition to thinking about the possible messages that could be in the log, it is also important to think about where the RUN command falls in the code. Let's revisit an example from *Chapter 7, Designing and Developing ETL Code in SAS*, when we read in some variables from the BRFSS, as mentioned earlier, which is an annual health survey administered by phone in the US. We did the following data step transformation steps:

1. We read in native variable _AGE80, which is age in years, and recoded a grouping variable called agrp, which put _AGE80 into three categories.

2. We read in native variable VETERAN3, which indicates whether the survey respondent was a veteran or not. We created a cleaned-up version of VETERAN3 called vetgrp, and created a two-state flag called vetflag.

3. We read in the native variable SLEPTIM1, which is average hours of sleep per night. We found some suspicious values that were very low (1 to 3 hours) and very high (15 to 24 hours), but we decided to keep them. But there were some unallowable values in SLEPTIM1 in the form of codes 77 and 99, so we transformed it as SLEPTIM2, dropping those codes.

4. Although the administrative native variables FMONTH and _STATE were in the dataset, we did not transform them. FMONTH refers to the numeric month of the survey (1 through 12), and _STATE refers to a two-digit code corresponding to the state where the respondent is from.

To begin to talk about debugging, let's start by reading in a copy of the SAS dataset we used in *Chapter 7, Designing and Developing ETL Code in SAS*, called Chap9_1. We will place it in the directory, map to LIBNAME X, and run PROC CONTENTS with the VARNUM option:

```
LIBNAME X "/folders/myfolders/X";
data brfss_a;
    rownum = _n_;
    set X.Chap9_1;
RUN;
PROC CONTENTS data=brfss_a VARNUM;
RUN;
```

Let's look at some features of the code:

1. We start by using a data step to copy dataset X.Chap9_1 into WORK and name it brfss_a.

2. However, as we are copying it, we create the variable rownum, which means row number. We set it equal to _n_, which is a way of making SAS place the row number of each record in that column. We can then use this rownum as an **index** for the table if we want.

Because we created the rownum variable in the transformation step, it is now the first variable in the dataset, as we can see from the PROC CONTENTS output:

Variables in Creation Order				
#	Variable	Type	Len	Format	Informat
1	rownum	Num	8		
2	_STATE	Num	8	BEST12.	BEST32.
3	_AGE80	Num	8	BEST12.	BEST32.
4	VETERAN3	Num	8	BEST12.	BEST32.
5	SLEPTIM1	Num	8	BEST12.	BEST32.
6	FMONTH	Num	8	BEST12.	BEST32.

Figure 9.1 – PROC CONTENTS output

In *Chapter 7, Designing and Developing ETL Code in SAS*, we worked out our transformation steps separately:

1. We built modular code that recoded agrp from _AGE80.

2. We also built modular code that created vetgrp and vetflag from VETERAN3.

3. We built more modular code that recoded SLEPTIM1 to SLEPTIM2.

4. We checked each of the transformed variables for accuracy.

Now, we could theoretically put all of our transformation steps together. Let's do that here (skipping the check for accuracy):

```
data brfss_b;
    set brfss_a;
    agrp = 9;
    if _AGE80 le 34
            then agrp = 1;
    if _AGE80 ge 35 and _AGE80 le 64
            then agrp = 2;
    if _AGE80 ge 65
            then agrp = 3;
    vetgrp = VETERAN3;
    if VETERAN3 in (7, 9, .)
            then vetgrp = 9;
    vetflag = 0;
    if VETERAN3 = 1
            then vetflag = 1;
    SLEPTIM2 = SLEPTIM1;
    if SLEPTIM2 gt 24
            then SLEPTIM2 = .;
RUN;
```

Notice how long the code became when we put it together. Also, notice that there is only one RUN command at the end. Since we built the code step-by-step, we know it should all run. Therefore, we put the RUN command at the end of all of these transformation steps.

But imagine we were trying to build this code from scratch. Let's say we typed all this code, but we made a mistake and forgot the semi-colon after the last line of the `_AGE80` recode, which is after `then agrp = 3`. This is a very typical mistake to make, but it is very hard to troubleshoot when looking at the log file. Here is an example of what the log file looks like with this mistake:

```
80          data brfss_b;
81          set brfss_a;
82          agrp = 9;
83          if _AGE80 le 34
84          then agrp = 1;
85          if _AGE80 ge 35 and _AGE80 le 64
86          then agrp = 2;
87          if _AGE80 ge 65
88          then agrp = 3
89          vetgrp = VETERAN3;  ←———  Error starts here
            ___
            22
ERROR 22-322: Syntax error, expecting one of the following: !, !!, &, *, **, +, -, /, ;, <, <=, <>, =, >, ><, >=, AND, EQ, GE, GT,
              IN, LE, LT, MAX, MIN, NE, NG, NL, NOTIN, OR, ^=, |, ||, ~=.

90          if VETERAN3 in (7, 9, .)
91          then vetgrp = 9;
92          vetflag = 0;
93          if VETERAN3 = 1
94          then vetflag = 1;
95          SLEPTIM2 = SLEPTIM1;
96          if SLEPTIM2 gt 24
97          then SLEPTIM2 = .;
98          RUN;

NOTE: The SAS System stopped processing this step because of errors.
WARNING: The data set WORK.BRFSS_B may be incomplete.  When this step was stopped there were 0 observations and 10 variables.
WARNING: Data set WORK.BRFSS_B was not replaced because this step was stopped.
NOTE: DATA statement used (Total process time):
      real time           0.01 seconds
      cpu time            0.00 seconds      Notes here indicate system stopped processing
                                            and datasets are incomplete.

99
100         OPTIONS NONOTES NOSTIMER NOSOURCE NOSYNTAXCHECK;
112
```

Figure 9.2 – Log from running code with error

Omitting a semi-colon is a common problem in SAS programming, and it takes a disciplined eye to interpret the log:

1. First, it is important to always start interpreting errors at the first error encountered in the log. That is because that error will cause other errors, so if you fix that one, the others might go away.

2. Next, it is important to remember that the log alerts you to an error in the process that happens *after* you omit a semi-colon, *not* the process that happens *before* the semi-colon that was omitted. So, when we see `vetgrp` processing is erroring out in the log, we should not look at `vetgrp` for troubleshooting if our semi-colons are in the right place. Rather, we should look one step before the `vetgrp` transformation, at `agrp`.

This situation is why code is usually developed modularly, and then later, can be assembled into these long ETL files. The fewer RUN statements we have, the greater efficiency we gain in processing. However, when developing the ETL code, we have to conduct modular tests to make sure the code will run. That's how we build code into these long RUN statements.

As mentioned before, ETL code in a SAS data warehouse lives a long life and is constantly being groomed, manicured, and upgraded. Therefore, very long data steps like this one, which develop over time, may need to be edited at some point, and this can cause some challenges, so it is nice to know some troubleshooting strategies for data steps.

Troubleshooting strategies for data steps

Debugging data steps is notoriously tricky, and since the early days of SAS, programmers have creatively used the PUT command to help them debug data step code. As described by Beatriz Garcia and Alberto Hernandez in their SAS white paper, *Basic Debugging Techniques* (listed in the *Further reading* section), PUT can be used to print values to the log as SAS runs its code. These values can help the programmer troubleshoot.

Let's start with a simple example from the white paper that does not require a dataset to run. Let's say we knew an individual who weighed 65 kilograms and was 1.51 meters tall. We know the formula for **Body Mass Index** (**BMI**) is weight divided by height squared, and we want to use SAS to calculate this individual's BMI. Using the following code, we could make SAS act like a calculator and calculate BMI, outputting the answer to the log file:

```
DATA _NULL_;
    Weight=65;
    Height=1.51;
    BMI= Weight/(Height*Height);
    PUT BMI;
RUN;
```

Let's examine this code:

1. Our first statement is DATA _NULL_. Normally, this would specify the name of an output dataset. However, _NULL_ in SAS specifies a dataset with no rows or columns. It is used under different circumstances, and one of these circumstances is one we have, which is we just want to make a calculation, not output a dataset.

2. Notice that there is no set statement. Since we are outputting a dataset with no rows or columns, we do not need to worry about importing any or creating any.

3. The next two lines of code look like we are creating variables in a dataset. Because we are using _NULL_, we are basically setting temporary variables to a value. These variables will go away when the data step is over. The variables we are setting are Weight=65 and Height=1.51.

4. The next line, BMI= Weight/(Height*Height), executes the BMI formula on our two temporary variables, Weight and Height. The value of the result of the formula is stored in the temporary variable BMI.

5. Now, the data step is almost over. Since we used _NULL_, all these variables, including BMI, are going to go away when it ends. Before the RUN statement, if we add PUT BMI, then we can ask SAS to print the value of BMI to the log before the data step is over and BMI goes away.

Because this is a data step, there is no output. But since we added PUT BMI, we should see the result of the BMI calculation in the log. Let's look at it:

```
1               OPTIONS NONOTES NOSTIMER NOSOURCE NOSYNTAXCHECK;
72
73              DATA _NULL_;
74              Weight=65;
75              Height=1.51;
76              BMI= Weight/(Height*Height);
77              PUT BMI;
78              RUN;

28.5075216
NOTE: DATA statement used (Total process time):
        real time             0.00 seconds
        cpu time              0.01 seconds


79
80              OPTIONS NONOTES NOSTIMER NOSOURCE NOSYNTAXCHECK;
92
```

Figure 9.3 – Log from the data step code with _NULL_ and PUT

To point out where the result of the BMI calculation was reported to the log, a red box was placed around the number 28.5075216. This demonstration shows how variables with values can be held in SAS's memory (and not in any dataset) and can be displayed in the log using the PUT statement in a data step. One source of confusion when troubleshooting data steps is the values that variables are taking at various times during processing. Since processing is sequential, a PUT statement can be added at different points to report information and then reviewed in the log later.

> **Using the _NULL_ dataset in SAS:**
>
> A blog post by SAS explains that _NULL_ is a reserved keyword in SAS that specifies a dataset with no observations and no variables (link to blog post in the *Further reading* section). Common cases when _NULL_ is used are when you want to use a data step that displays a result, defines a macro variable, writes a text file, or makes calls to the EXECUTE subroutine.
>
> The blog post describes six ways that the _NULL_ dataset is used in SAS. The first is the way we used it in the demonstration – as a calculator. We are also using it in the sixth way described, which is as a debugging tool. _NULL_ can also be used to display characteristics of a dataset and create a macro variable from a value in a dataset, as well as to create a macro variable from a computational result. Additionally, it can be used to edit a text file or ODS template on the fly.

PUT can also be used when troubleshooting during data transformation. Let's look at an example of this. You may remember when we were using the same dataset in *Chapter 7, Designing and Developing ETL Code in SAS*, we were surprised that the variable for the average number of hours of sleep per night, SLEPTIM1, contained 1 for some respondents. We had trouble understanding how someone could report sleeping an average of 1 hour per night.

Let's say we wanted to identify records of respondents who reported sleeping only 1 hour per night and do further research on them. One way to do that would be to figure out the value of each rownum where SLEPTIM1 = 1, and then further investigate those records. Let's do a data step that copies brfss_a into brfss_b and include a PUT statement:

```
data brfss_b;
    set brfss_a;
    if SLEPTIM1=1 then PUT "SLEPTIM1=1 for rownum: " rownum=;
RUN;
```

Let's look more closely at the third line of code:

- We start by using an if/then statement with if SLEPTIM1 = 1 then.

- Normally in a data step, we'd ask SAS to update data as a result of meeting this criterion. In this case, we ask SAS to write whatever is in the PUT command to the log whenever a record in the data meets the criterion.

- The next part of the line of code says "SLEPTIM1=1 for rownum: ". This indicates to SAS that each time a record meets the criterion, SAS is supposed to print this string to the log: SLEPTIM1=1 for rownum:.

- The last part of the line of code says rownum=. This tells SAS to print the current value of rownum after the string in the log. Therefore, if the current rownum were 10, our custom log message would be SLEPTIM1=1 for rownum: rownum=10.

Let's run the code and look at the log file:

```
1          OPTIONS NONOTES NOSTIMER NOSOURCE NOSYNTAXCHECK;
72
73         data brfss_b;
74         set brfss_a;
75         if SLEPTIM1=1 then PUT "SLEPTIM1=1 for rownum: " rownum=;
76         RUN;

SLEPTIM1=1 for rownum:  rownum=482
SLEPTIM1=1 for rownum:  rownum=502
SLEPTIM1=1 for rownum:  rownum=1442
SLEPTIM1=1 for rownum:  rownum=1568
SLEPTIM1=1 for rownum:  rownum=1660
```

Figure 9.4 – Log from transformation code with PUT

This is just the top of the log file, which is actually three pages long. It shows the first five rows of numbers that met the criterion as printed to the log file from our PUT command.

As is shown here, the PUT strategy is very powerful and can be creatively leveraged to insert code into data steps to help the programmer put messages in the log for troubleshooting purposes. On the other hand, effectively using the PUT strategy for troubleshooting data steps requires a lot of experimentation and prior knowledge, and can be quite tedious.

Since coding errors tend to fall in certain categories, it seemed that there should be a more structured way to debug a data step. Therefore, data step debugging tools have been developed, and they are especially useful when debugging do loop code, so we will discuss them in the next section.

Debugging the do loop code

A debug function was developed for data steps in SAS that was particularly helpful for troubleshooting the do loop code, but it was difficult to use so it was not that popular. Recently, much of this functionality has been built into SAS Enterprise Guide, which is much more usable, especially for debugging do loop code. Both of these topics will be covered in this section.

Using the original data step debugger

S. David Riba presented his SAS white paper *How to Use the Data Step Debugger* at the SAS Users Group International 25 in 2000, which benchmarks the time when this capability became widely available to the SAS community (link in the *Further reading* section). In his white paper, Riba is actually referring to a data step option called /DEBUG. This option is called at the beginning of the data step on the data line.

The /DEBUG option can only be called on a data step. When it is called, it allows the user to interactively step through the execution of the data step, which can help immensely with troubleshooting. /DEBUG is not supported in SAS University Edition, so no example is presented here from that environment. However, in his white paper, Riba describes how in release 6.11 of SAS, the debugger launches an interactive environment with a **source window** and a **log window**. Here is a diagram:

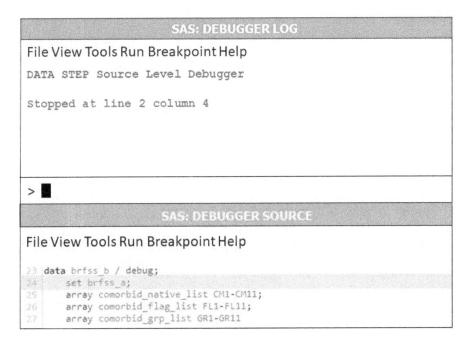

Figure 9.5 – Diagram of debugger windows

Also in his white paper, Riba explains that the debugger itself has certain commands. First, there are commands that control the program's execution:

- GO starts or resumes the data step execution.
- JUMP changes the point at which to resume the execution of the data step.
- STEP executes the data step one step at a time.

There are also commands in the debugger that manipulate data step variables:

- CAL or CALCULATE evaluates expressions and displays the results.

- DES, DESC, or DESCRIBE displays the attributes of the variable.

- E, EX, or EXAMINE displays the current value of a variable, or of a list of variables. It can also cause a variable to be displayed in different SAS formats.

- SE or SET assigns a new value to a variable.

Some commands control debugger requests:

- B or BREAK sets a breakpoint that causes SAS to stop execution at an executable statement.

- WATCH or W suspends execution when SAS sees that the value of a variable has changed.

- DELETE or D deletes breakpoints set by BREAK and variables set with WATCH.

- TRACE or T displays a record of program execution in the log window.

- LIST produces a list of items requested. The items can be BREAK (or B) for current breakpoints, WATCH (or W) for current watch points, FILES (or F) for current external files written out, and INFILES (or I) for current external files read. LIST can also request DATASETS or D, which will list all the input/output datasets. To list all items possible in a LIST command, the programmer can use LIST _ALL_.

Now that we have used data steps and do loops, we can see how useful these commands would be. However, it is also clear that simply learning how to use the debugger properly—even with the help of a window—was rather challenging. It is not clear how popular this debugging approach was.

The power in this debugging process is being able to experience a data step step by step. When programming do loop code, what often goes wrong is the execution of one of the loop iterations. But when a whole data step runs, it is hard to tell exactly what went wrong during the loop.

Imagine you asked SAS to loop through a series of variables to recode them, and you erroneously included a variable that did not work with the code. When the loop got to that variable, it would hit an error, and it would be hard to diagnose exactly what caused that. We'd love to know what the value of i was, for example, when the loop errored out. We'd also like to know the name of the source variable that was not cooperating with the loop code, and what went wrong when the code was executed on it.

All of that information could theoretically be obtained by using the /DEBUG function, but it was somewhat challenging to do. More recently, the /DEBUG function has been built into the **SAS Enterprise Guide**, and now it is much easier to use.

Using the data step debugger in SAS Enterprise Guide

At the SAS Global Forum 2017, Casey Smith of SAS explained how he worked to build the debugger functionality into SAS Enterprise Guide so it could be more easily used (link to the video in the *Further reading* section). He gave the example of trying to graph a rate. He was to read in a file, do some data step transformations, then display the graph, however, it did not display. He demonstrated using the debug function in SAS Enterprise Guide to troubleshoot this.

> **Note:**
>
> The data step debugger in SAS Enterprise Guide cannot be used to debug PROC SQL, PROC IML (which implements SAS/IML language for matrix programming), or SAS macro programs. It also cannot be used to debug data steps that read data from CARDS or DATALINES commands. See the SAS blog post *Using the Data Step Debugger in SAS Enterprise Guide* for more information (link in the *Further reading* section).

As can be intuited from watching Smith's demonstration, many of the /DEBUG commands from the original function are available, it's just that they are easier to deploy using the GUI provided by SAS Enterprise Guide. Here is a diagram of the interface:

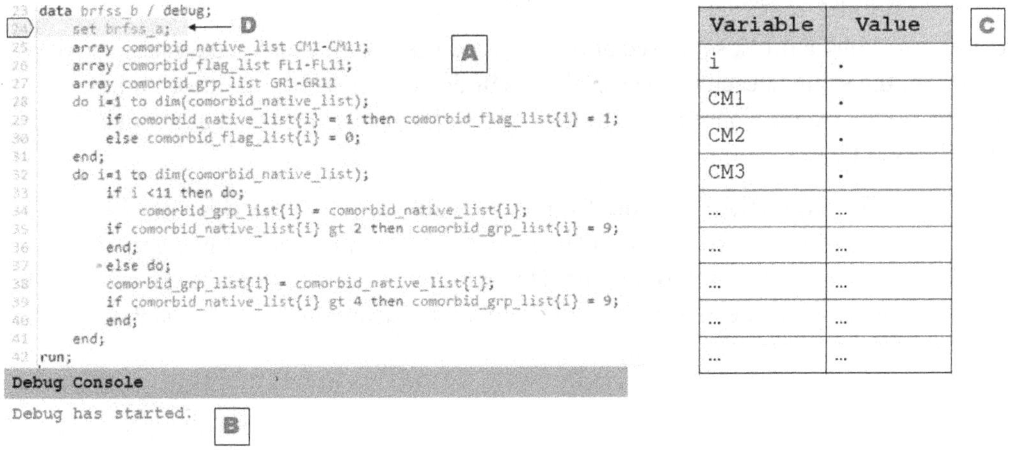

Figure 9.6 – Diagram of the debugger in SAS Enterprise Guide

Notice how the debug window opens in SAS Enterprise Guide. In the diagram, **A** refers to the window showing the code, and **B** refers to the debug console window. The window setup looks similar to the setup used in earlier versions of SAS. However, what is new is the panel on the right labeled **C**. This panel displays a list of all the variables in the code, and the value they take on at the current step. In the diagram, **D** refers to the current step, which is indicated by a yellow placeholder.

Although menu commands are available, essentially, this version has the functionality of many of the original /DEBUG commands. For example, BREAKS can be still be set. If you set a breakpoint in a loop, each time you advance the loop, you can review the value in the variable panel, as shown in this diagram:

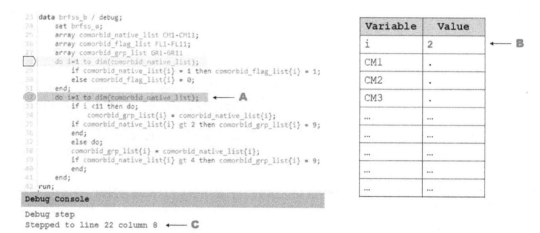

Figure 9.7 – Diagram of using breaks when debugging in SAS Enterprise Guide

The diagram shows a break set in red at the point marked **A**. At that break, as each step is advanced, the value of each of the variables is updated as shown in **B**. Information about the exact location where SAS is in the code and dataset is reported in the debug console labeled **C**.

It is also possible to directly edit values from the variable panel. In the video, the demonstrator uses this approach after he realizes that his graph is not showing up because some value was erroneously set to missing. He fixes this problem by editing the value in the panel, successfully debugging the program, and causing the graph to display.

As described earlier, while the SAS Enterprise Guide debugger expands options for debugging data steps, it cannot help us debug macros, which is the topic we will discuss next.

Debugging SAS macros

As has been said before, it is very difficult to edit and debug SAS macros. One way to make this work easier is to use a step-by-step design process to develop macros. That way, at each point in the step, if something goes wrong, it is easier to troubleshoot. This can also be helped by the `%PUT` command. These approaches will be described here.

Avoiding errors through the design process

Let's revisit *Chapter 8*, *Using Macros to Automate ETL in SAS*, for a moment, where we developed a macro called `%make_flag` that created a series of two-state flag variables based on a series of grouping variables that were all coded similarly. In *Chapter 8*, *Using Macros to Automate ETL in SAS*, we mentioned that we should use a step-by-step process for designing a macro that starts with creating a working data step with one of the variables and then converting parts of this data step to macro variables. We said we would show the details of that process here.

Let's use a copy of the dataset we used in *Chapter 8*, *Using Macros to Automate ETL in SAS*, for demonstration, which is a SAS dataset from the BRFSS named `Chap9_2`. Let's place this dataset in the directory mapped to `LIBNAME X` and run this code:

```
LIBNAME X "/folders/myfolders/X";
data brfss_a;
    set X.Chap9_2;
RUN;
PROC CONTENTS data=brfss_a VARNUM;
RUN;
```

The code copies the dataset into WORK, naming it brfss_a, and runs PROC CONTENTS on it with the VARNUM option. Here is the list of variables:

				Variables in Creation Order	
#	Variable	Type	Len	Format	Informat
1	_STATE	Num	8	BEST12.	BEST32.
2	ASTHMA3	Num	8	BEST12.	BEST32.
3	CVDINFR4	Num	8	BEST12.	BEST32.
4	CVDCRHD4	Num	8	BEST12.	BEST32.
5	CVDSTRK3	Num	8	BEST12.	BEST32.
6	CHCSCNCR	Num	8	BEST12.	BEST32.
7	CHCOCNCR	Num	8	BEST12.	BEST32.
8	CHCCOPD1	Num	8	BEST12.	BEST32.
9	HAVARTH3	Num	8	BEST12.	BEST32.
10	ADDEPEV2	Num	8	BEST12.	BEST32.
11	CHCKDNY1	Num	8	BEST12.	BEST32.
12	DIABETE3	Num	8	BEST12.	BEST32.

Figure 9.8 – List of variables from PROC CONTENTS

Except for _STATE, all of the other variables refer to different diseases the respondent reported having. All are coded such that 1 = **Yes**, so we made a macro that would take in a native variable and output a flag variable that was coded 1 where the disease variable was coded as Yes (and 0 in all other cases). This table plans out the names of all the flag variables we plan on making based on the disease variables in the dataset:

#	Variable Name	Definition or Question	Coding	Flag Name
1	_STATE	Code for state.	12 = Florida (FL), 25 = Massachusetts (MA), 27 = Minnesota (MN)	not applicable
2	ASTHMA3	(Ever told) you had asthma?	1 = Yes, 2 = No, 7 = Don't know/Not sure, 9 = Refused, blank = missing.	asthma_flag

#	Variable Name	Definition or Question	Coding	Flag Name
3	CVDINFR4	(Ever told) that you had a heart attack also called a myocardial infarction?	1 = Yes, 2 = No, 7 = Don't know/Not sure, 9 = Refused, blank = missing.	HA_flag
4	CVDCRHD4	(Ever told) you had angina or coronary heart disease?	1 = Yes, 2 = No, 7 = Don't know/Not sure, 9 = Refused, blank = missing.	CHD_flag
5	CVDSTRK3	(Ever told) you had a stroke?	1 = Yes, 2 = No, 7 = Don't know/Not sure, 9 = Refused, blank = missing.	stroke_flag
6	CHCSCNCR	(Ever told) you had skin cancer?	1 = Yes, 2 = No, 7 = Don't know/Not sure, 9 = Refused, blank = missing.	skinCA_flag
7	CHCOCNCR	(Ever told) you had any other types of cancer?	1 = Yes, 2 = No, 7 = Don't know/Not sure, 9 = Refused, blank = missing.	othCA_flag
8	CHCCOPD1	(Ever told) you have chronic obstructive pulmonary disease (C.O.P.D.), emphysema, or chronic bronchitis?	1 = Yes, 2 = No, 7 = Don't know/Not sure, 9 = Refused, blank = missing.	COPD_flag

#	Variable Name	Definition or Question	Coding	Flag Name
9	HAVARTH3	(Ever told) you have some form of arthritis, rheumatoid arthritis, gout, lupus, or fibromyalgia?	1 = Yes, 2 = No, 7 = Don't know/Not sure, 9 = Refused, blank = missing.	arth_flag
10	ADDEPEV2	(Ever told) you have a depressive disorder (including depression, major depression, dysthymia, or minor depression)?	1 = Yes, 2 = No, 7 = Don't know/Not sure, 9 = Refused, blank = missing.	dep_flag
11	CHCKDNY1	Not including kidney stones, bladder infection or incontinence, were you ever told you have kidney disease?	1 = Yes, 2 = No, 7 = Don't know/Not sure, 9 = Refused, blank = missing.	kid_flag
12	DIABETE3	(Ever told) you have diabetes?	1 = Yes, 2 = Yes, but female told only during pregnancy, 3 = No, 4 = No, prediabetes or borderline diabetes, 7 = Not sure, 9 = Refused, blank = missing.	diab_flag

Table 9.2. Native variables and intended flag names in the Chap9_1 dataset

In the table, we can see that each native disease variable is named under the **Variable Name** column. Under the last column, **Flag Name**, we see the name of the two-state-flag we want to make using the macro.

In *Chapter 8, Using Macros to Automate ETL in SAS*, we made a macro that automatically coded the flag variable for each native variable as 1 or 0, depending upon the input variable value (with 1 indicating they had the disease and 0 indicating all other answers). We found we could run it for every variable to automate this coding task. Remembering back to *Chapter 7, Designing and Developing ETL Code in SAS*, we will recall that if we added up all the flag variables, we could create a **disease index**. This was the reason for wanting to automate the coding task.

In *Chapter 8, Using Macros to Automate ETL in SAS*, we did not go through all the steps in developing the macro. We instead focused on the final macro, with the intention of revisiting the development of this macro in the current chapter. This diagram shows the original data step code for developing this macro at the top and the final code for the macro at the bottom:

Original data step code

```
data brfss_b;
    set brfss_a;
    asthma_flag = 0;
    if ASTHMA3 = 1
        then asthma_flag = 1;
PROC FREQ data=brfss_b;
    tables asthma_flag * ASTHMA3 /list missing;
RUN;
```

Final macro code

```
%MACRO make_flag5(native_var=, flag_var=, data_in=, data_out=);
data &data_out;
    set &data_in;
    &flag_var = 0;
    if &native_var = 1
        then &flag_var = 1;
PROC FREQ data=&data_out;
    tables &flag_var * &native_var /list missing;
RUN;
%MEND make_flag5;
```

Figure 9.9 – Original data step code compared to the final macro code

Let's look at the figure and think about the steps it took to get from the original code to the final macro code:

1. In the original code, the output dataset is `brfss_b`, and the input dataset is `brfss_a`. The native variable ASTHMA3 (indicating asthma) is recoded into the flag variable `asthma_flag`. Then, a two-way PROC FREQ is run on ASTHMA3 versus `asthma_flag` to check the variable.

2. The first step was to turn the entire snippet of code into a macro. The command `%MACRO make_flag1` was added before the code and `%MEND make_flag1` after the code. Then the macro was tested, and it ran.

3. In the next step, the macro was renamed `make_flag2`, and the macro variable `native_var` was added. It was added to the `%MACRO` statement, and ASTHMA3 was replaced in the code with a reference to `&native_var`. Then, the macro was tested setting `native_var=ASTHMA3`, and it ran.

4. In the next step, the macro was renamed `make_flag3`, and the macro variable `flag_var` was added, both to the `%MACRO` statement, and to the code, replacing `asthma_flag`. Then, using the macro variable values in the original code, it was tested, and it ran.

5. In the next step, the macro was called `make_flag4`, and the input dataset, `brfss_a`, was converted to a reference to the macro variable, `data_in`. The code was tested and it ran.

6. Finally, in the last step, the macro was called `make_flag5`, and the output dataset, `brfss_b`, was replaced with the macro variable `data_out`. The code was tested and it ran.

7. After the macro was completed, it was run successfully on all the variables included in the dataset. The results were studied to ensure it worked on all of them.

Once the final macro is built, it can be renamed and published without a number, but it is helpful to keep a record of the code used to build the macro.

> **Note:**
> A common problem—especially with nested loops—is ensuring that every `%DO` statement is paired with an `%END` statement. This problem can be greatly reduced through careful indentation of code blocks so `%DO` and `%END` statements can be located quickly visually. There is also a link to a SAS white paper in the *Further reading* section that uses an automated process for matching `%DO` and `%END` statement pairs.

As mentioned before, PUT is a command that can be used to print messages to the log during a data step. A comparable command that can be used with macro variables is called %PUT.

Using %PUT to display values of macro variables

If we wanted to troubleshoot the functioning of our macro, we might want to know the value of the macro variables at various processing steps. This could theoretically be achieved using the %PUT command during processing to ask SAS to print the value of the macro variable at the current point in processing to the log. However, in SAS University Edition, this output cannot be seen.

Therefore, for example, we will refer to the SAS documentation page for the %PUT command (available in the *Further reading* section). Here is a diagram that shows example code on the left beside the log output on the right for comparison:

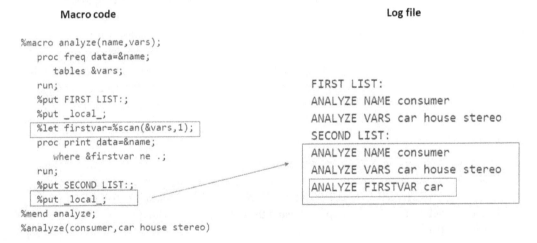

Figure 9.10 – Example of macro code and a log file using the %PUT command

As can be seen by the code on the left, in the blue box, a %LET statement defines a macro variable called firstvar. Next, there is a red box around a %PUT _local_ command. This command asks SAS to write all the **local variables** to the log with their current values. In SAS, local variables are variables that are only available while the macro in which they are defined is in the process of executing. Using the %PUT _local_ command gets around having to recall and list the names of all the macro variables used in the macro—this statement will print all of them automatically.

On the right side of the code in the log file, there is a red box around the part of the log. This is the part responding to the code in the red box on the left side of the diagram. Essentially, it has a response to the %PUT _local_ command. We see that the value of several macro variables is printed in the log. Imagine we were troubleshooting the evaluation of the variable firstvar, which is defined in the %LET statement with the blue box around it on the left side of the diagram. In the blue box on the right, we see the value of firstvar is printed in the log, allowing us to troubleshoot whether or not the variable is being evaluated correctly in the macro.

SAS is a powerful program in many languages. SAS programmers must, therefore, become sophisticated and adept at using many different ways to debug and troubleshoot SAS code across these different languages.

Setting system options to help with debugging macros

In *Chapter 8, Using Macros to Automate ETL in SAS*, we discussed how setting **system options** in SAS impacts the SAS session's environment, and how these options can be adjusted strategically during the SAS session to help us with macro programming. Here, we will talk about a few more system options that can help us with troubleshooting macros, specifically MLOGIC, SYMBOLGEN, and MPRINT. These options and others are the topic of the SAS white paper, *SAS System Options: The True Heroes of Macro Debugging*, by Kevin Russell and Russ Tyndall (link available in the *Further reading* section).

> **Note:**
> The SAS white paper, *SAS System Options: The True Heroes of Macro Debugging*, also covers other system option settings that are useful for macro debugging, including MFILE, MLOGICNEST, MPRINTNEST, MAUTOLOCDISPLAY, and MCOMPILENOTE. Examples of using the options are provided in the white paper.

As suggested in *Chapter 8, Using Macros to Automate ETL in SAS*, setting system options to troubleshoot macro errors falls along the strategy of getting information to print in the log file that provides clues as to what is causing a problem with the macro. Following this strategy, here is a summary of what the options MLOGIC, SYMBOLGEN, and MPRINT do:

- MLOGIC: This option is turned on by placing the code options mlogic; before the code that builds the macro. With MLOGIC on, messages are printed to the log that show steps in the execution of the macro.

- SYMBOLGEN: This option is for showing the values to which macro variables are resolving during processing. This option is turned on by placing the code options symbolgen; before the code with the macro variables.

- MPRINT: This option is turned on by placing the code `options mprint;` before the code that builds the macro; although, if using multiple options, they can be listed together (for example, `options symbolgen mprint;`). As you may remember from *Chapter 8, Using Macros to Automate ETL in SAS*, SAS's macro processor takes in macro code and rewrites it as SAS code before executing it. MPRINT causes this rewritten SAS code to appear in the log.

As with using the debugging capabilities in SAS Enterprise Guide, becoming adept at using system options for macro troubleshooting takes practice. Used together, these system options can be very helpful in troubleshooting large macros that have been developed and evolved over many years.

Summary

This chapter gave an overview of alternatives to debugging and troubleshooting SAS code, specifically data step code. First, we discussed the importance of well-formed and well-formatted code, especially with respect to ETL protocols in a data warehouse that might live on for a long time. We demonstrated how to leverage messages in the log for guidance when troubleshooting data steps, and how to use the PUT statement to write values to the log during a data step. Next, we looked specifically at ways to troubleshoot do loop code and went over debugging functions that SAS has built into Enterprise Guide. Finally, we covered approaches to debugging and troubleshooting macros.

Because the SAS code is so complicated, building data step code that processes big data usually turns into a big project with a lot of code. With more code, there are more opportunities to need to edit the code, especially to accommodate improvements in the data warehouse. This chapter gave a practical guide as to how to guide the development of this bank of code in such a way that it stays manageable when it gets big. It also provided tools and strategies for editing pieces of such a big data bank without causing unwanted ripple effects in other parts of the data warehouse.

While this chapter focused on the details of debugging SAS code, the next chapter switches the topic to think about how to serve the needs of SAS data warehouse users. It talks about strategies for serving the needs of different user groups, including providing data curation files to programmers, as well as managing a data stewardship committee to support developers. It also provides practical guidance for creating useful cross-walking variables for programmers using longitudinal data stored in the warehouse.

Questions

1. True or false: Because data step code formatting doesn't impact how the program runs, it doesn't matter how the programmer formats data step code files.

2. Imagine you develop a transformation step, but when you run it, you see many errors. What are the initial troubleshooting steps you can take?

3. How can using a PUT statement in a data step help a programmer troubleshoot code?

4. You realize you need to design 10 transformation steps for a new source dataset. How might you approach this task to minimize data step troubleshooting?

5. What does BREAK do if set on a loop command in the data step debugger?

6. Imagine you need to make a macro involving two loops and four macro variables, and you are in a hurry. Is it a good idea to skip using the step-by-step process, and just try to make the macro code from scratch since you are in a hurry?

7. Imagine you decide to slow down and use a step-by-step process to build the macro that involves two loops and four macro variables. How can the %PUT _local_ statement help you with this project?

Further reading

- SAS white paper *Errors, Warnings, and Notes (Oh My) A Practical Guide to Debugging SAS Programs* by Lora D. Delwiche and Susan J. Slaughter – available here: https://www.lexjansen.com/wuss/2016/168_Final_Paper_PDF.pdf

- SAS white paper *How SAS Thinks or Why the Data Step Does What it Does* by Neil Howard, available here: https://support.sas.com/resources/papers/proceedings/proceedings/sugi28/189-28.pdf

- SAS blog post *Converting Variable Types – Use PUT() or INPUT()?* by Sunil Gupta, available here: https://blogs.sas.com/content/sgf/2015/05/01/converting-variable-types-do-i-use-put-or-input/

- SAS white paper *Basic Debugging Techniques* by Beatriz Garcia and Alberto Hernandez – available here: https://www.pharmasug.org/proceedings/2012/PO/PharmaSUG-2012-PO17.pdf

- SAS blog post *Six Ways to use the _NULL_ Data Set in SAS* by Rick Wicklin – available here: `https://blogs.sas.com/content/iml/2018/06/11/6-ways-_null_-data-set-sas.html#:~:text=In%20SAS%2C%20the%20reserved%20keyword,no%20observations%20and%20no%20variables.&text=The%20_NULL_%20data%20set%20is,calls%20to%20the%20EXECUTE%20subroutine`

- SAS white paper *How to Use the Data Step Debugger* by S. David Riba – available here: `https://support.sas.com/resources/papers/proceedings/proceedings/sugi25/25/btu/25p052.pdf`

- SAS video *Demo: Data Step Debugging in Enterprise Guide* – available here: `https://video.sas.com/sasgf17/detail/videos/general-sessions/video/5383313949001/demo:-data-step-debugging-in-enterprise-guide?autoStart=true`

- SAS blog post *Using the Data Step Debugger in SAS Enterprise Guide* by Chris Hemedinger – available here: `https://blogs.sas.com/content/sasdummy/2016/11/30/data-step-debugger-sas-eg/#:~:text=It%20can't%20be%20used,data%20from%20CARDS%20or%20DATALINES`

- SAS white paper *Matching %DO-%END Pairs - A Macro Debugging Technique* by Zizhong (James) Fan – available here: `https://www.lexjansen.com/nesug/nesug07/cc/cc10.pdf`

- SAS documentation of `%PUT` macro statement: `https://documentation.sas.com/?docsetId=mcrolref&docsetTarget=n189qvy83pmkt6n1bq2mmwtyb4oe.htm&docsetVersion=9.4&locale=en`

- SAS white paper *SAS System Options: The True Heroes of Macro Debugging* by Kevin Russell and Russ Tyndall, available here: `https://www.lexjansen.com/nesug/nesug10/ff/ff10.pdf`

Section 3:
Using SAS When
Serving Warehouse
Data to Users

This section focuses on ways to use SAS in data warehousing so as to meet user requirements. First, we consider the different user classes of a SAS data warehouse, and what their needs are. To fulfill those needs, we cover setting **data stewardship policies**, creating **data curation files**, and serving up specific types of **warehouse variables** that could be particularly helpful to analyst users.

Next, we discuss interconnectivity between SAS and other systems, such as SQL. We cover packaging SAS data for export, as well as using an **open database connectivity (ODBC)** protocol with the **SAS/ACCESS** component to connect to another live data environment.

Finally, we cover how visualizations currently use the **output delivery system (ODS)** in SAS, and how serving SAS data up for print format is different than serving it up on the web. We review an example of an **interactive online dashboard** in SAS, and discuss SAS's new analytics suite, **SAS Viya**.

This section comprises the following chapters:

- *Chapter 10, Considering the User Needs of SAS Data Warehouses*
- *Chapter 11, Connecting the SAS Data Warehouse to Other Systems*
- *Chapter 12, Using the ODS for Visualization in SAS*

10
Considering the User Needs of SAS Data Warehouses

Maintaining a data warehouse entails so much work that it is important to step back once in a while and remind ourselves that the point of all this work is to serve data warehouse customers or users. This chapter focuses on how to serve the needs for the two main classes of SAS data warehouse users: **analysts** and **developers**.

First, we examine the roles of analysts and developers, and consider how their needs differ. Next, we discuss how to serve the needs of analyst users, through providing data access and serving up useful **foreign keys** and **crosswalk variables**. After that, we focus on serving the needs of warehouse developers through policies and procedures overseen by a *data stewardship committee*, and through providing and maintaining standardized curation and other support.

We will learn the following in this chapter:

- The difference in needs between the two main classes of data warehouse users, which are analyst users and developer users

- How, as a manager, you can adapt SAS or supplement it to meet the needs of both these classes of users

- Tools and resources that managers and programmers can use to leverage the power of a SAS data warehouse

- How to empower a data stewardship committee to help create policies that promote flexibility and cross-training across SAS data warehouse teams

Technical requirements

The multiple datasets used for demonstration in this chapter are available online on GitHub: `https://github.com/PacktPublishing/Mastering-SAS-Programming-for-Data-Warehousing/tree/master/Chapter%2010/Data`.

The code bundle for this chapter is available on GitHub here: `https://github.com/PacktPublishing/Mastering-SAS-Programming-for-Data-Warehousing/tree/master/Chapter%2010`.

Needs of data warehouse users

Tim Mitchell, a business intelligence architect who writes on many data science topics, published an article titled *Why Data Warehouse Projects Fail* (a link to the article is available in the *Further reading* section). He noted that many data warehouse projects fail—some have estimated the rate of failure to be as high as 50%. Since Mitchell has experience of rescuing data warehouse projects, he listed several reasons why he thinks data warehouse projects fail:

- **The data warehouse does not have an objective**: In *Chapter 7*, *Designing and Developing ETL Code in SAS*, we described a hypothetical data warehouse with the objective of studying the **quality of life** (**QoL**) of United States veterans after they leave the service. However, Mitchell points out that in a surprising number of cases, leadership failed to answer the question of *why* the organization was building a data warehouse. Some would say they assumed their organization needed one simply because *everyone else has one*.

- **Coding starts before necessary management structures are in place**: We recognize throughout the book that data warehouse projects need to be well-managed. Mitchell points out that in data warehouse projects, programmers should not jump into coding. A data warehouse, he says, is *first and foremost a business project, not a technical one*. To prevent failure, Mitchell recommends ensuring that the correct personnel with the appropriate mix of expertise are involved to architect, build, and test the warehouse solution.

- **Too little attention is paid to ETL protocols**: Throughout this book, but especially in *Chapter 7, Designing and Developing ETL Code in SAS*, we emphasize the research, design, and documentation of our ETL protocols. ETL protocols form the foundation of what value is added to the raw data by having it housed in the warehouse. Hence, high-quality protocols lay the foundation for a high-quality warehouse.

- **Too little attention is paid to the needs of stakeholder groups**: Mitchell points out that warehouses need regular maintenance, training for essential personnel, and coordination between an overall vision and the regular production of deliverables. Mitchell observed that technical staff can get disconnected from other stakeholders. This leads to the question of who the **stakeholder groups** of the data warehouse actually are, and how to serve their needs.

Let's conceive of data warehouse stakeholders as individuals in different user classes, then consider the needs of these different user classes.

Considering classes of data warehouse users

When designing a SAS data lake or warehouse, it is important to consider the different classes of stakeholders or users. The word **users** may seem simple, but there are times when the concept of users becomes complicated. On one hand, analysts who gain permission to access data in a SAS data warehouse or lake are definitely in one category of users, because they are being served by the system. On the other hand, as described in *Chapter 4, Managing ETL in SAS*, **developers** fall into distinct user groups, depending upon their levels of access to data within the data warehouse or lake. Therefore, we have at least two classes of users whose needs the system must serve: analysts and developers.

Developers fall into subclasses, as described in *Chapter 4, Managing ETL in SAS*, depending upon what function they serve with the data. Analysts also fall into subclasses, which are described in this table:

Class	Subclass (Function)	Group Functions
Developers	Raw data ETL	Individuals involved in ETL processing from raw data. At least one senior programmer and one leader should be in this group. This should be a very small group.
Developers	Processed data ETL	Individuals involved in ETL processing of already-processed data as a part of maintaining the warehouse. This will be a very large group and may have subgroups within it that are specialized to support particular warehouse services.
Developers	Archival	Individuals involved in archiving data and documentation off of live warehouse resources. At least one senior programmer and one leader should be in this group. This should be a very small group.
Developers	Curation	Individuals involved in maintaining documentation, policies, and standardization in the data warehouse. In a healthy data warehouse structure, this group will be 25%–50% of the size of the Processed data ETL group. In other words, in a healthy data warehouse structure, if the Processed data ETL group consisted of ten people working at 100% full-time equivalent (FTE), there should be at least two to five people assigned as 100% FTE to the curation group.

Class	Subclass (Function)	Group Functions
Analysts	Raw data analysts	These are individuals who have permission to obtain raw datasets or data tables from the data warehouse. The expectation is that they will make analytic datasets and do independent analyses. They need access to the datasets to which they have permission, resources to work with the data, and curation.
Analysts	Application-based analysts	These are individuals who have permission to access the data in the data warehouse through an application interface that controls their access. The expectation is that the application will allow them to use the data to do their analysis, and they are not removing the data from the application to do an independent analysis. Individuals in this user group need access to data warehouse applications, with permissions set accordingly. They also need support for accessing and using the application (which may involve access to additional resources and curation).

Table 10.1 – Data warehouse user classes and subclasses

As described in *Chapter 4, Managing ETL in SAS*, and reflected in the table, although a warehouse may have more granular levels of developers, you might divide developers into functions associated with four subclasses:

- Conducting **extract, transfer, and load (ETL)** on raw data.
- Conducting ETL on already-processed data.
- Archival duties including taking data off of the system.
- Curation duties including viewing data and creating documentation.

Note that according to the table, the largest subclasses of developers should be *Processed data ETL* and *Curation*, with the *Curation* group about 25%–50% of the size of the *Processed data ETL* group. In some situations, the *Curation* group may be larger, and possibly even equal in size to the *Processed data ETL* group. This can be seen in the situation where a data lake hosts many minimally-processed, longitudinal datasets from different sources because these features increase the need for documentation.

Please also note that analysts fall into roughly two subclasses: **raw data analysts** and **application-based analysts**. In reality, analyst user subclasses will fall into subcategories based on the functions offered by the system to serve up data. Imagine a system where a small group of analyst users are provided permission to obtain minimally processed data from a data lake. Let's say that on that same system, a large group of other users are provided the opportunity to log into a data warehouse through an application hosting a processed version of the same data. In that case, both analyst user subclasses classes would be being served by the same data warehouse system, just in different ways. Because they would be served in different ways, they would have different sets of needs.

In the next section, we will think about serving user needs by considering the different classes of users, and how their needs may be different.

Considering the needs of each class of users

This section will first discuss the needs of analysts using the data warehouse. Then, we will consider the needs of developers involved in managing and maintaining the warehouse.

Considering the needs of analysts

Large SAS systems can function as **data warehouses** and **data lakes** at the same time, in theory. This is because in order to develop a data warehouse, you need a lot of independent data tables from different systems related to the same entity (such as an organization). These tables can be served up in a minimally-processed, federated manner, as in a **data lake**, or they can be assembled into a structure and served up as a part of a **data warehouse**.

Although one data system could theoretically function as both a data lake and a data warehouse at the same time, this rarely happens in practice. This is because the needs of analyst users of data lakes and data warehouses differ. If the data tables are served separately, as in a data lake, analyst users are usually provided a data extract in a somewhat raw format (for example, the transfer of a file in the format of `*.sas7bdat` or `*.csv`). Analyst users receiving data from a data lake have a particular set of needs surrounding the documentation and meaning of the data, and actually using the data in the analysis.

However, if tables are served up as a part of a data warehouse structure, they are usually incorporated into an application (or more briefly, *app*), and this app controls access to the data, including queries, and whether or not data or reports can be downloaded or otherwise extracted from the system. The app may even control the analysis and visualization of the data. Users who access a data warehouse are usually doing it through some sort of app, so these users have a different set of needs than data lake users. Logically, the primary need of data warehouse users tends to be application support.

Normally, a system is designed specifically to function as either a data lake or a data warehouse, but not both, even though it is theoretically possible to function as both at the same time. This diagram summarizes the continuum of needs experienced by analyst users of SAS data lakes and warehouses:

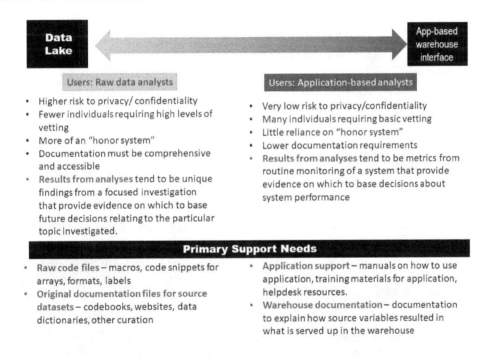

Figure 10.1 – Continuum of subclasses of analysts and their primary support needs

As can be seen from the diagram, raw data analyst users tend to need raw code files and other files that would support these (such as SAS format files), and original documentation files for source data, since the data they are using is minimally processed. On the other hand, app-based analysts predominately require app support. However, they also require documentation. Since the data is highly processed, they require documentation that pertains both to the source datasets, as well as to the variables and other data added through warehouse processing.

To ensure that analyst users gain value from using the data lake or data warehouse, it is important to consider the value that is added by simply processing the source data into the lake or warehouse:

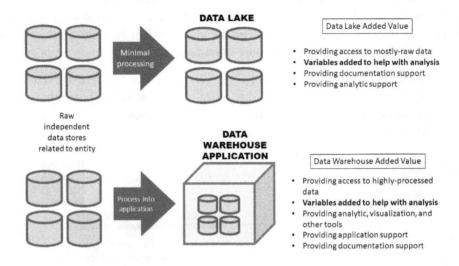

Figure 10.2 – Differences between a data warehouse and a data lake with respect to added value

As can be seen from the diagram, data lakes and data warehouses may start with the same source data. However, the value that is added by having them in a lake or warehouse is different depending upon whether it's a lake or a warehouse. In a data lake, data is minimally processed. This means that the data in the different tables may be challenging to join. It also means that the analyst will really need to understand the underlying source data in order to assemble an analytic dataset and conduct an analysis. Therefore, the value added comes in documentation support and analytic support. It also comes with variables that can be added to help with assembling analytic datasets and conducting an analysis.

A data warehouse is a completely different scenario in terms of data processing. In a data warehouse, data tends to be highly processed, and is housed in an application that provides the means for data access. Because data warehouses tend to be apps, these apps can provide analytic, visualization, and other tools to add value to the source data. In other words, by simply logging into the application, the application user can have access to some terrific resources for analyzing, visualizing, or otherwise working with the data housed in the warehouse. Of course, these application features cannot add value if they are not well supported—both in terms of *app usage* and in terms of the *documentation* of the native and transformed variables and tables. Because data is more highly processed in data warehouses than in data lakes, usually, more variables (and data tables) are generated from the source data by the warehouse when compared to a data lake. These can add value as well, so long as they are well-documented and served up in a way that analysts can use them.

Specific examples of how to provide support to these different analyst user groups with documentation will be provided in the *Data stewardship for serving warehouse users* section. Next, let's move on from thinking about analysts to considering the needs of developers of data warehouses and data lakes.

Considering the needs of developers

Developers are not always considered as a user group when thinking of a data lake or data warehouse. But in a practical sense, as described in *Chapter 4, Managing ETL in SAS*, developers must be partitioned into user groups to facilitate electronic permissions to control data access anyway. From the standpoint of a manager, it is useful to carry the analogy forward and treat developers like a user group with its own special needs. That way, the manager can ensure that the needs of developers are met so they can efficiently carry forward with their work on the data system.

It is safe to say that most of the documentation developed to serve analyst users will also be highly used by the developer user group as well. However, the developers have specific, enhanced needs for support and documentation, specifically when it comes to understanding source data from other systems. While analyst users are only exposed to the data served up by the warehouse, developers are often tasked with actually studying and understanding the source data, even data that does not eventually get served up in the warehouse.

In fact, as demonstrated in *Chapter 7, Designing and Developing ETL Code in SAS*, developers are often responsible for much of the ultimate design of the data warehouse, especially the added variables. Developers must decide what variables should be added to meet the needs of both developer and analyst users, and how to code and document these variables. Therefore, it makes sense that developers would need a lot more documentation from source datasets than analysts users. It also makes sense that developers might create documentation strictly for other developers maintaining the data system that focuses on ETL and other maintenance tasks. Specific examples of such documentation will be described in the *Data stewardship for serving warehouse developers* section.

To summarize, we went over the different types of data warehouse users and compared and contrasted their different needs. In the next section, we will focus on analyst users, and how to specifically support their needs.

Data stewardship for serving warehouse users

This section talks about conducting data stewardship for analyst users. First, we will cover setting up data access policies in order to provide access to all classes of users. Next, we will talk about how the act of structuring data in a warehouse or lake causes analysts to have needs for certain types of variables. We will discuss adding variables called **foreign keys** to the warehouse, and this is followed by a discussion of the development of **crosswalk variables**.

Providing data access

Restricting access to data to only those who should be approved is very important in a data system, regardless of whether it is a data lake or data warehouse. This section will describe how to set up an administrative process to grant access to analyst users while ensuring that data in the system is safeguarded.

Providing access to a data warehouse app

As mentioned earlier, analysts who access data through a data warehouse were classified as app-based analysts. In reality, in a large data warehouse app, there may be subclasses of app-based analyst users. Imagine a healthcare system sets up a data mart. App-based analysts working for the pathology laboratories in the healthcare system should be granted access to the data tables and tools they need to conduct their analyses. But they probably should not be granted access to radiology records, because they do not work for the radiology department.

Depending upon the **business rules** set up in the data warehouse application, subgroups of app-based analysts should be created, such as the laboratory subgroup and the radiology subgroup. Different sets of policies would need to be crafted. First, it would need to be determined how a person could qualify to be in either of the subgroups—laboratory or radiology. Next, a process would need to be set up by which a person can be granted access. This process would require at least the following elements:

- Demonstration of the person's eligibility for the group

- Assurance that the person gaining access has been adequately trained on policies and procedures, especially with respect to data privacy

- Clear instructions for the technical workers who will adjust permissions so as to grant the person access to the system

- Documentation of approval by necessary leaders

Policies surrounding granting access to the database app are often written up as a part of a larger set of administrative policies. The policies often include flow charts, explaining how access is granted to the different levels of the system (including the developer levels) and official forms that are to be used. The policies and processes are typically set up by the **Data Stewardship Committee**, which will be described later in the chapter, under *Managing a data stewardship committee*.

Providing access to data outside of an app

In a data lake scenario, typically, an app is not used to control access. Instead, minimally processed datasets are placed on servers, and access is controlled through being given access to the server. In terms of providing access to data stored this way, just as with an app, a process is needed for approval that includes the four elements listed previously.

However, something should be said here about the difference between providing access to a server full of data versus being provided access to an application that controls data access. There is much more of an *honor system* required of analysts accessing data in data lakes as opposed to analysts accessing data as a part of applications. This means the granting of access of analysts to data lakes should be given a very high level of scrutiny. Access at that level puts the analysts on similar ground as developers.

For this reason, data lakes may choose not to allow analysts access to the data lake. They may instead arrange for a subclass of developers whose role is to make analytic datasets for analyst users who are not granted access to the data lake server but are instead given data extracts for analysis. In that case, the following applies:

1. The developer would work with the analyst user using documentation from the data lake to specify the development of analytic datasets needed by the analyst.

2. The developer would then assemble the analytic datasets specified by the analyst, and transfer them to the analyst for analysis.

This is clearly the safest model with respect to data privacy. However, it requires a lot of support:

1. First, developers in this role need to have extremely good communication and customer service skills, and be adept programmers while also being knowledgeable about the datasets in the lake.

2. Next, developers need to continuously support the analyst after the transfer of analytic datasets, because they may need to be remade as the analyst uses them and learns more about the data. Also, the analyst is likely to have questions as they complete their data project.

3. Finally, both the analyst and the developer in this model would require extensive source documentation and data curation in order to fulfill their roles.

In earlier chapters, we have talked about variables of the data lake or data warehouse that can add value for analysts. In *Chapter 6, Standardizing Coding Through the Use of SAS Arrays*, we talked about adding **index variables**, and in *Chapter 7, Designing and Developing ETL Code in SAS*, we talked about adding grouping variables, flags, and cleaned-up versions of variables. The next section talks about needs imposed on users by housing data in the warehouse structure and different variables that can be added that meet these needs.

Serving needs created through the warehouse structure

A discussion of serving the needs of data warehouse users is really only helpful in the context of an example of a data warehouse or lake that collects datasets along a theme over time for the purpose of meeting an objective. In *Chapter 7, Designing and Developing ETL Code in SAS*, we described a hypothetical data warehouse built to explore the QoL experienced by US veterans after they leave military service. Throughout the book, we have been working with the 2018 **Behavioral Risk Factor Surveillance System (BRFSS)** dataset for our examples. This data comes from an anonymous health survey done by phone annually in the US. We discussed how the BRFSS would be a perfect dataset to host in such a data warehouse for several reasons:

- First, the dataset is very large (over 400,000 records), and it includes a question where the respondent states their veteran status. This allows the warehouse to be able to use BRFSS data to make estimates about both veterans and non-veterans.

- Next, the survey is conducted annually and, by design, the questions do not change much from year to year. This provides the ability for long-term trends to be established.

- Third, the dataset is known to be valid and reliable, and it is well-characterized, so there is a lot of documentation. It also has many variables that could help analysts study veteran QoL, such as whether or not respondents have various disease conditions, respondent average hours of sleep per night, and respondent use of medical marijuana.

- Finally, the dataset includes **foreign keys**, or indexes that allow it to be *hooked up* to other datasets. Specifically, it includes a code for the geographic variable for state (_STATE) that could be used to connect other state-level data into the warehouse.

> **Considering geographic variables in a data warehouse**
>
> Geographic variables are especially important to consider hosting in data warehouses, as they can greatly improve the possibilities for data analysis. The BRFSS includes data identifying the respondent's residence at the state level (_STATE) identified by a public identification number called a **FIPS number**. The BRFSS data also includes a variable for a smaller level of geographic grouping within the states, called counties, that are also identified with a FIPS number. To keep the demonstration simple, we are handling data at the state level, but in a real scenario, a better decision would be to also include the variable for lower-level geographies, such as county and city, if they are available.

These characteristics provide opportunities for serving the needs of data warehouse analyst users through creating and providing additional data in the warehouse:

- **Serving up related data**: The BRFSS data includes the respondent's state of residence. This variable is coded such that other data about the same state (such as the number of veterans) could be hooked on from other datasets and added to the data warehouse. The code for state could be used as a foreign key.

- **Serving up crosswalk variables**: The BRFSS data includes a few variables that have changed over time that are not comparable across years. For these situations, the data warehouse could add variables that create one standardized grouping variable that can be used over time, called a **crosswalk variable**.

We will continue to demonstrate with examples using our hypothetical veterans data warehouse and the 2018 BRFSS dataset. The dataset we will use for this demonstration is in SAS format and is called Chap10_1. This dataset should be placed into the directory mapped to LIBNAME X, and the following code run:

```
LIBNAME X "/folders/myfolders/X";
RUN;
PROC CONTENTS data=X.chap10_1 VARNUM;
RUN;
```

This code maps LIBNAME X to the directory of your choice, then runs PROC CONTENTS on the dataset, using the VARNUM option, so that the variables are printed in creation order. There are only four variables in the dataset, so rather than looking at the PROC CONTENTS output, let's instead review some information about the four variables taken from the codebook presented here:

Order	Variable Name	Question/Description	Potential Answers
1	_STATE	State identified by FIPS code	12 = Florida (FL) 25 = Massachusetts (MA) 27 = Minnesota (MN)
2	VETERAN3	Have you ever served on active duty in the United States Armed Forces, either in the regular military or in a National Guard or military reserve unit?	1 = Yes 2 = No 7 = Don't know/not sure 9 = Not asked or missing
3	MARIJAN1	During the past 30 days, on how many days did you use marijuana or hashish?	Not documented
4	USEMRJN2	During the past 30 days, which one of the following ways did you use marijuana the most often? Did you usually… [Please select one. Did you…]	1 = Smoke it (for example, in a joint, bong, pipe, or blunt) 2 = Eat it (for example, in brownies, cakes, cookies, or candy) 3 = Drink it (for example, in tea, cola, or alcohol) 4 = Vaporize it (for example, in an ecigarette-like vaporizer or another vaporizing device) 5 = Dab it (for example, using waxes or concentrates) 6 = Use it some other way 7 = Don't know/not sure 9 = Refused

Table 10.2 – Variables in the Chap10_1 dataset

Let's examine each of these variables:

- _STATE is a two-digit code for state. The code, called a FIPS code, is used in other datasets with state-level information. Therefore, _STATE makes a good foreign key to use to hook on information from other datasets related to the same state.

- VETERAN3 provides us with information on who in the BRFSS is a veteran and who is not.

- MARIJAN1 and USEMRJN2 are newer variables in the BRFSS. MARIJAN1, which asks how often the respondent has used marijuana in the last 30 days, is not documented in the codebook for some reason.

- USEMRJN2, the next marijuana variable, asks the most frequent way the marijuana was consumed. This question is only asked of respondents reporting marijuana use and has many categories of responses.

The _STATE variable will be used to demonstrate how to hook in other data to the warehouse in order to better serve user needs. The marijuana variables will be used to demonstrate how to create a crosswalk variable to serve the needs of users doing longitudinal analysis.

Adding, using, and serving up foreign keys

Let's borrow some terminology from the informatics field that refers to specific types of **indexes** in data tables: **primary keys** and **foreign keys**. A primary key is a column in a table that uniquely defines each row in the table. For example, in *Table 10.1*, where we list our four variables in our example dataset, we have an **Order** column. This column is the primary key for that table because the value on each row is unique to the row. Therefore, specifying a particular primary key in that table, such as two, specifies a unique row, which is the row for VETERAN3. As a practical matter, a primary key can be added to any table by simply adding a column like the **Order** column that numbers each row uniquely.

Using foreign keys

A foreign key, then, is a number stored in a column in a table that uniquely refers to another record in another table (typically by referring to the primary key in the other table). Imagine a table that had two columns, and in the first column, all the FIPS codes for each of the US states were listed (which is the code that is stored in _STATE, such as 12). In the second column, which was named **Description**, the name of the state was spelled out (for example, Florida). Then, the _STATE variable could be seen as a foreign key to a FIPS table that decodes states using the primary key in the FIPS table, because a FIPS code of 12 uniquely identifies the **Description** Florida.

Serving up foreign keys from the public domain, like FIPS codes, can be helpful by providing useful variables in a data warehouse. These variables allow an analyst to enrich the data in the warehouse by adding more information from different sources. For example, if we had other information from the different states about veterans, such as their income levels, we could use the FIPS code to patch the data into the warehouse and combine it with our BRFSS data. Let's do an exercise like this in the next section.

Adding data to the warehouse using foreign keys

Since our hypothetical data warehouse is about US veterans and QoL after leaving service, it might be helpful to see how their income compares after they leave service. The US census has established state-level measurements of veteran income 1 year, 5 years, and 10 years after leaving service. They present the 25th, 50th (median), and 75th percentile in their data, which is downloadable from the internet in $*$.csv format (see link in the *Further reading* section). Since we are interested in QoL after leaving service, we are most interested in data 5 and 10 years after the veteran leaves service. Even though we have the 2018 BRFSS file, in the census data, the closest estimate we have to 2018 data in the census file is 5-year post-service estimates calculated in 2010.

As we review the data dictionary for the census dataset, we find that there is a FIPS code for the state in a column called geography, and there is the 75th percentile measurement of veteran salary called y5_p75_earnings. We obtain these two variables just for 2010, the most recent year they are available in the dataset. This demonstration dataset is provided and is named vet_2010_5yrEarnings.csv. Here is a shortened version of what this dataset looks like:

geography	y5_p75_earnings
1	68280
2	73590
4	63330
5	44310
6	59580
8	57570
9	60030
10	
11	103400
12	54610
...	...

geography	y5_p75_earnings
25	66100
26	57800
27	54100
...	...
53	65920
54	52890
55	52920
56	58560

Table 10.3 – Selection of records from vet_2010_5yrEarnings.csv

Remember, the **geography** column refers to the FIPS code, so that the value for 12, which refers to Florida, corresponds to a 75th percentile annual salary (**y5_p75_earnings**) of $54,610. We also have the values for the other two states represented in our demonstration dataset, which are Massachusetts (**geography** = 25, **y5_p75_earnings** = 66100) and Minnesota (**geography** = 27, **y5_p75_earnings** = 54100). Using the FIPS code in our BRFSS dataset as a foreign key, we could *hook on* the value **y5_p75_earnings** and add it to our data.

Let's examine how we would do that. First, we would have to convert the dataset vet_2010_5yrEarnings.csv to SAS. Let's place this dataset in the directory mapped to LIBNAME X and run this PROC IMPORT code on it to convert it to a SAS dataset in WORK named vet_a, followed by PROC CONTENTS to verify the data imported successfully:

```
FILENAME REFFILE '/folders/myfolders/X/vet_2010_5yrEarnings.
csv';
PROC IMPORT DATAFILE=REFFILE
    DBMS=CSV
    OUT=WORK.vet_a;
    GETNAMES=YES;
RUN;
PROC CONTENTS DATA=vet_a;
RUN;
```

Let's examine the code:

- As described in *Chapter 2, Reading Big Data into SAS*, PROC IMPORT obtained the vet_2010_5yrEarnings.csv file from LIBNAME X, then converted it to SAS dataset vet_a with the same headings as in the *.csv file (GETNAMES = YES), and placed it in the WORK directory (WORK.vet_a).

- Because the data in vet_2010_5yrEarnings.csv was displayed in *Table 10.3* earlier, we will not look at the results of the PROC CONTENTS code here. Instead, we will assure ourselves from viewing the log file that the data was imported successfully, and the two columns available are geography (state FIPS code) and y5_p75_earnings (the value of the 75th percentile of earnings in that state by veterans in 2010).

To add the variable y5_p75_earnings to our BRFSS dataset, our strategy will be to follow these steps:

1. Start by converting the veteran's earnings dataset from *.csv to *.sas7bdat by using a data step and reading it into WORK, naming it vet_a, as we did earlier.

2. Read our BRFSS dataset X.Chap10_1 into WORK and name it brfss_a using a data step.

3. Use a data step including a merge function to merge the earnings variable y5_p75_earnings from the vet_a dataset into the brfss_a dataset.

As described earlier, we will use the FIPS code as the index to hook the two data tables together, but this is more easily done in SAS when the index variables have the same name. For that reason, before we do our merge, we will rename the FIPS variable in vet_a from geography to the name it has in our BRFSS dataset, which is _STATE. We will use a data step to do this, and output the vet_b dataset with the renamed variable:

```
data vet_b;
    set vet_a (RENAME = (geography = _STATE));
RUN;
```

As can be seen in the code, the vet_b dataset is output to WORK, and the geography variable has been renamed _STATE through the RENAME command.

Next, as described with using indexes in *Chapter 1, Using SAS in a Data Mart, Lake, or Warehouse*, because we will be using _STATE as the index variable to connect the two datasets, we need to first sort both datasets by _STATE. This code uses a data step to copy X.Chap10_1 into a dataset named brfss_a into WORK, then sorts the two datasets—vet_b and brfss_a—by _STATE:

```
data brfss_a;
    set X.Chap10_1;
PROC SORT data = vet_b;
    by _STATE;
PROC SORT data = brfss_a;
    by _STATE;
RUN;
```

Now, both datasets—vet_b and brfss_a—are in the WORK directory and are sorted by _STATE. This means we are ready for our merge.

In our particular merge, we do not expect any change in the number of rows in the dataset. In fact, if there were a change in the number of rows in the dataset, that would indicate an error. Imagine there was an error in the vet_b dataset where the FIPS code of 12 was assigned to both Florida and Georgia, and the one for Georgia was missing. This would mean that if we connected our dataset with FIPS codes of 12 into vet_b containing the error, we would get more records in the resulting dataset for FIPS code = 12 than we put in. This is because certain records in brfss_a would hook to more than one record in vet_b. In that case, vet_b would not have FIPS serving as a primary key as it is technically defined, which is what would cause this error.

To monitor for that error, it is a good practice to run PROC FREQ on the original dataset before the merge (brfss_a), and another instance of PROC FREQ on the dataset after the merge (brfss_b), to ensure that records were not erroneously added on or subtracted. Let's run a one-way PROC FREQ first on brfss_a here, by _STATE:

```
PROC FREQ data = brfss_a;
    tables _STATE;
RUN;
```

Here is our output table:

The FREQ Procedure

_STATE	Frequency	Percent	Cumulative Frequency	Cumulative Percent
12	15242	39.18	15242	39.18
25	6669	17.14	21911	56.33
27	16990	43.67	38901	100.00

Figure 10.3 – PROC FREQ of _STATE before merge

We observe in the output that for _STATE = 12, there are about 15,000 records; for _STATE = 25, there are over 6,000 records, and for _STATE = 27, there are almost 17,000 records. We will monitor these numbers as we merge our datasets in transformation steps to ensure we do not erroneously pick up or lose any records.

For the next step, we will craft our merge code to achieve our goal, which is to use a data step to create dataset brfss_b as a copy of brfss_a with the y5_p75_earnings variable included from vet_b. How we do the merge operation will ensure that the datasets are joined with the appropriate value corresponding to the _STATE of each record. When thinking of SAS merges, SQL programmers find it helpful to reflect on *joins* in SQL. Others will likely find the SAS white paper *Merging vs. Joining: Comparing Data Step with SQL by Malachy J. Foley* helpful as an explanation of SAS merges (available in the *Further reading* section).

Here is our merge code:

```
data brfss_b;
    merge brfss_a (IN = In1) vet_b (IN = In2);
    by _STATE;
    if (In1 = 1 and In2 = 1) then output brfss_b;
RUN;
PROC CONTENTS data = brfss_b VARNUM;
RUN;
PROC FREQ data = brfss_b;
    tables _STATE;
RUN;
```

Let's look at this code line by line:

1. After beginning the data step with `brfss_b` data, the next statement is the `merge` statement. The statement could simply include the names of the two datasets to be merged, which are `brfss_a` and `vet_b`. But instead, `IN` options are included in parentheses after each dataset. This is a function in SAS. The `IN` variable is a temporary variable that is set to `1` if the record is in the dataset, and is set to `0` if it is not. This variable goes away after the data step is completed. But within the data step, specifying `(IN = In1)` after stating `brfss_a` is a way of telling SAS to keep track of the records in `brfss_a` using a temporary variable called `In1`, and to set that temporary variable to `1` for the records in `brfss_a`, and `0` for the ones that are not in `brfss_a`. Because `(IN = In2)` is stated after the `vet_b` dataset, SAS learns that a parallel temporary variable will be kept about the records in `vet_b` under the name `In2`.

2. In the next line, we have the by statement, which is by `_STATE`. Because we sorted both datasets by `_STATE` before doing this `merge` operation, and because both datasets have a variable of the same type and coding named `_STATE`, we do not get an error. Also, because it is specified as the by variable, `_STATE` is the index variable used to determine the values of `In1` and `In2`. Since records in `brfss_a` only have values of `12`, `25`, or `27` in `_STATE`, `In1` only equals 1 for those values of `_STATE`. In `vet_b`, where we have all the states, other values of `_STATE` (such as the value for Michigan, which is `26`) have an `In2` equal to 1, while their `In1` is equal to 0.

3. The last code in the data step is `if (In1 = 1 and In2 = 1) then output brfss_b`. Please note that an alternative way to state this would be `if (In1 & In2)`. First, it is important to observe the criteria in the parentheses, which refer to temporary variables `In1` and `In2`. The criteria indicate that if the record (as identified by by variable `_STATE`) is in `brfss_a` and also in `vet_b`, then the record should output into the `brfss_b` dataset. Second, it is important to note that the `output` option can output multiple datasets, but all of them have to have been previously named on the first line of the data step. Since the first line of our data step was `data brfss_b`, the `output` option in this code will successfully output a dataset with the name `brfss_b`. This dataset should only contain records that have `_STATE` in both `brfss_a` and `vet_b`, which would be those with `_STATE` = 12, 25 or 27.

4. After the data step is complete, PROC CONTENTS is run on the newly merged dataset brfss_b using the VARNUM option (which sorts the output in the order of variables made). The purpose of this operation is so the analyst can quickly glance at the bottom of the output and verify that the new variables from the merged dataset (in this case variable y5_p75_earnings) successfully merged. We will assume this operation successfully happened and will skip viewing this output.

5. Finally, the last operation is a one-way PROC FREQ on the _STATE variable conducted on the newly merged dataset brfss_b for the purposes of comparing the distributions to the PROC FREQ run prior to the merge to ensure that the frequencies listed did not change. For the purposes of this demonstration, we will assume they did not change, and skip viewing the output.

This diagram can help visualize how the IN temporary variable works in the merge in the data step:

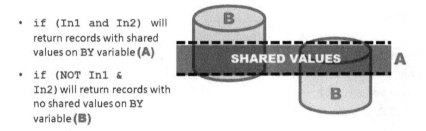

- if (In1 and In2) will return records with shared values on BY variable (A)

- if (NOT In1 & In2) will return records with no shared values on BY variable (B)

Figure 10.4 – Diagram contrasting merges with shared versus unshared BY variables

As shown in the figure, the dataset on the left represents a dataset with some values of the BY variable that are shared with the dataset on the right. If specifying in the merge if (In1 and In2) (assuming we called the temporary variables In1 and In2 as we did in the demonstration code), we are asking for the set of records designated in the diagram by the letter **A**, which represent records with shared values on the BY variable. The following code would do this with brfss_a and vet_b, resulting in the other_option dataset:

```
data other_option;
    merge brfss_a (IN = In1) vet_b (IN = In2);
    by _STATE;
    if (In1 and In2) then output other_option;
RUN;
```

In SQL, this would be called an **inner join**. By contrast, if we specify `if (NOT In1 and In2)` in the merge, we are asking to output records labeled **B** in the diagram, which are records with no shared values on the `BY` variable (which in SQL would be an **outer join**). The following code demonstrates this with `brfss_a` and `vet_b` creating a dataset like this, naming the output the `no_match` dataset:

```
data no_match;
    merge brfss_a (IN = In1) vet_b (IN = In2);
    by _STATE;
    if (NOT In1 and In2) then output no_match;
RUN;
```

To complete the analogy, a diagram including code examples is provided so SQL programmers can understand how to execute **left joins** and **right joins** using a data step:

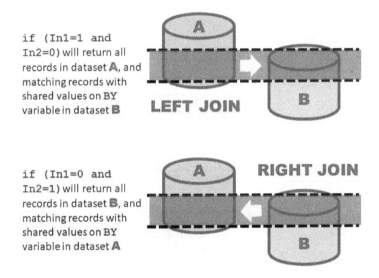

if (In1=1 and In2=0) will return all records in dataset **A**, and matching records with shared values on BY variable in dataset **B**   **LEFT JOIN**

if (In1=0 and In2=1) will return all records in dataset **B**, and matching records with shared values on BY variable in dataset **A**   **RIGHT JOIN**

Figure 10.5 – Diagram showing analogs in a data step to SQL left and right joins

As can be seen from the diagram, manipulating the `IN` temporary variables during a merge is necessary in order to achieve both left joins and right joins from SQL.

Typically, data warehouse ETL that includes merges usually focuses on **inner joins**, like the one we did between `vet_b` and `brfss_a`. In that case, we joined two datasets on a common variable, `_STATE`, which was present in both native datasets. Sometimes, we have state-level data in different datasets, but we are not lucky enough to have state variables formatted in the same way in each dataset that could help us merge the datasets as we did with `_STATE`. In those situations, we may have to deal with multiple foreign keys.

Dealing with multiple foreign keys in the warehouse

Before we leave the topic of using foreign keys, let's contemplate adding one more set of data to our warehouse, which could theoretically come from the **American Hospital Directory (AHD.com)**. This website compiles data reported quarterly to the US government about hospitals by state (see web link under *Further reading*). Here is a shortened summary of some of the data about hospitals and states available in the public domain and presented on their web page:

State_Desc	Number_Hospitals	Staffed_Beds
AK - Alaska	10	1235
AL - Alabama	89	15335
AR - Arkansas	52	7872
AS - American Samoa	1	0
AZ - Arizona	73	13416
CA - California	344	74724
CO - Colorado	55	8181
CT - Connecticut	34	8798
DC - Washington D.C.	7	2117
DE - Delaware	8	2072
FL - Florida	215	56245
...	...	...
MA - Massachusetts	73	15192
MD - Maryland	53	10802
ME - Maine	19	2887
MI - Michigan	105	23681
MN - Minnesota	55	10504
...	...	...
WA - Washington	62	10322
WI - Wisconsin	77	11080
WV - West Virginia	34	5396
WY - Wyoming	14	1261

Table 10.4 – Summary of hospital data from AHD.com

As can be seen from the hospital data, we run into a problem because FIPS code is not included. The only foreign key we could use to hook the data onto our BRFSS data would be the variable called `State_Desc`. This would require attaching a crosswalk dataset that included the FIPS code to `State_Desc`, as shown here in a truncated version:

State_FIPS	State_Abbrev	State_Desc
1	AL	AL – Alabama
2	AK	AK – Alaska
4	AZ	AZ – Arizona
5	AR	AR – Arkansas
6	CA	CA – California
8	CO	CO – Colorado
9	CT	CT – Connecticut
10	DE	DE – Delaware
11	DC	DC - Washington D.C.
12	FL	FL – Florida
...	...	...
25	MA	MA – Massachusetts
26	MI	MI – Michigan
27	MN	MN – Minnesota
...	...	...
53	WA	WA – Washington
54	WV	WV - West Virginia
55	WI	WI – Wisconsin
56	WY	WY – Wyoming
60	AS	AS - American Samoa
66	GU	GU – Guam
69	MP	MP - Northern Mariana Islands
72	PR	PR - Puerto Rico
78	VI	VI - Virgin Islands

Table 10.5 – State_Desc to FIPS crosswalk

Let's review the columns in this crosswalk table:

- **State_FIPS**: This column contains the FIPS code, which reflects the same value contained in `_STATE`. However, in the crosswalk dataset, this variable is named `State_FIPS`, not `_STATE`.
- **State_Abbrev**: This is the two-digit state abbreviation. This is a foreign key used in many government datasets, but has not been used in any of the datasets demonstrated so far in this chapter.
- **State_Desc**: This is another foreign key for state. This is the one reflected in the AHD data.

As with the veteran data, we could theoretically patch on the hospital data to our brfss_b dataset, and make brfss_c. But the issue with missing the crosswalk variable makes us have to execute this in a series of discrete steps:

1. First, import the AHD data (AHD.csv) into a SAS dataset (ahd_a).
2. Next, import the crosswalk data (State_xwalk.csv) into a SAS dataset (xwalk_a).
3. In xwalk_a, the FIPS code is named State_FIPS. Rename this in a data step to _STATE, and output the xwalk_b dataset.
4. Sort both xwalk_b and brfss_b by _STATE, then use a merge function as we did with vet_b to attach the crosswalk variables and output the brfss_c dataset.
5. Now that we have the by variable named State_Desc that we will use with ahd_a that was patched on from the crosswalk dataset, we can now sort ahd_a by State_Desc, sort brfss_c by State_Desc, and merge them into brfss_d.

Let's do each of these steps one at a time. Let's start by importing the ahd.csv dataset from the directory mapped to LIBNAME X into a SAS dataset in WORK, and naming it ahd_a:

```
FILENAME REFFILE '/folders/myfolders/X/AHD.csv';
PROC IMPORT DATAFILE=REFFILE
    DBMS=CSV
    OUT=WORK.ahd_a;
    GETNAMES=YES;
RUN;
PROC CONTENTS DATA=ahd_a;
RUN;
```

This code imports the dataset into WORK and names it ahd_a. Although we included a PROC CONTENTS on ahd_a in the code, we will not show the results here, because an example of the data is already shown in *Table 10.4*.

Now, we will do the second step, which is to import the crosswalk file, State_xwalk.csv, into WORK as a SAS dataset and name it xwalk_a:

```
FILENAME REFFILE '/folders/myfolders/X/State xwalk.csv';
PROC IMPORT DATAFILE=REFFILE
    DBMS=CSV
```

```
        OUT=WORK.xwalk_a;
        GETNAMES=YES;
RUN;
PROC CONTENTS DATA=xwalk_a;
RUN;
```

This code places `State_xwalk.csv` into `WORK` and names it `xwalk_a`. It also runs `PROC CONTENTS` on `xwalk_a`, but we will not show the results here, because the data in `xwalk_a` is shown in *Table 10.5*.

The name of the FIPS code in `xwalk_a` is `State_FIPS`, not `_STATE` as we would like in order to make our merge easier in SAS. Therefore, we next use a data step to rename the `State_FIPS` variable to `_STATE` in the crosswalk dataset:

```
data xwalk_b;
    set xwalk_a (RENAME = (State_FIPS = _STATE));
RUN;
```

The data step outputs the `xwalk_b` dataset with the renamed variable, `_STATE`.

Now, we move on to sorting each dataset, then merging the crosswalk dataset onto the BRFSS dataset to create the final output dataset `brfss_c`:

```
PROC SORT data = xwalk_b;
    by _STATE;
RUN;
PROC FREQ data = brfss_b;
    tables _STATE;
RUN;
data brfss_c;
    merge brfss_b (IN = In1) xwalk_b (IN = In2);
    by _STATE;
    if (In1 = 1 and In2 = 1) then output brfss_c;
RUN;
PROC CONTENTS data = brfss_c VARNUM;
RUN;
PROC FREQ data = brfss_c;
    tables _STATE;
RUN;
```

Let's review this code:

1. First, we see the crosswalk dataset, xwalk_a, being sorted by _STATE, as well as brfss_b being sorted by _STATE, as preparation for the merge.

2. Next, we see a merge that will result in an inner join between xwalk_a and brfss_b. Because we only expect the records with _STATE = 12, 25, or 27 in both datasets to merge, we do not expect that the dataset resulting from the merge, brfss_c, will have a different number of observations than brfss_b. However, it will have the extra two columns from the crosswalk merged, which are State_ Abbrev and State_Desc, as shown in *Table 10.5*.

3. Finally, we run PROC CONTENTS and PROC FREQ on brfss_c to verify that the two variables from xwalk_a, State_Abbrev and State_Desc, were successfully merged, and that the number of rows in the dataset did not change from brfss_b to brfss_c.

Finally, we get to our last step, which is to merge the ahd_a data onto brfss_c:

```
PROC SORT data = ahd_a;
    by State_Desc;
RUN;
PROC SORT data = brfss_c;
    by State_Desc;
RUN;
PROC FREQ data = brfss_c;
    tables _STATE;
RUN;
data brfss_d;
    merge brfss_c (IN = In1) ahd_a (IN = In2);
    by State_Desc;
    if (In1 = 1 and In2 = 1) then output brfss_d;
RUN;
PROC CONTENTS data = brfss_d VARNUM;
RUN;
PROC FREQ data = brfss_d;
    tables _STATE;
RUN;
```

Let's review the code:

1. First, both the datasets to be merged, which are `ahd_a` and `brfss_c`, are sorted by the index that will be used for the merge, which is now `State_Desc`. We patched `State_Desc` onto `brfss_c` as a foreign key, and now we can use it to hook on data from `ahd_a` using a `merge` command. In order for us to use `State_Desc` as the merge index, both datasets need to be sorted by this variable first.

2. Next, we run a one-way `PROC FREQ` of `_STATE` in `brfss_c` so we can compare these numbers to the ones in the merged dataset and verify that the frequencies did not change.

3. In the next set of code, we include the data step with the merge, where we merge `brfss_c` and `ahd_a` by `State_Desc` using an inner join. The output dataset is `brfss_d`.

4. To verify that the hospital data was added to `brfss_d`, and that the number of rows in `brfss_d` did not change in the merge, we end by running `PROC CONTENTS` and `PROC FREQ` on `brfss_d`.

Now that we have finished adding the salary data from `vet_b`, the crosswalk variables from `xwalk_b`, and the hospital variables from `ahd_a`, let's look at the `PROC CONTENTS` on `brfss_d` that was output from the code:

Variables in Creation Order					
#	Variable	Type	Len	Format	Informat
1	_STATE	Num	8	BEST12.	BEST32.
2	VETERAN3	Num	8	BEST12.	BEST32.
3	MARIJAN1	Num	8	BEST12.	BEST32.
4	USEMRJN2	Num	8	BEST12.	BEST32.
5	y5_p75_earnings	Num	8	BEST12.	BEST32.
6	State_Abbrev	Char	2	$2.	$2.
7	State_Desc	Char	20	$20.	$20.
8	Number_Hospitals	Num	8	BEST12.	BEST32.
9	Staffed_Beds	Num	8	BEST12.	BEST32.

Figure 10.6 – PROC CONTENTS from brfss_d

In the output, we can see our four native variables, followed by the `y5_p75_earnings` variable from `vet_b`, the `State_Abbrev` and `State_Desc` variables from `xwalk_b`, and the `Number_Hospitals` and `Staffed_Beds` variables from `ahd_a`.

There were many steps and much effort associated with attaching the hospital data from the AHD dataset. This was because we needed to attach a crosswalk dataset. But this provided us with an improvement overall to the data warehouse. Now, if we wanted to attach data from another dataset that had a state variable that was not formatted as a FIPS code, we'd have more flexibility. We could use either `State_Abbrev` or `State_Desc` as an index to merge on other data. In fact, other analysts using the data in the warehouse could request these indexes and be able to hook more data on from other datasets outside of the warehouse. Therefore, hosting many foreign keys, especially to geographic regions, can greatly add value to a data system and help support the needs of analyst users.

In this demonstration, we practiced how to use crosswalk datasets to connect variables from different datasets. In the next section, we will talk about creating **crosswalk variables** to help make longitudinal analyses go more smoothly.

Crosswalking data over time

The BRFSS is an example of a dataset that is available every year. Since it is a cross-sectional survey, different people are surveyed each year, but they are asked largely the same questions. Therefore, most variables are the same from year to year. Although the BRFSS is a survey, there are administrative datasets that are available on an annual basis that are comprised of data from live business systems. One example can be seen in the **Healthcare Cost and Utilization Project (HCUP)** datasets (the link is in the *Further reading* section). These datasets contain data from US healthcare facilities. Because the file specifications rarely change over the years, new annual HCUP data files tend to be in the same structure as previous files.

However, variables do change over time, both in surveys and in business environments, because the environment evolves. For HCUP, shifts in the healthcare environment may require shifts in the exact variables HCUP collects from healthcare facilities. And in the BRFSS, changes to health and healthcare could cause changes to the questions being asked.

A perfect example can be seen in the questions about marijuana use in the BRFSS. In the US, marijuana was illegal at the state level in most states in the early 2000s, but as time went on, states began to legalize marijuana purchase and use in various forms. Because marijuana was approved for medicinal use in many states, eventually, the BRFSS had to wrestle with adding new questions regarding marijuana use.

A review of the codebooks from previous BRFSS annual datasets reveals that marijuana questions were not included in the core survey until 2016. Here is a review of five marijuana-related questions included in the BRFSS in the 2016, 2017, and 2018 datasets:

Variable name	Question	2016	2017	2018	Comment
MARIJANA	During the past 30 days, on how many days did you use marijuana or hashish?	X	X		
MARIJAN1	Not documented in codebook			X	Coded the same as MARIJANA
USEMRJNA	During the past 30 days, how did you use marijuana? [Please tell me all that apply. Did you....]	X			Accepts multiple answers
USEMRJN1	During the past 30 days, what was the primary mode you used marijuana? [Please select one. Did you...]		X		Accepts one answer. Answer list is the same as USEMRJN2.
USEMRJN2	During the past 30 days, which one of the following ways did you use marijuana the most often? Did you usually... [Please select one. Did you...]			X	Accepts one answer. Answer list is the same as USEMRJN1.

Table 10.6 – Marijuana variables in BRFSS 2016, 2017, and 2018 files

Let's make a few observations about the variables in the table:

- Notice that the variables we have in our 2018 dataset, which are MARIJAN1 and USEMRJN2, are reflected in the table. According to the table, these variables were not used previously.

- In 2016 and 2017, the variable that measured whether or not the respondent used marijuana was worded, *During the past 30 days, on how many days did you use marijuana or hashish?* and called MARIJANA. If the person responded with any number of days, they were then asked the next question, about mode of use.

- In 2018, there was no comparable variable to MARIJANA documented in the codebook, but there was one that was not documented that was named MARIJAN1 that appeared to be coded the same way.

- USEMRJNA, USEMRJN1, and USEMRJN2 are all variables about the mode of marijuana use, but they are all coded differently and were all asked in different years (2016, 2017, and 2018).

- USEMRJNA is different from the other two, USEMRJN1 and USEMRJN2, because it accepts multiple answers, whereas the other two only ask for one answer.

- USEMRJN1 and USEMRJN2 use the same answer list, but the questions are worded differently.

Imagine the data warehouse wanted to host data in the past, even before these variables were available. It also wanted to make it so analyses run on all years of data hosted in the data warehouse could make use of the marijuana variables. Because the variables first did not exist, and then changed in coding over a few years, a way to make such an analysis possible is to make crosswalk variables. We will start by making categorical crosswalk variables for the marijuana variables in BRFSS.

Crosswalking categorical variables

Let's focus on developing a crosswalk for the marijuana use variables. Let's start by examining the variable that was collected in 2016, called USEMRJNA. Remember, this variable supposedly holds multiple answers to this question. This is how USEMRJNA was coded according to the codebook:

- 1 = **Smoke it (for example, in a joint, bong, pipe, or blunt)**
- 2 = **Eat it (for example, in brownies, cakes, cookies, or candy)**
- 3 = **Drink it (for example, in tea, cola, or alcohol)**
- 4 = **Vaporize it (for example, in an ecigarette-like vaporizer or another vaporizing device)**
- 5 = **Dab it (for example, using waxes or concentrates)**
- 6 = **Use it some other way**
- 7 = **Don't know/not sure**
- 9 = **Refused**
- Missing (indicated by period) = **not asked**

When reviewing the contents of the actual variable in the 2016 dataset, it was revealed how one variable could hold multiple answers. All of the answers were crammed into the same variable. For example, if a respondent answered 1 for **Smoke it** and 2 for **Eat it**, their answer was recorded as 12. Even though the number 12 is not a numerical measure, and simply indicates the codes 1 and 2 together, it also can be seen as the minimum number of all the answers containing multiples. Therefore, records with variables coded as greater than 12 represent respondents who gave multiple answers to the question.

On one hand, data stored this way is messy and can be hard to analyze. On the other hand, this question produced very rich data about the respondents. Most respondents only answered one method, so when there were multiple answers, they were very revealing. However, because both the question and the answers behind the variable USEMRJNA are so different from the ones behind USEMRJN1 and USEMRJN2, we can already see it might be difficult to crosswalk these variables in meaning over time.

Now let's examine the coding of USEMRJN1 and USEMRJN2. Actually, the coding is identical to USEMRJNA, except that multiples are not accepted. Remember, the only difference between USEMRJN1 and USEMRJN2 is in how the question is worded, not in the possible answers. Because the question is worded almost the same way in USEMRJN1 and USEMRJN2, it is tempting to actually store these two variables in one variable. After all, they use the same coding system.

While it is tempting to do this, it is a bad practice:

- Even though USEMRJN1 and USEMRJN2 share the same coding system, they are actually technically different variables, because they are answers to different questions.

- The coding system used by USEMRJN1 and USEMRJN2 handles all the codes in those variables, but it does not take into account the coding of multiples that took place in the USEMRJNA variable. There is no code for multiples in the coding system used by USEMRJN1 and USEMRJN2.

- Also, the coding system in USEMRJN1 and USEMRJN2 uses missing (represented by a period) to indicate the question was not asked (because the respondent did not say they used marijuana in the last 30 days). But there is no code to indicate that the variable did not exist at the time of the dataset, which is the code that would need to be used in old BRFSS data, before 2016.

The solution is to create a brand new coding system that handles all these problems:

- The new coding system should have a code that indicates that multiple answers were given. That code can be used for records in the 2016 dataset with answers of 12 or greater in USEMRJNA.

- The new coding system should also have a code that indicates that the question did not exist at the time of the dataset.

Since we are making a new coding system, let's improve upon the native one. The original coding system had too many levels to be useful for analysis. Therefore, let's collapse some of the levels:

- Studies using PROC FREQ on the variables in the different years revealed that the most common level by far is 1, which refers to smoking it.

- Studies also revealed that levels 2 and 3 (referring to eating it and drinking it, respectively) were rarely mentioned, and could reasonably be collapsed together. These could be placed in a *Consumed* category.

- In addition, it was realized that levels 5 and 6 (referring to dabbing it or using it another way, respectively) were also low frequency, so these could be collapsed together into an *Other* category.

- Because the multiple answers in the 2016 dataset were also rare, we can add them to the *Other* category by expanding it into *Other or multiples*.

- Since 7, 9, and missing all refer to unknown values, these could be collapsed under an *Unknown* category.

- Finally, we could add a category *Not available in time period* to accommodate coding the variable in all the records in datasets prior to 2016 before the variables existed.

Let's call our new coding system USETYPE:

- 1 = **Smoked**
- 2 = **Consumed**
- 3 = **Vaporized**
- 4 = **Other or multiples**
- 8 = **Not available in time period**
- 9 = **Unknown**

Now that we have designed our categorical crosswalk variables, we can amend our ETL protocol to accommodate generating USETYPE. For the datasets before 2016, USEMRJNA will need to be recoded into USETYPE using this coding:

USEMRJNA	USEMRJNA Desc	USETYPE	USETYPE desc
1	Smoke it (for example, in a joint, bong, pipe, or blunt)	1	Smoked
2	Eat it (for example, in brownies, cakes, cookies, or candy)	2	Consumed
3	Drink it (for example, in tea, cola, or alcohol)	2	Consumed
4	Vaporize it (for example, in an ecigarette-like vaporizer or another vaporizing device)	3	Vaporized
5	Dab it (for example, using waxes or concentrates)	4	Other or multiples
6	Use it some other way	4	Other or multiples
7	Don't know/not sure	9	Unknown
9	Refused	9	Unknown
12 and higher	Multiple responses (concatenated)	4	Other or multiples
(blank)	Unknown (also "not asked")	9	Unknown
Not applicable	This level of USETYPE is for BRFSS datasets before 2016 that may reside in the warehouse, when marijuana questions were not asked	8	Not available in time period

Table 10.7 – Crosswalk of USEMRJNA to USETYPE

As can be seen from the table, in the processing of the 2016 dataset, ETL code can be crafted that creates USETYPE based on the native coding of USEMRJNA. The following table shows how in the 2017 and 2018 datasets, USEMRJN1 and USEMRJN2 can be crosswalked into USETYPE:

n	USEMRJN1 and USEMRJN2 desc	USETYPE	USETYPE desc
1	Smoke it (for example, in a joint, bong, pipe, or blunt)	1	Smoked
2	Eat it (for example, in brownies, cakes, cookies, or candy)	2	Consumed
3	Drink it (for example, in tea, cola, or alcohol)	2	Consumed
4	Vaporize it (for example, in an ecigarette-like vaporizer or another vaporizing device)	3	Vaporized
5	Dab it (for example, using waxes or concentrates)	4	Other or multiples
6	Use it some other way	4	Other or multiples
7	Don't know/not sure	9	Unknown
9	Refused	9	Unknown
(blank)	Unknown (also "not asked")	9	Unknown
Not applicable	This level of USETYPE is for BRFSS datasets before 2016 that may reside in the warehouse, when marijuana questions were not asked	8	Not available in time period

Table 10.8 – Crosswalk of USEMRJN1 and USEMRJN2 to USETYPE

> **Note:**
>
> In *Chapter 6, Standardizing Coding Using SAS Arrays*, we discussed using conditions in arrays, and in *Chapter 8, Using Macros to Automate ETL in SAS*, we discussed using conditions in macros. Imagine you were setting up ETL code to process the BRFSS datasets from prior to 2016, and 2016, 2017, and 2018. You could theoretically make code using conditions in both arrays and macros to handle the differences between the code files in coding USETYPE. For the files before 2016, the code would just add USETYPE = 8, whereas for 2016, the coding for USETYPE would be based on the USEMRJNA native variable, and for 2017 and 2018, the coding for USETYPE would be based on the native coding of USEMRJN1 and USEMRJN2. Using conditions in arrays and macros could automate this ETL protocol.

The following is an example of ETL code we could use on brfss_d to create USETYPE in our 2018 BRFSS data extract:

```
data brfss_e;
    set brfss_d;
    USETYPE = 9;
    if USEMRJN2 = 1
          then USETYPE = 1;
    if USEMRJN2 in (2, 3)
          then USETYPE = 2;
    if USEMRJN2 = 4
          then USETYPE = 3;
    if USEMRJN2 in (5, 6)
          then USETYPE = 4;
RUN;
PROC FREQ data = brfss_e;
    tables USEMRJN2 * USETYPE / list missing;
RUN;
```

The code inputs brfss_d and outputs brfss_e, and during the data step, it codes USETYPE based on the values in USEMRJN2 as described in the table. Finally, we run a two-way PROC FREQ between USEMRJN2 and USETYPE to ensure the coding is correct.

In this section, we talked about crosswalking the categorical variables that referred to mode of marijuana use into a standard system USETYPE. However, we did not focus on the continuous variables MARIJANA and MARIJAN1, which are the number of days (within the past 30 days) that the respondent used marijuana. The next section discusses crosswalking continuous variables that change over time.

Crosswalking continuous variables

In the case of MARIJANA and MARIJAN1, as a result of our research, we suspect that both variables refer to the same question and the same set of answers. We suspect that the only thing that changed from 2017 to 2018 is BRFSS renamed the variable from MARIJANA to MARIJAN1. In a case like this, it is tempting to try to overcome this problem by renaming one of the variables. For example, if we processed the 2017 file in 2017 with the variable named MARIJANA, we might be tempted to simply rename the native MARIJAN1 variable as MARIJANA in the 2018 file.

The problem with this approach is that native variables should not be edited during ETL. This means that they should not be recoded or renamed. This is because that native variable may need to be used in other ETL, and needs to stay frozen as a standard measure. If we want to combine these two variables into the same variable for crosswalking purposes, we can simply create a new variable name and code both into that one.

Imagine we invent a new variable and name it `MARDAYS`:

- When doing ETL on the `MARIJANA` variable in the 2017 file, we simply create a new variable called `MARDAYS` and copy `MARIJANA` into it.

- When doing ETL on the `MARIJAN1` variable in the 2018 file, again, we simply create a new variable called `MARDAYS` and copy `MARIJAN1` into it.

`MARIJANA` and `MARIJAN1` present a simple case of crosswalking, because only the variable name changed. When it comes to other continuous variables, crosswalking can get more complicated:

- **Financial variables**: Variables indicating money, such as salaries, home values, and business loan levels, need to be recalculated each year to adjust for inflation in order to be comparable across years. It is not unusual to see crosswalk variables that refer to values in a year (for example, value in 2016 dollars, value in 2017 dollars, and value in 2018 dollars).

- **Variables derived from formulas**: Some continuous variables, such as mortality indexes in healthcare, are created from formulas. Sometimes, these standard formulas undergo a change. In that case, business rules will dictate what the data warehouse should do. If both formulas are still being used, then the warehouse should calculate both variables and make them available. If the old formula is being phased out, then only variables using the new formula should be served up.

> **Note**
>
> As mentioned in Tim Mitchell's article, there can be a problem in a data warehouse where developers who are building and maintaining the warehouse can become detached from the needs of analyst users. Management is encouraged to view analyst users as customers being served who could provide input on how to be served better. When considering getting new datasets, or changing the processing of datasets, focus groups of analyst users and developers could be brought together to co-design the new processing and variables.
>
> When data warehouses host datasets in a longitudinal manner as described with the BRFSS, there will be many analysts who become regular users of such data. This user group can be the subject of market research surveys that can ask about their needs for crosswalk variables, as well as other features in the warehouse, such as analytic tools, documentation, visualizations, and reports.

This section described data stewardship for serving analyst users of the warehouse. This type of stewardship is also helpful for warehouse developers because it helps them serve user needs. However, developers have their own needs for data stewardship that should be acknowledged and are discussed in the next section.

Data stewardship for serving warehouse developers

Although there are many tasks associated with data stewardship for warehouse developers, this section will focus on two important functions: managing a **data stewardship committee** that provides oversight of the data warehouse, and maintaining and providing **data curation files** specifically aimed at developers rather than analysts. These two functions will be discussed here.

Managing a data stewardship committee

There are several different **stakeholder groups** surrounding a data lake or data warehouse. These go beyond the obvious stakeholder groups of analyst users and developers. They also include groups who have a stake in the data system, even if they are not directly connected to the project. Some stakeholders may never log in to the data system.

One set of stakeholders includes the leadership team that initiated the development of the data system. Other sets of stakeholders can be representatives of entities whose data is represented in the system. Imagine a data warehouse used for a health insurance company that includes claims data. People who work to adjudicate claims in the claims department may not use the data warehouse. Also, the people who make contracts with healthcare providers and facilities that are configured in the data warehouse may not use it, either. However, they would be stakeholders in the data warehouse because their data is included, and they would want it represented correctly.

Depending upon what the data system does and who it serves, there could be other stakeholder groups. Groups that make decisions based on reports from the data system could represent other stakeholders, as well as groups who develop reports from the system. Nevertheless, it is important to identify all the relevant stakeholder groups in order to set up a data stewardship committee. That is because, on such a committee, each stakeholder group needs to be represented.

In practice, data stewardship committees are often named something different, such as the *Data Committee*, or the *X Committee*, where *X* represents the name of the data system. The committee is typically chaired by a leader in the department that manages the data system, and this is often a permanent seat (with back-up chairs designated as other leaders from the same department). The other members of the committee are there by way of their position in the company. For example, for health insurance, the director of the Claims Department and the director of the Contracting Department could be tasked with serving on the committee. That way, whoever is filling those roles will also be expected to complete committee work.

Because data stewardship committees typically comprise busy leaders, meetings do not happen very often—typically once per month or per quarter. Issues with the data system that need to be dealt with by leadership are collected into an agenda and discussed at committee meetings. The chair's department manages the committee and keeps records of committee activity. Completing committee business typically involves crafting and disseminating new policies and procedures that are then implemented with the support of the committee.

Doing a good job of managing a data stewardship committee is a great way to serve the needs of developers. This is because developers require guidance in the form of well-considered policies approved by leadership. In fact, the data stewardship committee can provide a forum for developers to pitch ideas and ask questions that they want leadership to consider. So while data stewardship committees help improve the function of the data system, their work also serves to support data system developers.

Providing curation and other support

In addition to receiving support from the data stewardship committee, warehouse developers can also be supported by useful and high-quality **data curation files**. The bottom of the diagram in *Figure 10.1* mentioned several types of curation and support files that should be made available to support analysts and can also be useful for developers. Curation files will come from two sources:

- **Outside organizations**: Most curation from outside organizations that are hosted in the data system will consist of documentation referring to native datasets in the data system. However, outside organizations using the same native datasets may also produce curation hosted by the data system.

- **Internal sources**: Employees, consultants, and others working internally on the data system will also produce curation files. Although some of these may be about native datasets hosted in the data system, most of the internally authored curation will be to document how the data lake or warehouse is constructed, as well as policies and procedures for its maintenance and management.

This table provides a summary of different curation items, and lists the source and primary user group:

Item	Source	Primary Users	Comment
Original survey codebook	Organization providing data	Analysts Developers	Only pertains to survey data.
Miscellaneous source dataset documentation, such as data entry manuals, and documentation of business rules	Organization providing data	Analysts Developers	
Data dictionaries for native datasets	Organization providing data	Developers	If many native variables are served up, analysts may find use for data dictionaries for native datasets.
Data dictionaries for the data warehouse or lake	Data warehouse	Analysts Developers	This includes documentation of grouping variables, flags, index variables, and crosswalks added by the data system.
Code for labels and formats	Data warehouse	Analysts	
Macros used for ETL	Data warehouse	Analysts Developers	Both analysts and developers use these files to check against documentation such as data dictionaries.
Results of studies done of native data that led to the design of ETL code	Data warehouse	Developers	
Instructions for using ETL code	Data warehouse	Developers	This is the documentation behind ETL protocols. It can also include entity-relationship diagrams for tables included in the system.
Data warehouse application documentation	Data warehouse	Analysts Developers	This includes manuals, training materials, policy documents, and other guidance.

Table 10.9 – Common curation files

As can be seen in the table, many curation items, such as survey codebooks for native variables from survey datasets, and data dictionaries for warehouse-derived variables, are used by both analysts and developers. But a few items on the list—especially instructions for using ETL code, and the results of studies on which ETL code is based—represent curation files aimed primarily at developers.

Chapter 7, Designing and Developing ETL Code in SAS, described taking smaller SAS datasets and conducting studies to help inform the development of grouping and flag variables, and in this chapter, we act on the results of studies using the marijuana variables BRFSS in order to develop the USETYPE crosswalk variable. Documentation of these studies, along with instructions for using the ETL code, are valuable items of data curation aimed solely at developers. The table describes many different types of curation files that are to be created by the data warehouse. For guidance on how to create such files, you are referred to an online course titled *Data Curation Foundations* (the link is available in the *Further reading* section).

A final point that should be made about maintaining a data warehouse is that the less siloed the talent is, the stronger the data warehouse is. Different types of analysts within their analyst class should cross-train across subclasses so they can support each other, and different types of developers should do the same. Imagine a developer who normally works to de-identify data on a de-identification server cross-trained with a developer who worked on maintaining a cloud server. This means that if either developer needed to be away from their duties, another individual could take their place. But more importantly, having the developers cross-train would help them understand the challenge inherent in each other's job. This way, the developer on the de-identification server may modify their protocols to better serve customers after learning about different customer needs from the cloud service developer and vice versa.

Summary

The purpose of this chapter was to provide guidance to SAS data warehouse leaders on how to focus on the needs of their customers, which are categorized as users who are either analysts or developers. That is because improving user experience will encourage the onboarding of more users and the conduct of more analyses. The more people and analyses using a SAS data warehouse, the closer it strives to fulfill the business objective that led to setting up a SAS data warehouse in the first place.

This chapter first focused on serving the needs of analyst users through providing data access and providing particularly useful analytic variables such as foreign keys and crosswalk variables. Next, the chapter described how to serve the needs of warehouse developers through managing a data stewardship committee, and providing data curation and other support.

This chapter provided a template to readers as to how to manage a SAS data warehouse that is hosting data from many different sources, even if it is within a single organization. It also provided guidance on identifying stakeholders and setting up a data stewardship committee if one is not already organized. It explained the types of policies that might be made by this committee, such as policies around data access and hosting remote datasets from other organizations. It also showed how programming policies can be made that enhance the analyst user experience, such as revising the ETL protocols so they serve up useful foreign keys and crosswalk variables. These variables encourage analysts to connect warehouse data to other datasets and offer more flexibility for longitudinal analysis.

In the next chapter, we focus on interconnectivity between SAS data warehouses and data systems in different formats. We discuss serving SAS to other systems and connecting to non-SAS data storage.

Questions

1. Why is it helpful to consider both analyst and developer users in a data warehouse or data lake?

2. Why do analyst users of data lakes and developer users of data warehouses need extensive documentation on source datasets?

3. What are the pros and cons of having analysts access a data lake server directly?

4. How does having multiple foreign keys in a data warehouse make it more useful to analysts?

5. Imagine you are hosting an annual dataset in your data warehouse. One year when you receive the dataset, you learn that a new additional categorical variable is included that you find valuable, named ADJPRICE. For the next 2 years, you receive the dataset with ADJPRICE in it coded according to the same system, but the third year, you receive the dataset without ADJPRICE but with ADJPRICE2, which is coded slightly differently than ADJPRICE. If you were to make a crosswalk variable to handle ADJPRICE and ADJPRICE2 in datasets over all these years, what coding would it have to handle?

6. Why might SAS data warehouse developers need their own set of curation files that are not seen by analyst users?

7. Curation files, such as data dictionaries and crosswalk tables, can take a long time and a lot of effort to make. What should be taken into account when deciding whether or not to spend time and effort on making particular curation files?

Further reading

- *Why Data Warehouse Projects Fail* by Tim Mitchell, available here: `https://www.timmitchell.net/post/2017/01/10/why-data-warehouse-projects-fail/`

- Data about veteran salaries by state from the US census: `https://lehd.ces.census.gov/data/veo_experimental.html`

- SAS white paper, *Merging vs. Joining: Comparing the Data Step with SQL* by Malachy J. Foley, available here: `https://support.sas.com/resources/papers/proceedings/proceedings/sugi30/249-30.pdf`

- American Hospital Directory state hospital data available here: `https://www.ahd.com/state_statistics.html`

- Healthcare Cost and Utilization Project (HCUP) datasets: `https://www.hcup-us.ahrq.gov/databases.jsp`

- Online course, *Data Curation Foundations*, available here: `https://www.linkedin.com/learning/data-curation-foundations/data-curation-in-data-science?u=2125562`

11
Connecting the SAS Data Warehouse to Other Systems

When SAS was originally designed, it was meant to run entirely in its own SAS environment. Data editing, analysis, reporting, and visualization all happened within the same SAS environment. But as the world has evolved, we now have the internet, and SAS has evolved to include interoperability features.

Now, SAS data can be served to other systems, and the **SAS/ACCESS** component can now be used to connect to non-SAS data storage, such as other data warehouses in **SQL**. Reporting and visualization of SAS data can now take place outside of SAS using software optimized for these functions. This chapter will cover these improvements in the interoperability of SAS.

This chapter will teach the following skills:

- How to design a system where data from a SAS data warehouse is served up in a non-SAS application, such as serving up a **star schema** in **IBM Cognos**

- How to create a connection between a SAS data warehouse and a non-SAS data warehouse (such as one with data stored in SQL), and copy the data into the SAS data warehouse

- How to create a connection between a SAS data warehouse and data in the cloud to create interoperability

Technical requirements

There are no datasets for this chapter because the code that is presented is not runnable and is for explanation purposes only. The code bundle for this chapter is available on GitHub here: `https://github.com/PacktPublishing/Mastering-SAS-Programming-for-Data-Warehousing/tree/master/Chapter%2011`.

Serving SAS to other systems

SAS is known for its ability to conduct ETL protocols on very large datasets. While other products have this capability, SAS also has the ability to conduct advanced statistical analyses, such as **regressions**. When compared to data warehouse storage software from other companies, such as **Oracle's SQL**, SAS is superior when it comes to the capacity to perform analytic functions. Even creating a **contingency table** such as the one produced by `PROC FREQ` can be overwhelming in terms of processing effort when attempted by other data warehouse software.

However, even though SAS has evolved over time, the continued difficulty in programming efficient I/O in SAS through the sophisticated use of **data step language** continues to hamper SAS when it competes with other products. This is why, traditionally, SAS data warehouses were only constructed to serve analysts who were planning on analyzing the data using SAS as the analytic software. That way, the only I/O costs would be in placing the data on the server and conducting ETL. Analysts would not experience any significant I/O costs.

But in today's world, there are so many products that can be used for data presentation and visualization, so it makes sense that SAS data warehouses would consider serving data to other systems. Perhaps the COVID-19 worldwide pandemic presents a relevant use case. Most state health departments in the US collected and analyzed data on infectious disease outbreaks routinely in SAS before COVID-19. After the COVID-19 outbreak started, these departments were asked to send regular data extracts with COVID-19 outbreak data to the federal government for tracking, analysis, and visualization purposes. Because the products used for these functions by the federal government included SAS as well as other tools, the data scientists essentially found themselves serving SAS data to several different systems. In today's world, if you collect any large dataset in SAS, you should expect that at some point, you may need to serve all or some of it to another system.

For our first example of serving SAS data to a non-SAS system, let's assume we do not want to create a dynamic connection between another system and our SAS system. Let's instead imagine we are like the US state health departments and want to serve up a SAS data extract to a non-SAS system by extracting a dataset and transferring it to the other system. Serving SAS data to a non-SAS system can be seen as happening in two different steps:

1. The SAS data warehouse creates a data extract and transfers it to non-SAS system developers.

2. The non-SAS system conducts ETL to incorporate the SAS data into the non-SAS system.

With respect to these steps, the handling of **identifiers** in the data is of utmost importance. This is because if sensitive personal identifiers are included in the extract in the first step, they will be transferred to the new system as a part of the data transfer in the second step. While this might be necessary, it must be considered carefully, given the potential risks to privacy and confidentiality.

Since dealing with personal identifiers in SAS data transfers is an issue in serving SAS data to non-SAS systems, this section will cover this topic. In addition, as implied by the second step, the SAS data that is received into the non-SAS system will typically be reshaped into a new structure through ETL. One common warehouse structure, the **star schema**, will also be discussed in this section.

Implementing de-identification policies

As described in *Chapter 10*, *Considering the User Needs of SAS Data Warehouses*, **foreign keys** are indexes that can be used to connect datasets. For example, in the US datasets that we have been using in this book, we have been using a FIPS code as a foreign key for the state. We have been using examples that include the state of Florida, whose FIPS code is 12. In the same chapter, we also examined how to leverage a FIPS code in a dataset as a foreign key so that we could hook on other data about the state, such as salary data, or data about the number of hospitals.

In *Chapter 10*, *Considering the User Needs of SAS Data Warehouses*, it was also mentioned that foreign keys that are in the public domain, such as FIPS codes, are extremely important to keep in the data warehouse, as they allow analysts and warehouse managers to enrich information in the warehouse by hooking on other datasets using the foreign key. But special considerations have to be made when using **private or personal foreign keys**, such as a healthcare **medical records number (MRN)** or a government-assigned number, such as a **social security number (SSN)**, which is a personal identifier used in the US, as part of a SAS data warehouse.

On one hand, these personal foreign keys might be very useful to data warehouse users, in that a lot of data could be aggregated together about one person using the SSN as a primary key that connects to related datasets using SSN as a foreign key. It is easy to imagine why a healthcare data mart would like to store MRNs so that it could hook together all records for one patient based on the MRN.

But it would be risky to store such identified data in the data warehouse, especially once it is all connected together and accessible via the internet. It would be bad enough if a hacker were to steal one table including SSNs plus other columns of data about individuals represented in the data warehouse. Imagine the hacker were to query the warehouse and steal all the connected data about the different people represented in the warehouse! Not only would the hacker obtain their SSNs, but they would also know so much information about them from the other data that they could successfully impersonate them on the phone, thus thwarting safeguards against social engineering.

This is why personal identifiers should be crosswalked whenever possible in a data warehouse, while public domain identifiers (such as FIPS codes) can exist in their native format. Here, we will discuss how to crosswalk the identifiers of individuals represented in the data warehouse. That way, we can leverage the ability to hook records together for the same individual, while also not putting that individual's data at risk for hackers to breach and steal.

Making a crosswalk table

The first step in arranging to use personal identifiers as foreign keys but not store them in the live data warehouse is to create an identifier crosswalk table. A brief example is provided here:

SSN	MRN	WarehouseID
111-22-3333	123-456	24789
123-45-6789	234-567	24790
987-65-4321	345-678	24791
999-88-7777	456-789	24792

Table 11.1 – Example identifier crosswalk table

For brevity, this table only has four rows, representing four different individuals. Let's review the columns in this table:

- **SSN**: This is a personal identifier that we would not want to store in the live data warehouse. Imagine there is a person who was assigned the SSN 111-22-3333, as is represented in the first row. If this were a healthcare data mart, we may need their SSN in order to connect certain datasets that are not directly healthcare-related (such as information from death certificates). That is because those datasets will have an SSN, but not an MRN.

- **MRN**: Continuing with that hypothetical scenario, if this were a healthcare data mart, we would also need the person's MRN. So, the person with the SSN 111-22-3333 also has the MRN of 123-456. In medical records, this MRN would be present, but not the SSN. This table would allow us to hook all the data together about this person, regardless of whether the source data had an MRN or SSN.

- **WarehouseID**: This is a unique identifier generated by the warehouse. This table is the only table that decodes this. If after ETL, all of the MRNs and SSNs were replaced with this WarehouseID number, the SSNs and MRNs would not need to be stored in the live data. Yet, the data would still be connected at the individual level and would enable comparisons between individuals and groups of individuals. We could also re-identify the data in an emergency by using this crosswalk table.

Once this crosswalk table is designed conceptually, we can arrange analysts and workflows to use the crosswalk table during ETL to effectively de-identify and connect the data that will end up in the live warehouse.

Arranging analyst workflow and server setup

To process raw, identified data into the live data warehouse while replacing identifiers with a warehouse identifier (ID) along the way requires a specific physical setup of servers and analysts:

- **De-identification server**: If you are planning to de-identify data in the live warehouse, generally, the first set of transformation steps has to do with ETL to replace personal identifiers. These should be the minimal steps performed by one or a few developers on an independent server.

- **ETL server**: Most ETL transformation steps will come after the initial ETL to replace identifiers. These steps should be done on a separate server system in an environment conducive to large teams performing ETL remotely.

This can be seen more clearly in a pair of diagrams. First, this diagram provides an example of how to set up the de-identification server:

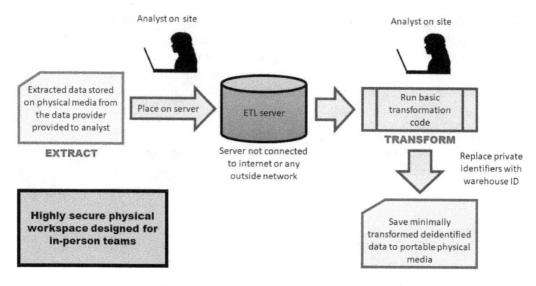

Figure 11.1 – De-identification server setup

Let's go over some features of the de-identification server setup:

- While most data warehouse functions can be performed remotely, de-identification is not one of them. The de-identification server should be placed in a physically safeguarded area. It should not have any connections to the internet, and analysts need to be on site in order to access the server.

- Physical media with raw, identified data from the data provider will also be in this location. This is another reason why this location needs to be physically safeguarded.

- The on-site analyst will place the raw data on the ETL server, and run basic transformation code that replaces private identifiers with the warehouse ID.

- The on-site analyst will finally place the minimally transformed de-identified data on portable physical media. These de-identified datasets can now be removed from this secure location and placed on the data warehouse network.

This next diagram shows how the de-identified data enters the data warehouse and highlights the functions of the data warehouse system:

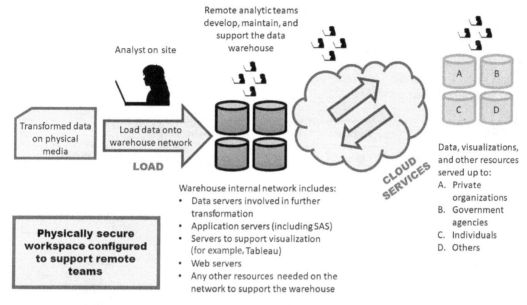

Figure 11.2 – Data warehouse functions

The diagram envisions a system that supports remote developers and analyst teams. The transformed de-identified data is loaded on this system, and developers connecting to the system on site or remotely conduct ETL and perform maintenance. Analysts connect to the system remotely and access resources.

Using this approach, although this system is on the internet and accessed remotely, no personal identifiers are stored. This ensures that if the data system were breached, identifiers could not be stolen.

Serving up a star schema

In our discussion of de-identifying data, we talked about how de-identified data undergoes ETL and is served up to users. However, we did not specify the final shape or structure of the data warehouse into which data is loaded, although using our discussion of keys, we seemed to be implying a relational structure. Many data warehouses use a typical **relational structure** comprising tables that are held together through primary and foreign keys. These could be diagrammed in an **entity-relationship diagram**, or **ERD** (also called a **schema**):

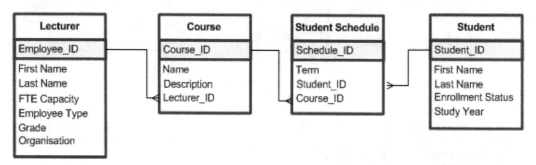

Figure 11.3 – Example ERD from a student registration database at a college. Image by Jon Cook, CC-BY-SA-3.0 (https://commons.wikimedia.org/wiki/File:Example_ERD2.png)

This shows a simple example of an ERD for a student registration database at a college. Each box in the diagram represents a data table with a title; the leftmost one is Lecturer. Each box lists the variables in the table, so in the Lecturer table, there are variables pertaining to the lecturer, including the **primary key** of Employee_ID. In the diagram, we see Employee_ID is used as a foreign key in the Course table, so the Course table can keep track of which lecturer in the Lecturer table is assigned to the course.

Notice the shape of the line connecting the Employee_ID primary key from the Lecturer table to the Lecturer_ID foreign key in the Course table. The endings of the line indicate that this is a **one-to-many relationship**, meaning for every lecturer in the Lecturer table, there are many courses related to them in the Course table. Data warehouses using this design may have very complex and extensive ERDs comprising many tables. Generally, managers will visualize parts or *neighborhoods* of their data warehouse with an ERD only showing tables associated with a particular function. This can help with troubleshooting.

Most production databases in business also share a relational structure. But because those databases are customized to support a business process – such as the college registration of students – the tables and relationships will all be designed to optimally function to support that process. In a data warehouse, we design a different relational structure that serves our business process, which is supporting analysts using transformed data for analyses. It is true that we perform a lot more ETL to load data in a data warehouse environment compared to the ETL we do when managing a production database in a business environment. But it is also true that many people who have worked in environments supporting business databases will find the warehouse work somewhat different but very familiar, given their ETL experiences in a business setting.

Although the relational structure is flexible, in that relational structures can be designed to support any business function, including data warehousing, there is another structure called the **star schema**, or sometimes called a **cube structure**, which is specifically optimized to be used in data warehousing. The conceptual process of serving up a star schema data structure from SAS data will be described here.

Comparing data warehouse structures conceptually

Relational databases, such as business databases with a SQL backend, are optimized for both editing data and querying data. That is because both of these activities are necessary for the business process. However, in a data warehouse, at least conceptually, the work being done by analyst users is only querying data, not editing it. Most, if not all, of the data editing would have been done during ETL by developers when in a totally different technological environment. Under this basic idea, the **star schema** was developed as an approach to restructuring warehouse data to completely optimize it for querying, with no consideration made for data editing.

To understand a star schema, it is first necessary to think logically about the context of a data mart in active use. In this book, for our coding examples, we have used core data from the health survey conducted by phone annually in the US called the **Behavioral Risk Factor Surveillance System (BRFSS)**. The BRFSS publishes yearly data files, and at the time of writing, the 2018 data file is the most recent one available. Since this book is being written in 2020, it means all of the data in 2018 – in all businesses, all countries, all organizations, and the BRFSS – have already happened. Though this may seem like a silly and obvious point, it is important to recognize that no new 2018 data will ever be produced from now on.

Because all of the data in 2018 has already happened, certain metrics that we have formed a habit of tracking in terms of changes over short periods of time – such as yesterday's maximum temperature outside, or yesterday's closing price of a stock we are following – have also already happened. For every time period in 2018, we know the maximum temperature outside in every location being measured, and the closing price daily of all stock being tracked.

Imagine you worked at a stock brokerage, and just after the markets closed, you wondered about the closing price of a certain stock. You would probably query your organization's business database that would look up that metric. The database you would be querying would probably have a table with all the companies with stock listed on it, and that might be related to a table about daily stock prices. This relational query would take some time to execute.

Now, imagine we create a data mart for your organization by taking copies of tables from your business database, and conducting ETL to load them into a warehouse structure optimized for querying. To grasp the concept, think of this overly simplistic scenario. Imagine that we took data about every stock your organization followed in 2018, and we created one huge table in an Excel spreadsheet. This one huge table represented every single day in 2018, every single company that had stock that your organization was following, and the closing price for each of these stocks every single day.

Since all of this information would be in one huge table, querying it would not take much time to execute. No table joins have to take place, so no primary or foreign keys would need to be used. No formulas need to be executed, and no complex criteria would need to be applied. These are the kinds of operations that bog SAS and other database software down when executing queries. Hence, in the scenario described, the query runs very fast, as it is essentially executing a lookup in a lookup table.

Of course, even the casual observer will realize that a huge table such as this – especially about all stocks, and all days in 2018 – will likely take up a huge amount of space. And we were just talking about the closing price of the stock. What about the average stock price for the day? We were also just talking about days. What about weeks, months, and quarters? These different time periods would definitely need their own tables. So, immediately, even using this simplistic scenario, we realize that if we want to optimize queries for all types of analysts using this method, we will need to make many different tables – which will take up even more space.

A star schema is not exactly what was depicted in our scenario, but it follows some of the principles described in the scenario:

- It contains really huge, long tables that are efficient to query.

- It minimizes the use of joins in queries.

- Formulas for time periods (such as maximum, average, and so on) are already calculated. Formulas are not run on the fly during queries. Instead, already-calculated metrics by different time periods are hardcoded, and just need to be looked up.

- Time periods are very important. Different calculated data is needed for different time periods. If a star schema wants to aggregate metrics by day, month, quarter, and year, it will need four sets of calculations – one for each time period.

To better illustrate the work that a SAS shop would need to do to reshape data into a star schema, we will consider an example scenario.

Designing a star schema warehouse

To conceptualize what an actual star schema might look like, let's design one together. We will use the hypothetical data warehouse described in *Chapter 10, Considering the User Needs of SAS Data Warehouses*, which includes data to facilitate the study of the **quality of life (QoL)** of US veterans after they leave the military. In the same chapter, we described including the following data in the warehouse:

- **Annual BRFSS health survey data**: The BRFSS dataset from the core survey is published online every year. The dataset includes the veteran status, information about health conditions and behaviors, and the FIPS code for the state.

- **Census data on salary**: Data from the US census on veteran salary by state is also available online and could be included in the data warehouse.

- **State hospital data**: Data about hospitals from the **American Hospital Directory (AHD)** in each state is available online, and could be included to shed light on the ease of healthcare access in different states.

One of the goals we stated for our hypothetical data warehouse about the QoL of US veterans was to be able to compare differences between states because states provide different levels of support to veterans. The datasets we plan to include all have a state variable. This suggests that analyst users of our data warehouse will likely do a lot of state-level queries. Because we believe this, we can start building our star schema assuming that analysts will be interested in retrieving state-level estimates.

Because we believe there will be a lot of state-level queries, let's create a **state fact**. A **fact** in a star schema represents a central table comprised entirely out of foreign keys (except for the primary key for the fact table itself). The fact table has a relationship with other tables, called **dimensions**. In our case, the different dimension tables related to the state fact table would need to reflect the data from the BRFSS table, the data from the census table, and the data from the AHD hospital table. Therefore, we know we will need three dimension tables, one for each of these datasets. Let's name the fact table `state_fact`, and the three dimension tables `brfss_dim` for the BRFSS data, `census_dim` for the data from the census table, and `ahd_dim` for the data from AHD.

> **IBM Cognos**
>
> A popular piece of software for serving up star schema data is called **IBM Cognos**. Theoretically, it is possible to use SAS to transform data into fact and dimension tables that are served up in the IBM Cognos application. However, although this has been done at times in practice, it is more common to use data software other than SAS to prepare star schema data for IBM Cognos. This is because ETL in SAS is such a time- and cost-intensive operation due to I/O challenges compared to data software such as Oracle SQL.
>
> If ETL is conducted in a SAS data warehouse, usually the data is also served up in SAS, as this avoids I/O issues. Because ETL in SQL and other programs may be faster and easier, it is not unusual for IBM Cognos implementations to conduct ETL in SQL and then load the resulting data directly into Cognos. For this reason, Cognos is actually seen as a competitor to SAS, when in actuality, both of them could be used on the same data project.

Now that we have designated our fact and dimension tables, we can visualize a star schema like this:

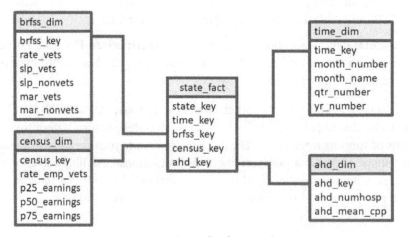

Figure 11.4 – Example of a star schema

Let's look at some features of this star schema design:

- Each of the boxes in the diagram represents a table in the star schema. The title of the table is at the top and is followed by a list of the names of the variables in the table.

- The state fact table, state_fact, only has keys in it for variables, as you can see by the names.

- In the time dimension table, which is time_dim, we see variables about month (month_name), quarter (qtr_number), and year (yr_number). But we do not see any variables about weeks, days, or shorter time periods.

- The census and AHD tables (census_dim and ahd_dim) are intended to provide summary statistics about geographic areas, such as statistics about veteran salary by state (variables ending in _earnings), and the number of hospitals per state (ahd_numhosp). Notice in the diagram that those types of metrics are listed among the variables in their tables.

- We realize that the BRFSS is a survey, so when we see a summary statistic in the BRFSS dimension table (brfss_dim) named rate_vets, it is necessary to realize that these are summary statistics that were calculated from the person-level BRFSS dataset.

Imagine this data was served up in an application designed for serving up star schemas, such as IBM Cognos. Let's specify a query we could use in the application based on the star schema we designed:

- Let's say we wanted to look at the percentage of veterans in the state of Florida, along with the number of hospitals in Florida.

- We would need to specify a time frame for our query. Given the structure of the database, we would only be able to see results by month, quarter, or year. Imagine we select a year.

Now that we have specified our query, the application would conduct the following functions:

- Using the state_fact table, it would filter rows that pertained to the state of *Florida* and the timeframe of *years*.

- Because the data on rates of veterans per state is in the brfss_dim table, and the number of hospitals is in the ahd_dim table, only those keys would be used in joins from the state_fact table. Since we do not want data from the census_dim table, that join would not be executed.

This simplistic example shows how the star schema can really speed up the processing time for queries. To be clear, SAS is not designed to serve up a star schema the way IBM Cognos is. However, SAS can conduct ETL on big data to create the huge tables that are served up in star schema software. This represents a scenario in which SAS may be used to transform data and serve it up to other systems, such as applications that provide access to query tools for star schemas.

Up to now, we have talked at length about conducting the ETL of data in SAS, and then providing it to other SAS or non-SAS systems by way of copying it to static, portable media, then loading it onto another network. But what if we want to make a dynamic connection between our SAS environment and a non-SAS data environment, such as SQL?

Connecting to non-SAS data storage

For much of SAS's early life, the SAS environment was very exclusive. Raw data brought into the environment largely comprised `*.txt` and `*.csv` files loaded to a SAS server from static media. The idea of connecting dynamically to another database housing data – in SAS or any other format – would have been a dream.

This dream became reality when SAS built in its **open database connectivity (ODBC)** connection capability. Using the ODBC, a programmer could run SAS, open an ODBC connection to another database in another environment, and query as well as extract data from this other environment into the SAS environment. The component that allows this capability is called **SAS/ACCESS** and can be added to Base SAS and SAS STAT when purchasing SAS. However, because this feature is not available in SAS University Edition, the examples we provide here will not be presented in runnable code.

Those of you who are familiar with SQL will recognize some of the functionality of the SAS ODBC. Remember, the main goal of opening an ODBC connection in SAS to another database is to be able to either run – in SQL terms – **select queries** on the data in the other database so that you can decide which data you want to extract, or **create table queries** that extract data from the native data system into your SAS environment. In SQL, a select query is one that provides the programmer with a customized view of either raw data or aggregated data but does not actually save the results of the query as a data table. In *Chapter 5*, *Managing Data Reporting in SAS*, we saw that SAS's PROC REPORT option produced output that was reminiscent of a select query in SQL.

> **Note**
>
> In this section, we will focus on using PROC SQL in SAS for connecting with non-SAS data storage and passing tables back and forth. However, there are other SAS approaches to doing this. As an example, in their SAS white paper *Transporting Data in the SAS Universe*, John Malloch and Jayne May Miller describe using PROC DBLOAD to transfer data from SAS to Microsoft SQL (see the link in the *Further reading* section). For the sake of brevity, this chapter will only cover methods of SAS data transfer using PROC SQL.

In SAS, when we run a report using PROC REPORT, we can filter the output based on placing criteria on the underlying data. However, we do not actually save a new dataset. If we run PROC REPORT on a dataset that contains data from three US states, but in PROC REPORT, we filter by only one of those states, the underlying SAS dataset still contains all the information about all three states. In SQL, there is a step in between viewing a filtered report and actually saving the results in the report as a new, filtered dataset. This step is called a **view**.

Understanding SQL views

In SQL, opening a connection to another remote SQL database is a common task. Also, in SQL, there is already a set of commands for creating views. Views are select queries that do not modify the underlying data but instead provide a view of it.

Imagine I am a SQL programmer at the US census, and I open an ODBC connection between my SQL database and another SQL database because I want to transfer data from their environment into my environment. I would complete the following steps:

1. **Establish a connection**: First, I would establish a connection to the remote SQL database environment, and through using my SQL interface and navigating, I would find the tables in the remote environment that I want to query. These are the tables that contain the data I want to transfer to my SQL environment.

2. **Create a select view to stage a query for data transfer**: Next, I would use my SQL interface to create **select query views** (similar to what we saw in the PROC REPORT demonstration) of the data in the other SQL system that essentially stage the results for the transfer to my SQL system. For example, imagine the remote SQL system included a table that had data from all US states, but because I work in Florida, I would only want to transfer the records that pertain to Florida. Also, let's say I only wanted a few columns from the source table in the remote system.

I could set up and save a view in my SQL interface that would show me dynamically what the dataset I'd be downloading actually would look like if I filtered the rows and the columns from the source data by my row and column criteria. Each time I connected and ran that view, it would show me the data in the source system in its current state. By looking at my intended data extract through views before actually downloading it, I could estimate how much effort it would be to copy the results of this view into my SQL environment and modify criteria in the view to suit my needs.

3. **Use a create table command to copy the data from the view from the remote SQL environment into a hardcoded table in my SQL environment**: Because views provide us with a view of the data without actually saving a dataset to disk, running a view takes very little processing time. That's why using SQL select queries to stage views in remote SQL systems before copying any data is preferable. The programmer limits the data transferred to what records are returned in their view, so once the optimal view is set up, a `create table` command is used to copy the actual data represented in the view from the remote system into a table in the local SQL system.

Although we will be using SAS to do this same operation, the steps that SQL programmers take to copy remote tables into their local system are roughly the same.

Using SAS to copy data from a remote data system

In terms of connecting to a remote system and transferring tables, the SAS programmer can do all of the things that the SQL programmer can do if they use `PROC SQL`. In fact, these steps are described in detail in Jeff Magouirk's SAS white paper, *Using PROC SQL and ODBC to Manage Data Outside of SAS* (available in the *Further reading* section):

1. **Configure an ODBC connection**: In SAS, you have to actually set up a connection so that it can be configured and parameters can be set for the environment to which you will be connecting.

2. **Create libnames that refer to the remote environment so that you can refer to remote tables in your code**: So far in this book, we have used `LIBNAME` to create an alias to directories on a server. `LIBNAME` can also be used to create an alias to a location in a remote environment. That way, if the remote server has a SQL database with a table in it, we can use the libname to refer to the specific table.

3. **Create registrations**: In the white paper, how to connect a user to a SQL database in SAS to retrieve tables is described, and therefore, it only covers the first two steps. However, if you are setting up connections for multiple SAS users on a server, you will need to do the third step of creating registrations. Registering a user, an ODBC server, and an ODBC library are tasks that are done in SAS. These efforts are mainly made to help SAS work with the remote data store, and are explained in the server SAS support documentation.

Once the connection between SAS and the other database environment is set up, the connection can be used to access the data in the remote environment. To do that, the programmer would take the following steps:

1. **Connect to the remote environment using ODBC**: The first step in doing the comparable operation in SAS compared to SQL using the ODBC would start with establishing an ODBC connection. Although ODBC connections can be established between SAS and many other data environments, for the purpose of demonstration, let's assume we are using a SAS ODBC connection to connect to a remote SQL server. During this task, you will create a libname that refers to the remote environment. For the purpose of this example, let's say we use the `other` libname for *the other database*.

2. **Use the libname established in the connection to refer to tables in the remote database environment**: Imagine we were connecting to a healthcare system database. We already established that the `libname` for this environment is `other`. Let's say the remote database has a table in it called `facilities`. We can query this table by referring to it using the syntax for libnames – in SAS code, that would be `other.facilities`.

3. **Use PROC SQL or data step commands to copy data from the remote environment into the SAS environment**: Although data step commands can be used, `PROC SQL` commands tend to be easier to use because they were designed specifically for processes such as copying tables from data stores across networks.

Imagine we wanted to copy the entire `facilities` table into our SAS environment into the `WORK` directory (keeping the name of the table as `facilities`). After establishing the ODBC connection and mapping the remote environment to the `other` libname, we could copy the entire `facilities` table from the database we mapped to the `other` libname into our `WORK` directory with the following data step:

```
data work.facilities;
    set other.facilities;
RUN;
```

Using PROC SQL, the same operation could be achieved with the following code:

```
PROC SQL;
    CREATE TABLE work.facilities as
        SELECT * FROM other.facilities;
QUIT;
```

In a healthcare database, a facilities table might be large. However, a table containing patients would be much, much larger. This would also be true of a table keeping track of billing. In those cases, we would want to be more selective about the data we transferred. This activity can be aided by using PROC SQL views.

Leveraging PROC SQL views for data transfer

Like SQL, PROC SQL in SAS has the ability to make views that can help with data transfer from a non-SAS system into a SAS system. As covered in the SAS white paper *Using SAS Views and SQL Views* by Lynn Palmer (link in the *Further reading* section), a **view** is like a stored query, so it contains no data. Instead, saving a view could be seen as storing an algorithm or formula. This saved algorithm or formula can be applied to an underlying table that will filter in data dynamically according to criteria when it is run.

Because views represent coding for an algorithm or formula (which is essentially a **query**), but not the actual data that would be filtered in by the query, storing a view on a SAS server takes up very little space compared to saving an actual data table, such as the hardcoded results of a query. Even so, results from views that are executed on the fly can act like tables, as they can be joined to or merged with tables and other views on the fly.

This is why storing PROC SQL views in a SAS data warehouse is a great way to organize regular data transfers from external databases in different formats into the SAS warehouse. Imagine you worked at a data warehouse that wanted to include monthly data collected from Florida healthcare facilities that are included in the datasets served up by the **Agency for Health Care Administration (AHCA)** in Florida (see the link in the *Further reading* section).

Let's assume a particular scenario. Say AHCA maintains a data warehouse in SQL, and your data warehouse is in SAS. Let's also assume you obtained permission from AHCA to connect to their system, and each month, your job is to log in to AHCA's system and transfer in the previous month's data of billing records from healthcare facilities. If it were February 2018, you would be logging in to AHCA and trying to transfer the data from January 2018.

In such a scenario, using an **ODBC connection**, PROC SQL views, and **SAS macros**, you could do the following set of tasks to transfer a designated month's worth of data from AHCA's non-SAS storage into your SAS environment:

1. Using the SAS/ACCESS component, run SAS code to establish an ODBC connection to the AHCA warehouse so that you can query tables inside AHCA's SQL database.

2. Call up a local SAS macro that allows you to set the value of variables to designate a specific time period. As discussed in *Chapter 8, Using Macros to Automate ETL in SAS*, variables can be established and their values set with the %LET command. For example, a xfer_month variable could be set to the value of the month of the data looking to be transferred (January would be 1), and xfer_year could be set to the year of the data to be transferred (such as 2018).

3. Once the SAS macro accepts the variable values, it should run the stored PROC SQL view on tables in the remote database at AHCA using the values of variables in the query criteria under the WHERE clauses. When querying a table in the AHCA database that includes dates, the month and year of the record dates could be filtered by criteria expressed by these %LET macro variables referred to in a WHERE clause in a PROC SQL command.

4. Using a CREATE TABLE command from PROC SQL on the results of the dynamic view can be used to copy just the data filtered in by the view into a data table in the local SAS data warehouse. We saw this process demonstrated before, but in that case, we copied an entire table from the SQL database into our SAS environment, not the results of a view.

Let's write some code that would accomplish this in our scenario. We will start by creating a view in our SAS database that will be applied to the other database. Let's pick up after we run code to set up an ODBC connection to the other database and set the libname to other. Next, we will set up a view that will save in our WORK directory, and call it recent_billing, to query the remote table named billing. And remember – we are setting macro variables for the month and year of the billing data we want to transfer, called xfer_month and xfer_year, respectively:

```
%LET xfer_month = 1;
%LET xfer_year = 2018;
PROC SQL;
CREATE VIEW AS WORK.recent_billing
    SELECT *
```

```
    FROM other.billing
    WHERE billing_month = &xfer_month and
    billing_year = &xfer_year;
QUIT;
```

Let's go over this code:

1. First, using %LET, we set the xfer_month macro variable to 1 for January and xfer_year to 2018. We will use these macro variables later, in the WHERE clause of the PROC SQL view we create.

2. Next, we call up PROC SQL. Notice that the PROC SQL part of our code starts with PROC SQL and ends with QUIT.

3. In our PROC SQL code, we create a view in our WORK directory called recent_billing. We do this by selecting all columns (SELECT *) from the billing table in the remote database called other.billing.

4. Notice how we express the WHERE criteria. Let's assume the other.billing table has two different variables: billing_month, which indicates a numeric month for the bill (such as 1 for January), and billing_year, which indicates the year (such as 2018). We can tell by the use of the ampersand (&) that when we set billing_month = &xfer_month and billing_year = &xfer_year, we are telling SAS to use the values from the macro variables we set in the %LET statement.

If we ran this code and created this view called recent_billing, it would allow us to set our macro variables and look at the data we want to transfer. But we can also use the view in the next step when we actually transfer the data represented in the view by referring to the view by name in subsequent PROC SQL code. Here is the code referring to the view we just made, named recent_billing, which we could use to copy the results of the view from the remote database to our SAS data warehouse. In this example, the table we create in our SAS data warehouse is placed in the WORK directory and is called billing_2018_01:

```
PROC SQL;
CREATE TABLE AS WORK.billing_2018_01;
    SELECT *
    FROM WORK.recent_billing;
QUIT;
```

As a result of running this code, the `2018_01_billing` table would be saved in the `WORK` directory and would represent the results of running the `WORK.recent_billing` view that had the macro variables set with the `%LET` statement to create a filter. Even so, tables extracted from a remote database using this method could be seen as raw data extracts. Though the use of `PROC SQL` views allows us to customize the transferred tables to some degree, they still will need to undergo ETL for incorporation into the SAS warehouse structure.

Data step views

Although this section focuses on creating views in `PROC SQL`, it is technically possible to also create views using SAS data steps. These views have several important drawbacks, including being more inefficient to run than data steps that manipulate hardcoded tables, not being able to support `PROC SORT`, and not being able to be saved outside of the SAS session. Therefore, `PROC SQL` views, which are extremely useful in data warehousing in SAS, are the focus of this section. For more information about data step and other types of SAS views, see the previously mentioned SAS white paper by Lynn Palmer in the *Further reading* section.

So far, we have seen an overview of how the ODBC and `PROC SQL` can be used from within a SAS environment to connect to a remote SQL data warehouse and transfer data into the SAS environment. Additionally, we have gone over how the ODBC and `PROC SQL` can be used to export data from a SAS data warehouse to non-SAS storage.

What we have not done is looked carefully at how to connect using the ODBC. We have also not considered how to export data out of our SAS environment to another environment. Both of these topics will be covered in the next section.

Exporting SAS data to non-SAS data storage

In their SAS white paper, *Transporting Data in the SAS Universe*, John Malloch and Jayne May Miller cover different scenarios where SAS tools can be used for transporting data (see the link in the *Further reading* section). Since we just went over the example of extracting data from a remote SQL database into SAS, let's cover the opposite scenario, where we want to connect to a remote SQL database and export data from SAS into the SQL database.

In their white paper, the authors cover two methods for moving data from SAS to Microsoft SQL: one using `PROC DBLOAD` and one using `PROC SQL`. Since the authors mention that SAS is moving away from `PROC DBLOAD`, we will only cover the `PROC SQL` scenario.

In their example, the authors use SAS PROC SQL code to connect to a remote SQL database. Next, they try to create a table in the remote SQL database using a CREATE TABLE command based on the SAS tables in their own database. However, as they note in their white paper, once you create the table in the remote SQL database with PROC SQL, you cannot edit the table structure with the ALTER TABLE command. Instead, you have to use an EXECUTE command, which invokes the **SAS/ACCESS pass-through facility**.

Let's take this opportunity to look into the details of setting up an ODBC connection, as well as to practice by making up some code based on the AHCA scenario. Imagine the remote AHCA database is named AHCA_database, and we set a libname to refer to it as other:

```
LIBNAME other ODBC NOPROMPT="DSN=AHCA_database";
```

Now, let's use PROC SQL to export a SAS data table named claims from our environment's WORK directory to AHCA's using the CREATE TABLE command. To demonstrate the SAS/ACCESS pass-through facility, once we export it, let's also use ALTER TABLE to add a blank, 10-character column to claims called blank_column:

```
PROC SQL;
    CONNECT TO ODBC (DATABASE= AHCA_database);
    CREATE TABLE other.claims as SELECT * FROM WORK.claims;
    EXECUTE (ALTER TABLE other.claims ALTER COLUMN blank_
    column CHAR(10)) BY ODBC;
DISCONNECT FROM ODBC;
```

Let's go through this imaginary code:

1. First, PROC SQL is invoked. Next, a connection is made using the ODBC. The ODBC could connect to a number of remote databases, so DATABASE= AHCA_database specifies the database to which the SAS programmer is connecting.

2. The next line with the CREATE TABLE command creates a table in the remote SQL database, which is mapped to the other libname. Therefore, telling SAS to create a table named other.claims is a way of telling SAS to create a table named claims in the remote database mapped to other.

3. The part of the CREATE TABLE command that says SELECT * FROM WORK. claims tells SAS that the table that is being created in the remote database is being drawn from the claims table in the WORK directory of the local SAS environment.

4. In the next line, an example of using the SAS/ACCESS pass-through facility is provided. This line starts with an EXECUTE command, which is a part of the SAS/ACCESS pass-through facility language. Then, in parentheses, language reminiscent of PROC SQL language is used to create an ALTER TABLE command that adds a 10-character blank column to the remote claims table named blank_column.

5. Notice that the EXECUTE command in parentheses is followed with BY ODBC, and this line is followed by a line telling SAS to disconnect from the remote database through DISCONNECT FROM ODBC.

This example shows that it is possible to export tables from a SAS database into a SQL database. But in general, it is more common for programmers to connect to a remote database to extract data into the local SAS database, rather than to push data out from the local databases to a remote database.

Up to now, we have discussed connections between SAS and non-SAS environments for the purpose of moving data back and forth from different environments. Before ending this chapter, let's discuss transforming and transporting SAS data to non-SAS environments and applications for the sole purpose of reporting.

Innovations in integrating SAS in reporting functions

When SAS was initially designed, network environments had to be enclosed. All functions – storage, ETL, and reporting – had to be done within the environment. Therefore, even though SAS was primarily focused on data management, it developed tools for reporting that could work within the SAS environment. However, those tools were not sophisticated, because SAS's objective was focused on data management, not data reporting.

As time went on, connectivity between systems became possible – and then standard – through the internet. This provided the opportunity for SAS to be used for exactly what it is good at – which is data storage and management – while having other products that are optimized for reporting be used to report SAS data. The next chapter, *Chapter 12, Using the ODS for Visualization in SAS*, will focus on modern uses of SAS tools for reporting and visualization.

Summary

In this chapter, we went over the different ways to serve SAS data to other systems, and different ways to connect SAS data warehouse environments to non-SAS storage. First, we discussed preparing data extracts to be served up to another environment asynchronously. We went over de-identifying data and implementing data de-identification in the warehouse workflow, and what would be needed to serve up data in a star schema. Next, we talked about connecting a SAS data warehouse to non-SAS data storage using SAS/ACCESS and an ODBC connection. We discussed how views work in PROC SQL, and how SAS and PROC SQL language can be used for data transfer. We looked at copying tables into a SAS database from a SQL database this way, as well as exporting SAS tables from our warehouse into a SQL database. Finally, we talked about how data transfer paradigms differ when being done for reporting purposes.

If you work at a data warehouse, you will likely be asked to do tasks such as the ones described in this chapter. You might be asked to prepare and transfer a SAS data extract to an external data warehouse, or you might be asked to set up an ODBC connection with a SQL server and remotely extract data. The exercises we went over showed you how you would approach these tasks depending on the circumstances. For regular data transfers from a SQL server, you might choose to use PROC SQL views and macro variables. But for serving up a star schema, you may simply conduct ETL to provide fact and dimension tables to another warehouse environment that will serve up the results in an application. Knowing the information in this chapter will empower you to consider these circumstances and make informed decisions.

The exercises we did in this chapter had to do with transferring data back and forth between live and asynchronous SAS and non-SAS environments. In contrast, the approach to the visualization of SAS data, both within SAS and outside of SAS, generally does not require this kind of data transfer. If SAS data is to be visualized in an external program, the SAS data needs to undergo ETL in SAS in order to be properly served up to the visualization software. This will be described in detail in the next chapter.

Questions

1. Why should de-identification take place on a server, rather than connected to the internet?

2. Why is it important to support remote work for analysts and developers?

3. Why do data warehouses configured in a star schema need a lot of space for data storage?

4. What does an ODBC connection using SAS/ACCESS allow the SAS user to do?

5. What are the advantages of using `PROC SQL` views to stage data transfers from non-SAS databases to SAS data warehouses?

6. SAS offers the use of an ODBC connection to connect to remote database systems in other formats (such as SQL), which we went over in this chapter. Under what circumstances might it be easier to transfer data into the SAS environment from a remote environment in a more manual way, without using an ODBC connection?

7. How could `PROC SQL` be used to export a dataset out of a SAS environment into a SQL database?

Further reading

- The *Understanding Star Schemas* website from the Oracle Data Mart Builder Population Guide, available here: `http://gkmc.utah.edu/ebis_class/2003s/Oracle/DMB26/A73318/schemas.htm`

- The SAS white paper, *Transporting Data in the SAS Universe*, by John Malloch and Jayne May Miller, available here: `https://www.lexjansen.com/nesug/nesug04/po/po06.pdf`

- The SAS white paper, *Using Proc SQL and ODBC to Manage Data outside of SAS*, by Jeff Magouirk, available here: `https://www.lexjansen.com/wuss/2004/data_warehousing/c_dwdb_using_proc_sql_and_od.pdf`

- The SAS white paper, *Using SAS Views and SQL Views*, by Lynn Palmer, available here: `https://www.lexjansen.com/wuss/2003/DatabaseManagement/c-using_sas_views_and_sql_views.pdf`

- Florida AHCA Office of Data Dissemination and Transparency: `https://ahca.myflorida.com/schs/DataD/DataD.shtml`

12
Using the ODS for Visualization in SAS

This chapter will focus on the basics of using the SAS **output delivery system (ODS)** for data visualization, both in print and on the web. Now, the ODS is embedded in **SAS reporting functions** and does not need to be called up separately.

This chapter covers how complex legacy code, containing **arrays**, **macro variables**, and **macros**, has been built and maintained for SAS reporting in print format. This SAS code is often continuously maintained, so migrating away from the SAS platform for this type of reporting poses a challenge. In contrast, SAS reporting to the web, which also deploys the ODS, poses its own challenges with **input/output (I/O)** and other limitations. We will cover two tools – the **SAS Enterprise Guide** and **SAS Viya** – that can help with these challenges. Finally, we will cover using other applications, such as **R** and **Tableau**, to visualize SAS data, and considerations that SAS warehouse developers must make when planning the direction of print and **web reporting** for their SAS data warehouse.

We will cover the following topics in this chapter:

- How reporting functions can be manipulated using SAS code
- How menus and features are used in the SAS Enterprise Guide and SAS Viya
- How to use the SAS Enterprise Guide and SAS Viya as interfaces to manage SAS visualizations and improve reporting to the web
- How to have SAS data visualized in software external to SAS, such as R and Tableau

Technical requirements

You will need the dataset in `*.sas7bdat` format used as a demonstration in this chapter, which is available on GitHub: `https://github.com/PacktPublishing/Mastering-SAS-Programming-for-Data-Warehousing/tree/master/Chapter%2012/Data`.

The code bundle for this chapter is available on GitHub here: `https://github.com/PacktPublishing/Mastering-SAS-Programming-for-Data-Warehousing/tree/master/Chapter%2012`.

The basics of using the ODS for data visualization

As was discussed in *Chapter 5, Managing Data Reporting in SAS*, SAS created the ODS in a very early version of SAS to simply output graphics. Since then, the ODS has become the basic function within SAS that enables graphical output from SAS. From the time the ODS was invented until about 2000, graphics were generally seen as **print graphics**, in that they were not anticipated to be output to the web or to be interactive. In fact, reports in general, regardless of whether they contained only tables, or graphics and tables together, were formatted for printing.

Even if those reports were never actually printed, they were consumed in a printed context, such as through a person reading a PDF or Word document online. We saw in *Chapter 5, Managing Data Reporting in SAS*, how the ODS could be used with plotting PROCs to output graphics files that could be saved in graphics format (for example, `*.jpg`). The approaches behind reporting directly to the web from a data store were just being developed in the early years of the web. At that time, SAS's ODS had been around for a long time, and many SAS shops had spent years developing intricate code for running extensive printed reports.

Theoretically, we can carefully manipulate code for SAS PROCs for graphics, or tabular formatting PROCs such as `PROC TABULATE`, to produce an ideal graphic. Alternatively (and perhaps more easily), we can output the minimum required visual information into a graphic using SAS, then use outside graphic editing software such as **Adobe Photoshop** or **Microsoft PowerPoint** to do the final editing. Therefore, some of these print approaches continue today with the addition of using modern software to finish the report.

Historically, to optimize I/O, SAS reports drew from data stores in SAS format. But as time went on, even SAS, a company that made its mark storing big data, started to struggle with storing big data. Although this may sound unbelievable, in recent years, it has begun to become clear that a data storage area (larger than SAS could handle) was needed to store huge datasets temporarily so that SAS could be used to process these huge datasets into a SAS data warehouse.

Generally, companies who have trouble storing big data solve this problem by turning to *cloud storage*. For SAS, this meant teaming up with the company **Snowflake**, which provides cloud data storage. Snowflake data stores can be accessed through SAS using an **open database connection** (**ODBC**) and the **SAS/ACCESS** component. We will talk about the development of SAS reporting code and using this code to report from Snowflake in this section.

Using macros in reporting

To better understand the challenges we see today in developing SAS code to report to both print and the web, it is helpful to examine how reporting in SAS has evolved over the years as SAS has adapted to keep up with the times. In *Chapter 5*, *Managing Data Reporting in SAS*, we went over the basics of using the ODS in reporting. We also talked about how as SAS evolved over the 1980s and 1990s, features were built into the software to improve the visual appeal of the output. PROC TABULATE was created to provide more attractive table output, and new sets of graphics PROCs were developed primarily to make use of the new capabilities of the ODS to output more communicative data visualizations.

After that, in *Chapter 6*, *Standardizing Coding Using SAS Arrays*, we talked about using SAS **arrays** in ETL, and in *Chapter 8*, *Using Macros to Automate ETL in SAS*, we discussed automating ETL using **macro variables** and **macros**. It is important to realize that all three of these features – arrays, macro variables, and macros – were also implemented to a great degree in reporting functions on datasets that would come directly out of a SAS data warehouse. Remember, when SAS data is kept in the SAS environment, the I/O is most efficient, so it was most efficient to report directly from SAS data using SAS reporting tools in a SAS environment.

As we talked about in *Chapter 8*, *Using Macros to Automate ETL in SAS*, as data anomalies would appear in the new data we would load, we would just add **conditions** to our macros. This would keep our code updated because then our macros would take that new anomaly into account, and keep running properly into the future. This approach made it so that the data warehouse could keep running and meeting its demands, and analysts could keep the code and reports up to date. However, this approach also had the unintended consequence of growing extremely complicated code over the years.

Over time, other data visualization software rose to prominence, such as **R** and **Tableau**, which we will talk about in the *Using SAS and R for visualizations* section. This meant that by the time this visualization software was available for SAS users to use on SAS data, SAS reporting code had become so complicated that there was a question about the cost benefit of rebuilding the reporting code into a new program.

Let's consider an analyst who knows both SAS and R who identifies a graph currently coming out of SAS reporting code that would look much more polished, modern, and professional in R. In a practical sense, it might take a lot of time to rebuild the code from SAS to R. Even if the analyst succeeds in this task, the analyst would then need to conduct validation studies to ensure that the reporting results coming out of R are identical to the ones that come out of SAS. Depending on the importance of the graphic, and considering the level of complication of SAS code, rebuilding the graphic from SAS into R might not be worth the hassle.

Let's do an exercise to see for ourselves how this can happen by reviewing some code provided in the SAS white paper *SAS Macro Programming for Beginners*, by Susan J. Slaughter and Lora D. Delwiche (available under the *Further reading* section), and writing our own based on what we observe. Even though this SAS white paper is about macros, you will notice that the paper focuses on talking about reports, not ETL. This is testimony to how widely macros have been used in reporting functions in SAS.

Let's illustrate using a simple example of hospitals in the US. If we were managing a healthcare data warehouse, we might want to include a table in our warehouse that was basically a list of all the states in the US, and a few columns of information about the hospitals in the state. A demonstration dataset named Chap12_1 is in *.sas7bdat format and includes public data about US hospitals taken from the internet. This should be placed in the directory mapped to LIBNAME X, and the following code run:

```
LIBNAME X "/folders/myfolders/X";
RUN;
PROC CONTENTS data=X.chap12_1 VARNUM;
RUN;
```

The code maps `LIBNAME X` to the directory specified and runs `PROC CONTENTS` on the `Chap12_1` dataset. To get a feel for the dataset, instead of looking at the `PROC CONTENTS` output, let's instead look at the first three rows of data in the dataset:

State_Desc	Number_Hospitals	Staffed_Beds
AK - Alaska	10	1235
AL - Alabama	89	15335
AR - Arkansas	52	7872

Table 12.1 – Top three rows of the Chap12_1 dataset

You may recognize that we used this dataset in *Chapter 10, Considering the User Needs of SAS Data Warehouses*, when we demonstrated crosswalks. Let's go over the columns:

- `State_Desc`: This is a description variable for the state. It is formatted by having the abbreviation (for example, `AK`) followed by a dash, and then the state name (for example, `Alaska`).

- `Number_Hospitals`: This is the number of hospitals in the state. For Alaska, it is `10`.

- `Staffed_Beds`: This is the number of staffed beds combined from all the hospitals in the state. So, we see that across Alaska's `10` hospitals, there is a total of `1235` beds available.

A **staffed bed** refers to a bed in a hospital with staffing assigned. This can change as hospitals add and remove resources. Therefore, it is reasonable to imagine running a health data warehouse and regularly loading new data over time about hospitals with staffed beds. The total staffed beds in a region – such as a state, county, or city – is seen as a measure of healthcare capacity in that region. Therefore, leaders may look to healthcare data warehouses to answer questions about current healthcare capacity in a region.

It is also common practice to run a series of reports after each ETL data load to review information about the data loaded, and how the load went. It is not unusual for the SAS developer to spend the first portion of their effort actually launching the ETL code, and then spend the next portion of their effort poring over reports about the processing to verify that the data was loaded correctly. Let's imagine we are programming a SAS report that is to come out among many reports that run after a data load to help the SAS developer ensure that the data loaded correctly.

Let's say that after ETL, we had a series of reports that went through the new data and reported the region with the maximum number of staffed hospital beds at each level:

1. The first set of code would run one report that would report the state with the most staffed beds from all the hospitals in the state combined. That is because **state** is our highest level in a region.

2. The next code would run a series of reports – one for each state. For each state, it would report the county with the most beds. This would be for the second level of the region, which is **county**.

3. If there was a smaller level of the region within the county for which to run the reports, this set of reports would come next.

If you are familiar with the material in the previous chapters, then you can appreciate how a description of these three types of reports implies how conditions, do loops, and macro variables could be used to accomplish these tasks in SAS. For brevity, let's examine doing only the first step, which is producing the report about the state with the most staffed hospital beds. Let's develop code that produces a report about the state with the most staffed beds in the Chap12_1 dataset:

```
PROC SORT DATA = X.Chap12_1;
    BY DESCENDING Staffed_Beds;
RUN;
DATA _NULL_;
    SET X.Chap12_1;
    IF _N_ = 1 THEN
    CALL SYMPUT("biggest", State_Desc);
    ELSE STOP;
RUN;
PROC PRINT DATA = X.Chap12_1 NOOBS;
    WHERE State_Desc = "&biggest";
    TITLE "State &biggest has the most hospital beds.";
RUN;
```

Let's unpack this code:

1. The first step is to run PROC SORT on the Chap12_1 dataset. This sorts the dataset by Staffed_Beds using the DESCENDING option. This rearranges the dataset to make it so that the first row now has the maximum value of Staffed_Beds.

2. The next step is to use _NULL_ to create and assign a value to a macro variable. As we discussed previously, in *Chapter 9, Debugging and Troubleshooting in SAS*, _NULL_ can be used for a variety of purposes, and this is one of them. We are doing a data step and appear to be making a dataset named _NULL_, but in reality, there is no dataset created to save in _NULL_. This maneuver is only used as a way to deploy data step code to set the value of a macro variable.

3. The code in the data step starts by identifying the row in Chap12_1 with the maximum value of Staffed_Beds. Because we sorted the dataset such that the first row now has the maximum value of Staffed_Beds, our intention is to take the first row and report on it. This explains the if statement in the data step. IF _N_ = 1 means if the observation number, or row number, which is indicated by _N_, is 1, then execute the code that follows. Therefore, the code that comes after IF _N_ = 1 will only be applied to the first record in the sorted dataset, which has the most staffed beds.

4. CALL SYMPUT is a command from the macro programming language that assigns a value produced in a data step to a macro variable. In this case, we are using CALL SYMPUT to make a macro variable called biggest, and assign it the same value as State_Desc. Remember, it only has one value of State_Desc to work with, and that is the value of the first record in the sorted dataset. That was ensured by the previous code of IF _N_ = 1. So, therefore, when we use CALL SYMPUT at this point in the code, SAS is going to assign the value of State_Desc from the top row indicated by _N_ = 1.

5. After the line with CALL SYMPUT, the code says ELSE STOP. This ends the loop started by the if statement. Of course, because IF _N_ = 1 selects only the first record, this loop ends after one record is processed, so this code runs very fast.

6. Notice how we immediately run PROC PRINT after this so that we can use the value we just assigned to our new macro variable, biggest. Notice that we are calling PROC PRINT on the original dataset, Chap12_1. By including the NOOBS option in the PROC PRINT statement, we are telling SAS not to print the number of observations on the output.

7. The next line is a WHERE clause that filters Chap12_1 by one criterion, which is State_Desc = "&biggest". This means that we are asking SAS to print all the records in the Chap12_1 dataset where the value of State_Desc matches the value of State_Desc stored in the biggest macro variable. By doing this, we select the top record with the most staffed beds, which we were able to dynamically select through creating the biggest macro variable.

8. In the next line, we add a title command and reuse our biggest macro variable in the title. Where we place the &biggest code, we will see SAS print the value of biggest on the output, which will be one of the values of State_Desc in the Chap12_1 dataset.

Let's run this code and view the output:

State CA - California has the most hospital beds.

State_Desc	Number_Hospitals	Staffed_Beds
CA - California	344	74724

Figure 12.1 – Report output from PROC PRINT

As we can see, in the Chap12_1 dataset, California is the state with the most staffed beds, with a grand total of 74 724 staffed beds (from a total of 344 hospitals). Notice how the value reported on the State_Desc output, which is CA - California, is reprinted in the title because this was the value stored in our macro variable, biggest.

More about CALL SYMPUT

When we discussed macros in *Chapter 8, Using Macros to Automate ETL in SAS*, we did not delve deeply into the topic. What the CALL SYMPUT command brings to mind is the fact that analysts and users are not the only sources of setting values in macro variables. There can also be the results that come out of data steps and PROCs that can be turned around and made into macro variables during the processing, as we did in the example code.

Thinking of SQL, when users set values to macro variables in SAS and then launch code, it is like what happens when a user launches a *stored procedure in SQL*. But when CALL SYMPUT is used in SAS, this is like using a *trigger in SQL*. Yunchao (Susan) Tian provides three examples of using CALL SYMPUT in complex macros in her SAS white paper, *The Power of CALL SYMPUT – DATA Step Interface by Examples*, which can provide further instructions (link available in the *Further reading* section).

Let's reflect on a few points about this short reporting exercise:

- This short and simple report required a lot of code. It required at least a data step, a PROC, and a macro variable.

- Even this short snippet of code was complicated, which means it will be hard to troubleshoot. Imagine that one time this report was run, the result looked illogical. We already know that the state of Alaska has only 10 hospitals. If Alaska showed up as the state with the biggest number of staffed beds on this report, we would assume it was an error. This would lead us to the need to troubleshoot this problem by considering this complicated code against this simple underlying dataset.

- We demonstrated the code at the top regional level of reporting, which is *state*. We did not demonstrate the next level, which would be county. However, we can imagine that running this report at the county level would add even more complex code after the code we already ran. We would probably have to use arrays to store the county codes, and we would likely have to use macro variables, macros, and do loops to ensure our reports come out accurately.

- If we used this code for reporting in a data warehouse that continued to load data regularly over time, eventually, we would encounter anomalies in new data that disrupt ETL. To solve this, we would add conditions to our macros, thus enabling the code to run, but making it even longer and more complicated.

The purpose of this demonstration was to illustrate how complicated SAS reporting code can become relatively quickly, and how overwhelming it can be to deconstruct it and reconstruct it in another program. However, this demonstration also serves to illustrate that once a complicated set of reports is built in SAS, it would be nice to be able to point the same code at other data that is stored same structure. It is also possible that other SAS data warehouses storing similar state hospital data would want to use this reporting code on their data.

This brings us to the next topic, which is the interoperability of SAS in the cloud. Specifically, let's look at using SAS to report from data stored in the cloud, in Snowflake.

Connecting to data in Snowflake

The early 2000s began a global expansion of databases, data warehouses, and data lakes, and this led to many SAS servers being set up or expanded in this era. These were usually physical servers set up in a server room on site at the organization hosting the warehouse:

Figure 12.2 – Example of a server room. As opposed to cloud servers, these servers are physically on site, and maintain data in this secure physical location. Photograph by Acirmandello, CC BY-SA 4.0 (https://commons.wikimedia.org/wiki/File:GIP_Servers.jpg)

As shown in the preceding photo, these **server rooms** are secure physical locations. These servers can hold any kind of data or applications, so a server room like this could be set up for a SAS data warehouse. However, the servers and the server room are finite, in that the room can only hold so much hardware, which means the room itself can only hold a finite amount of data. Many of the servers in a room such as this are also **application servers**, running applications, rather than **data servers**, storing data. This means that as time went on, many SAS data warehouses *filled up* with data and had nowhere to expand.

For anyone running a data warehouse, when the warehouse fills up, the first inclination is to find some way to use cloud storage to help. Usually, the first step is setting up the cloud storage and establishing a connection so we can use it as a **staging area**. Imagine we ran out of room in our SAS data warehouse, and we wanted to run the report we just coded. We might not have enough room in the actual warehouse to do these manipulations. But we could connect to cloud storage such as Snowflake, use a data step to move the data we want to work with to Snowflake temporarily, then run our reports off of that.

Our I/O would likely suffer from all the moving around, but if we are only extracting a small amount of data from our warehouse and reporting off of that, it may not be so bad. Admittedly, there would be little alternative if the data warehouse were actually full.

Let's imagine our SAS data warehouse was full, and so we wanted to connect to Snowflake, move some of our data over to Snowflake from our warehouse, and run the report we just programmed. Our first step would be to set up an ODBC connection to our Snowflake database. Here is some example code of how this could happen:

```
LIBNAME snow SASIOSNF
      server = "saspartner.snowflakecomputing.com"
      user = mwahi
      password = TAM813fla
      schema = HCDW
      database = USERS_DB
      warehouse = USERS_WH
      dbcommit = 10000
      autocommit = no
      readbuff = 30000
      insertbuff = 30000;
RUN;
```

Let's look at this code more carefully:

- Notice that this code is one long libname with a semi-colon at the end. This code maps the Snowflake database to the snow libname. The long code is about setting options and parameters for the LIBNAME command.

- Notice also that the engine used is SASIOSNF. This is the engine for accessing Snowflake.

- In this example code, the cloud database we are trying to access on Snowflake is on the server called saspartner.snowflakecomputing.com. The user is the author, mwahi, and their password to get into the Snowflake database is TAM813fla.

- All of the data stores are in the schema called HCDW (for healthcare data warehouse). The database that the user mwahi is accessing within the HCDW schema is called USERS_DB. It is likely that in this schema, there is a separate DEVELOPERS_DB database for developers. Within USERS_DB, there is a particular warehouse just for users, called USERS_WH, so warehouse = USERS_WH grants mwahi access to this.

- The dbcommit, autocommit, readbuff, and insertbuff options modify the LIBNAME connection in terms of how it behaves with transactions and memory while mwahi is using their connection. As there are many options that can be set on a Snowflake connection, you are encouraged to read the original SAS documentation for information about these options (link available in the *Further reading* section).

Let's assume that our LIBNAME code ran successfully. This means that now, the user mwahi can interact with tables in the Snowflake data warehouse using the snow libname before the table name. If our demonstration dataset named Chap12_1 was in USERS_WH, which is the data warehouse location mapped to snow, then we could refer to it by using the term snow.Chap12_1.

As described earlier, SAS reporting was historically geared toward making reports that would be served up in a printed format. Today, SAS has evolved, so now we can use the ODS to serve up data and reports to the web, which we will cover in the next section.

Serving SAS data to the web with the ODS

In *Chapter 5*, *Managing Data Reporting in SAS*, we talked extensively about how revolutionary the ODS was when it was invented because it allowed graphics to be output from SAS for the first time. In fact, the ODS continued to be updated and was eventually used to output graphics in HTM format, which fosters displaying graphics on the web.

The ODS is still the basic function that outputs graphics in SAS. It can still be directly called up and commands given directly, as we did in *Chapter 5*, *Managing Data Reporting in SAS*. But when programming SAS to display graphics on the web, the functions behind the ODS seem to fade into the background. This is because we use different tools, such as the **SAS Enterprise Guide** and **SAS Viya**, to help us develop programs in SAS that use the ODS, which we will talk about in this section.

But first, let's consider using SAS and the ODS to serve data and reports to the web by examining a web-enabled query interface that has been built using SAS products.

Interacting with SAS data over the web

When considering serving data to the web, we can imagine that this will need to involve several steps each time a user runs a report to the web from our SAS data. As a typical example, we could visit the online visualization platform for the **Behavioral Risk Factor Surveillance System (BRFSS)** data, which is built using SAS tools (link in the *Further reading* section). The BRFSS is an annual anonymous health survey done over the phone in the US. The data from the annual core survey is provided for download in `*.sas7bdat` format on the internet for those who want to analyze the raw data. The US government also provides this visualization platform for those who want to use tools to query the data online.

Throughout this book, we have used extracts from the 2018 version of this dataset in our examples. We have typically included data from only three states: Florida, Massachusetts, and Minnesota. Let's use the visualization tool and look at metrics for one of those states: Minnesota.

The interface provided is depicted by this diagram:

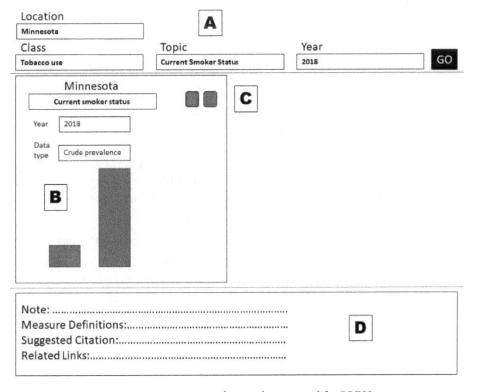

Figure 12.3 – Diagram of a visualization tool for BRFSS

Let's go over some features of this interface:

- The area labeled **A** in the diagram refers to a header area with overarching query parameters. In the diagram, when we selected **Tobacco use** under **Class**, the interface dynamically shifted to only allowing us the choice of limited topics. We chose **Current Smoker Status**. Under **Year**, we chose **2018**, then we clicked the **GO** button in the diagram.

- After clicking **GO**, we would have seen the area labeled **B** automatically update. Before clicking **GO**, it would have the old query parameters filled in. When the visualization page first loads, this graph area defaults to the selection **Alcohol Consumption**. This section labeled **B** does not update until **GO** is pressed, which can be confusing. But after it updates, there are more query parameters available. Under **Data type**, we chose **Crude prevalence**.

- The area marked **C** refers to two toggle buttons. The one on the left causes the panel labeled **B** to display as a bar chart, as shown (default). By clicking on the button on the right, the user can toggle to a different interface that displays a small table that provides the actual percentage and the confidence interval around it for the estimates (not shown, but a screenshot of the format of this output is shown in *Figure 5.16*).

- The area at the bottom marked **D** provides notes and other information. This section is also dynamic and updates when query parameters are updated.

As can be discerned from this example of serving SAS to the web, there are a lot of steps involved. Let's contemplate some of these steps:

1. First, we had to set some query parameters at the top of the page and then hit the **GO** button up in the section marked **A**. But queries were going on even as we were setting parameters because as we updated **Class** to say **Tobacco use**, the choices available under **Topics** also updated.

2. Clicking **GO** launched PROCs and data steps in SAS that took in the inputs from the web interface (for example, us selecting **Minnesota** and the other parameters), and returned results to the web that we could view (for example, a graph of the rate of current tobacco smokers in Minnesota from the 2018 dataset).

3. When the section labeled **B** was populated, we were offered the opportunity to update another parameter, **Data type**. If we were to update that, it would launch PROCs and data steps that would refresh this section.

There are probably many more events and pathways that the user could take with this interface; the steps outlined are only a few. But they are instructive, in the sense that even this simple interface can foster some complex user interaction.

In order to keep all of these steps organized, many programmers working to serve SAS data to the web like to use the SAS Enterprise Guide.

Using the SAS Enterprise Guide

Before we consider what the SAS Enterprise Guide does, let's think about what programming tools might be helpful to the SAS data warehouse programmer, especially when considering serving up data on the web:

- **Visualization of the sequence of events**: Thinking about the example, it would be helpful to have a visual pathway of what happens after a user clicks **GO**. There is a series of data steps and PROCs that result in the web interface being updated. Being able to visualize the steps in that pathway – which are essentially snippets of SAS code – would be helpful to understand and troubleshoot the overall process.

- **Visual menu for navigating disparate data stores**: When running a SAS data warehouse, the developer usually regularly connects to many SAS and non-SAS databases and data stores through SAS (using **SAS/ACCESS**). You can imagine that the BRFSS interface connects to datasets from many different years of the BRFSS. It would be nice to have them available as a panel or menu that is easily navigable.

- **Ability to call up other useful SAS windows when needed**: SAS software in its various formats already provides useful windows. The SAS editor, which color-codes programming as it is written, is helpful to the programmer who leverages these features. The SAS log file is extremely helpful for learning about data being processed and also about the SAS environment. SAS graphic output windows are also useful to developers behind the scenes. However, launching and navigating these windows can be clumsy in some versions of SAS, so it would be nice to have an interface that makes using these windows easier.

To solve these problems (and to also address other issues that affect other types of SAS users), SAS invented the **SAS Enterprise Guide**. Chris Hemedinger works with his colleague, Rick Wicklin, to maintain a chart of all the SAS Enterprise Guide releases over the years, which they have posted on SAS's blog (link available under the *Further reading* section). According to their chart (which is generated using SAS graphics), the first version of the SAS Enterprise Guide came out in 1999. This makes sense because this would be the time that SAS was forced to transition out of its own enclosed environments and connect to external environments, such as SQL databases and web interfaces.

To provide a quick idea of how the SAS Enterprise Guide can look, here is a diagram:

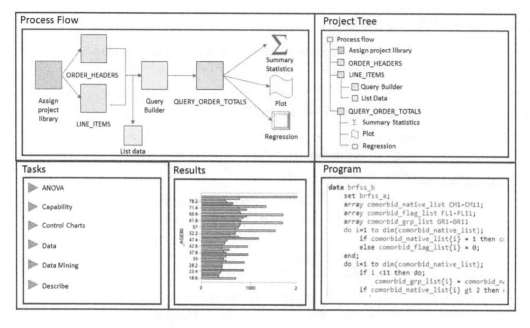

Figure 12.4 – Diagram of SAS Enterprise Guide windows

Let's revisit those features that we mentioned would be helpful for the programmer of a SAS data warehouse and see how they match up to the diagram:

- **Visualization of the sequence of events**: The **Process Flow** and **Process Tree** windows can help visualize these steps. There are other ways of visualizing the steps and interacting with the SAS Enterprise Guide to modify code and processing.

- **Visual menu for navigating disparate data stores**: Although not depicted in the diagram, the SAS Enterprise Guide provides a way of connecting and navigating to disparate data stores. These can be visualized in a SAS window. This makes it much easier for the SAS data warehouse developer to manage ETL on many different datasets at once.

- **Ability to call up other useful SAS windows when needed**: The diagram depicts a few windows that are available in the SAS Enterprise Guide. A **Tasks** window, a **Results** window, and a **Program** window can offer the programmer a multi-dimensional view of their current work, as well as programming shortcuts. Many other windows and visualizations are available, as the interface is customizable to the programmer's needs.

The SAS Enterprise Guide as an application

As can be seen by Chris Hemedinger's chart, the SAS Enterprise Guide has had many upgrades over the years and continues to be upgraded in 2020. It is important to note that the SAS Enterprise Guide versions represented in the chart are for the desktop client application, which works over many different SAS versions, and that is why it is updated so much. Hemedinger notes that there is a different SAS Enterprise Guide for the core engine of SAS that has been upgraded less frequently.

There are a few implications of this situation. Since the SAS Enterprise Guide is an application that was launched in 1999 to help manage SAS data projects and has been continuously supported, by 2020, it is now essential software in many SAS shops. This means that it will need to be continuously supported in the future, and future versions will need to be backward-compatible unless SAS develops a new approach to replace the current one.

The SAS Enterprise Guide was not developed as a web-enabled application in the beginning, because that was not possible in 1999. Current versions still have the look and feel of applications of that era. Yet, they need to continue to be supported. Therefore, SAS continues to invent and develop other products that can be more easily web-enabled, such as SAS Viya.

The SAS Enterprise guide allows the programmer to control many parts of the SAS code, including the ODS, with menus as well as code. Many programmers over the years have found the SAS Enterprise Guide helpful in managing data warehouses. However, newly minted programmers will note that the SAS Enterprise Guide seems somewhat old-fashioned and not web-enabled. Programmers interested in SAS's products designed for the web should look into SAS Viya.

Using SAS Viya

The idea behind **SAS Viya** was to upgrade SAS's analytics platform while also improving it, especially in terms of interfacing with the web. Tricia Aanderud covered the announcement of SAS Viya at the 2016 SAS Global Forum in her blog, *BI Notes* (link available in the *Further reading* section). She interpreted the announcement to mean that part of the SAS Visual Analytics and SAS Visual Statistics functions were converted into an HTML5 interface as part of Viya. After this announcement, other analytic functions have been built into Viya, and SAS continues to improve Viya.

SAS Viya ultimately serves as an analytics suite, the way **Amazon Web Services** (**AWS**) might be seen as an analytics suite for data loaded to Amazon servers. SAS Viya has many subcomponents that can be used for different tasks. All of them cannot be covered in this chapter, but since this book is about data warehousing, let's cover one subcomponent, called **SAS Data Preparation**.

Data warehousing involves a lot of ETL protocols, and SAS Data Preparation is developed specifically for ETL. As demonstrated in a SAS video by Mary Kathryn Queen (link available in the *Further reading* section), SAS Data Preparation can help with the following activities:

- **Profiling**: Queen demonstrates doing data discovery with SAS Viya profiling tools. The interface is similar to SAS University Edition, where results of data steps appear immediately in windows and are easily navigable. Other panes helpful for data profiling are also available, such as a graphing pane.

- **Preparing ETL**: Queen demonstrates a screen split into windows. Transformation steps can be run in one window, and the results viewed immediately in another, with a third panel allowing the steps to be navigable. This allows us to edit code interactively on the fly to ensure all transformed variables are appearing properly.

- **Scheduling ETL jobs**: After perfecting ETL code, Queen demonstrates how the programmer can use SAS Data Studio to develop plans for ETL jobs and schedule them. Being able to schedule ETL jobs in SAS is very useful because they can be scheduled to run when there are few users on the system to optimize I/O.

Earlier, we discussed the overhead associated with deconstructing SAS reports and reconstructing them in another software. There is a similar overhead associated with deconstructing SAS Enterprise Guide configurations and moving them over to SAS Viya. This is the primary consideration for an existing SAS shop when deciding whether or not to migrate some or all work currently done in the SAS Enterprise Guide to SAS Viya.

Reporting directly to the web with the SAS Enterprise Guide and SAS Viya

As described before, how we use the ODS now for reporting to the web is through using SAS code, either directly or through a platform such as the SAS Enterprise Guide or SAS Viya. For SAS shops running the SAS Enterprise Guide that want to make a web visualization like the BRFSS one we examined, please see the step-by-step process outlined in the SAS white paper *End-to-End Web Reporting Using SAS Enterprise Guide 3.0: Who Does What and When*, by Marje Fecht and Peter R. Bennett (link available in the *Further reading* section). For SAS Viya, there are different approaches to reporting SAS data directly to the web, depending on whether there is anticipated user interaction. Because SAS Viya is currently under continuous development, you are encouraged to access new information from SAS itself for guidance on reporting to the web directly from SAS data storage using SAS Viya.

Using SAS and R for visualizations

R is a software that can be integrated into reporting SAS data. With R, which is open source, it is possible to set up connections between SAS and R data. But the main difference between making plots and other visualizations in SAS versus doing it in R has to do with data handling. As we have seen with SAS, when using PROCs that create plots, such as `PROC UNIVARIATE`, SAS typically reads or calculates the relevant values from the entire dataset and plots them. In a scatter plot, this is necessary – but it is not necessary for all plots. Although some SAS PROCs have the ability to take in a summary dataset and visualize it, many SAS PROCs require processing the whole underlying dataset.

Let's think of a box plot for a moment. For a box plot, outliers aside, we technically only need to know five different points in order to create the image of the plot: the minimum, 25th percentile, median, 75th percentile, and maximum. If we were creating a time-series graph, showing how a rate changed over time, we would only really need the information about the x coordinate (the time the rate was measured) and the y coordinate (the value of the rate) for each data point.

Contrary to SAS, this is how R *always* assumes the programmer will prepare data for serving up in plots. R is also different from SAS in another way. While SAS has different **components** that can be added on, in R, the different add-on components are called **packages**. A famous package in R used for plotting is called `ggplot2`. Though **base R** has the capability of making plots, R programmers tend to prefer using `ggplot2` because it offers advanced plotting capabilities and features to edit the output of the plot. In both base R and `ggplot2`, the only data that needs to be inputted is the actual points that need to be plotted.

We have been talking about keeping a healthcare data warehouse. Let's imagine that we are including in that data warehouse information about hospitals, as well as information about US veterans and their access to healthcare. Imagine we are keeping data about three states – Florida (FL), Massachusetts (MA), and Minnesota (MN) – and we are keeping track of the proportion of veterans in the general population. Currently, the proportion of the population that is veterans in FL is about 0.07. In MA, it's about 0.06, and in MN, it's also about 0.06.

Let's say we were going to use our hypothetical SAS data warehouse to create a time-series plot of the rates of veterans in the population over time – from 2014 through 2018 – for each of the three states. SAS provides online examples for creating time-series plots because there are several different ways to approach the task and different PROCs that could be used (see the link in the *Further reading* section). Just looking at our options in SAS, we can see that if we used SAS for this task, the code would be complicated, and the output plot would look old-fashioned:

- We observe that we would first need to decide between different PROCs that would produce different outputs.

- Next, as with R, SAS may need the programmer to create a dataset in a certain format to facilitate plotting. This is where one set of I/O costs would occur.

- The other set of I/O costs would be when SAS actually generates and displays the plot.

One of the SAS examples given creates a time-series plot along with a moving average using PROC ARIMA. The code is quite extensive, with over 30 lines, and the output is difficult to read because it is so busy. By contrast, a similar time-series plot in R using ggplot2 took less than 10 lines of code.

Imagine we used SAS ETL approaches to generate the following SAS table (of fabricated data for demonstration purposes) from our data warehouse that we could export as *.csv:

state_abbrev	rpt_yr	vet_prop
FL	2014	0.09
MA	2014	0.05
MN	2014	0.06
FL	2015	0.08
MA	2015	0.05
MN	2015	0.05
FL	2016	0.07
MA	2016	0.04
MN	2016	0.06
FL	2017	0.08
MA	2017	0.05
MN	2017	0.06
FL	2018	0.07
MA	2018	0.06
MN	2018	0.06

Table 12.2 – Fabricated dataset optimized for plotting in R

Let's look at some highlights of this table:

- The table is very small and easy to transport. There are only three columns: `state_abbrev`, which is the state abbreviation, `rpt_yr`, which is the year of the measurement, and `vet_prop`, which is the proportion of veterans.

- Notice that the table could be said to be in the **entity-attribute-value** (**EAV**) format, in that each row designates a unique location and time. The first row, for example, designates the location of `FL` in `state_abbrev`, and the time of `2014` in `rpt_yr`. Then, in the third column, the actual metric to be graphed is presented, called `vet_prop`, which in our fabricated data for 2014 is `0.09`.

Once we generate the small summary dataset in the preceding table in SAS and export it as `*.csv`, we can import it into R and make the following plot using R's `ggplot2` package:

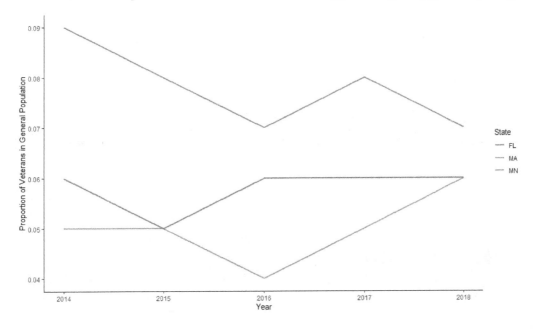

Figure 12.5 - Example time-series plot from R using ggplot2

The purpose of this demonstration was to show that it can be relatively easy to use R with SAS reporting functions without having to connect to a live environment, such as a SQL system. Instead, ETL can be done to create extracts that can be visualized in external software, as we saw here with R, which we will also consider next with Tableau.

Reporting SAS data in Tableau

Tableau is an online data visualization software that provides the opportunity to connect on the fly to different data stores for visualization. It allows a connection to a `*.sas7bdat` file so that SAS data can be visualized in Tableau (see the link in the *Further reading* section). The main function of Tableau's SAS integration tool is to import the file into Tableau so that data can be visualized in the Tableau interface (as we just did with R).

Currently, maintaining a live link between an updating data store in SAS and a visualization dashboard in Tableau would not be easily implemented. However, visualizing a dataset in SAS format obtained from a SAS data warehouse as we did with R would be possible using Tableau's connection resources.

Exporting extracts of data out of SAS and importing them into Tableau for reporting may seem like a simplistic idea, but there are greater purposes behind developing ways of reporting SAS data in other programs. SAS data reporting tools are limiting, but they have been around a long time. Therefore, many legacy reports have been designed in SAS, but these reports cannot leverage the power of Tableau, which easily generates colorful, multi-paneled, interactive, web-enabled reports, such as this one of global temperatures over the years created by Wikimedia Commons user Marissa-anna:

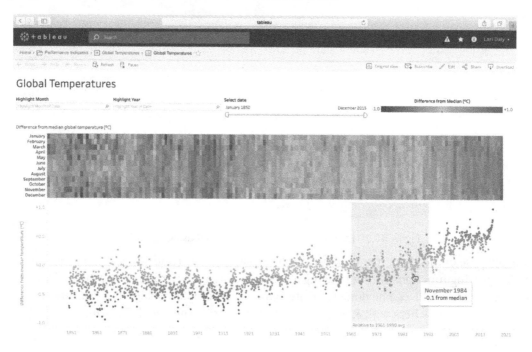

Figure 12.6 – Example of visualization in Tableau (https://commons.wikimedia.org/wiki/File:Global_Temperatures_Server.png)

Hence, nowadays, many SAS warehouses are considering rebuilding their historical SAS reports into another software that was specifically designed for reporting. Taking extracts from SAS and developing visualizations in external products such as R or Tableau can expand the possibilities.

Considerations when reporting SAS warehouse data

Today, anyone running a SAS data warehouse likely has the obligation of providing some sort of reporting. Perhaps they are producing a monthly set of finance reports using SAS reporting PROCs that are saved to PDF format and circulated among executives at a company. In addition, they may be supporting an online query tool such as the one we examined with the BRFSS. Therefore, leaders of SAS data warehouses have to make some reporting-related considerations:

- **Different considerations in print versus web reporting**: As described, the legacy of print reporting using arrays, macro variables, and macros in SAS is difficult to migrate away from, so they tend to continue to be supported, although plans can be made in this area. For web reporting, different considerations need to be made, the most prominent one being I/O.

- **I/O with SAS and web reporting**: The primary concern with SAS with any function is always I/O. When we are doing any type of reporting to the web, I/O becomes even more crucial. Interactive web interfaces such as the one from the BRFSS are much slower and clumsier than comparable ones built in R or Tableau. But visualizing data stored in SAS on the web interactively using R or Tableau might be challenging at the current time.

These basic considerations – print versus web reporting, legacy code versus implementing a new approach, and requirements to serve interactive web interfaces – must be weighed against the current status of the SAS data warehouse, and its strategic direction going into the future. Where and how data is stored will greatly influence reporting considerations, as well as the specific requirements for reporting, and the resources available to the project.

Summary

Although this chapter was specifically about visualization using the SAS ODS, it echoes two basic themes we see throughout this book. The first theme is that many believe that SAS's power is in its PROCs, its data step language, and its capabilities with arrays, macro variables, and macros. These features have been used over the years to develop extensive print reports in SAS using the ODS (not to mention complex ETL code). Now, the same basic features are being used to serve SAS data to the web. SAS is a company that has constantly innovated throughout its life, creating new products and capabilities, such as SAS/ACCESS, and the ability to connect to cloud storage in Snowflake. It is likely that SAS will continue on this route, developing new subcomponents and products to address new technological challenges.

A second theme that runs throughout this chapter and this book is that while SAS continues to innovate, the direction of its innovation seems misguided in some ways. As we discussed in *Chapter 1, Using SAS in a Data Mart, Data Lake, or Data Warehouse*, when SQL posed an early challenge to SAS's big data storage and I/O, SAS did not rebuild their core engine, and instead innovated new capabilities, such as the macro language and PROC SQL. When reporting code became overly complicated through the use of various novel SAS features, rather than redesigning the SAS ODS approach, SAS simply invented tools such as the SAS Enterprise Guide to manage the complexity.

When the visualizations in SAS began to look old-fashioned, rather than teaming up with another software company known for visualizations and developing an integration, SAS continued to innovate on their own ODS output. Now, SAS Viya seems to have built into it the same limitations that come with the basic design of the SAS data step language, macro language, and other languages that ran optimally in their day but are not optimized for the modern global computing environment. As an example, today, it is hard to imagine SAS developing visualizations and reporting tools that would surpass the functionality and appearance of those coming out of R and Tableau.

So, at the end of this chapter and this book, you are encouraged to ponder what SAS really is. As we learned in *Chapter 1, Using SAS in a Data Mart, Data Lake, or Data Warehouse*, SAS originally arose from the need to do complex statistical procedures, such as ANOVA, and the data step language arose from needing to format the data a certain way to run a PROC that would do ANOVA. Reporting capabilities through the ODS were an afterthought for SAS. In this chapter, we learned how SAS data warehouse reporting is done using the print method, as well as how to serve SAS data to the web. Both the extensive SAS reporting code, and SAS capabilities for serving data to the web, are not optimized for today's environment.

Also, as we saw in developing our simple report in this chapter, the data step language was not optimized for today's ETL. Rather, it was optimized for ETL in the environment when it was developed, in the 1970s and 1980s. In this chapter, we saw how these days, we may need to use Snowflake as cloud data storage overflow if our SAS data warehouse fills up and we need processing room. Because the data step language has not been optimized for today's environment, determining how to minimize I/O when moving data back and forth from the SAS environment to Snowflake will be its own challenge.

In closing, it is important for the modern data warehouse developer to realize that both reporting and ETL were an afterthought in SAS's development and that these core processes have not been rebuilt since they were originally conceived. Perhaps the most valuable component of SAS remains its analytics engine, which provides optimized statistical analyses, such as PROC FREQ for frequencies, and regression analyses. Data warehouses that do not use SAS now could probably leverage the best part of SAS through SAS Viya. On the other hand, data warehouses that are deeply entrenched with SAS currently should consider making strategic plans that allow the entire warehouse to optimize its use of all its warehouse software. Previously, we dreamed of a fully enclosed, physically secure, comprehensive SAS environment comprising data servers and analytics applications as a home for our big data. Today, we envision that the ideal data warehouse or data lake deploys an orchestra of SAS and other data warehouse products in order to optimally serve its developers, analysts, other users, and stakeholders most harmoniously into the future.

Questions

1. Why is it important to build in and review SAS reports about ETL after an ETL protocol is run?

2. We used a SAS query tool for the BRFSS data and found it to be very limiting in terms of interactivity and data display. What is the source of these limitations?

3. Imagine you ran a data warehouse where you had to connect to 15 different SQL databases and import data into SAS on a regular basis. How could the SAS Enterprise Guide help you?

4. Imagine you were advising a SAS data warehouse that had not implemented the SAS Enterprise Guide or SAS Viya, but they want to implement one of them now. Which would be the most appropriate?

5. Imagine you work at a SAS data warehouse that is very extensive and has been around for decades. Each year, they have upgraded their SAS software, and now have the most current visualization tools. Why might you nevertheless choose to visualize some of the SAS data in non-SAS software?

6. Imagine a SAS data warehouse comes to you asking for recommendations on whether or not they should rebuild the reports they are currently running in SAS into R or Tableau. What kind of information would you need in order to give them good advice?

7. Imagine you receive funding to start a new data warehouse today. What components of SAS would be the best to use?

Further reading

- The SAS white paper, *SAS Macro Programming for Beginners*, by Susan J. Slaughter and Lora D. Delwiche, available here: `https://support.sas.com/resources/papers/proceedings/proceedings/sugi29/243-29.pdf`

- The SAS white paper, *The Power of CALL SYMPUT – DATA Step Interface by Examples*, by Yunchao (Susan) Tian, available here: `https://support.sas.com/resources/papers/proceedings/proceedings/sugi29/052-29.pdf`

- SAS documentation for SAS/ACCESS ODBC connection to Snowflake: `https://documentation.sas.com/?docsetId=acreldb&docsetTarget=p19i7uzcbso1szn1pczxn88co3g1.htm&docsetVersion=9.4&locale=en`

- Online visualization platform for BRFSS data: `https://www.cdc.gov/brfss/brfssprevalence/index.html`

- The SAS blog post by Chris Hemedinger, *Through the Years: SAS Enterprise Guide Versions*, available here: `https://blogs.sas.com/content/sasdummy/2020/08/25/eg-versions-through-years/`

- The blog post, *SAS Visual Analytics 7.4: Modern Theme Vs Classic Theme*, by Tricia Aanderud, available here: `https://bi-notes.com/sas-visual-analytics-modern-theme-export-excel/`

- The SAS video by Mary Kathryn Queen on Viya Data Preparation, available here: `https://www.youtube.com/watch?v=sl8AtKwzn4M`

- SAS time series web examples: `https://support.sas.com/rnd/app/ets/examples/gplot/index.html`

- The SAS white paper, *End-to-End Web Reporting Using SAS Enterprise Guide 3.0: Who Does What and When*, by Marje Fecht and Peter R. Bennett, available here: `https://support.sas.com/resources/papers/proceedings/proceedings/sugi30/246-30.pdf`

- Instructions to connect to a `*.sas7bdat` file from Tableau, available here: `https://help.tableau.com/current/pro/desktop/en-us/examples_statfile.htm`

Assessments

Chapter 1

1. SAS manages data using the procedural data step language, so the programmer must spell out exactly how queries will execute. SQL is a declarative language, so the programmer declares the query results they want, and an optimizer program determines how the query will execute.

2. Setting criteria using the WHERE clause will cause SAS to skip over reading records that do not meet the criteria. If the criteria are set on an IF rather than a WHERE clause, SAS will read the whole record first and evaluate the criteria afterward.

3. SAS/ACCESS.

4. It is a high-cardinality variable that is used repeatedly in WHERE clauses in processing.

5. It is best to store data independently of SAS code for privacy reasons, and it is usually easier to enter the data into a spreadsheet and then read it into SAS. However, it is possible to use SAS to enter data, and then save the resulting dataset as the live data.

6. By using all SAS components, you can achieve the fastest I/O possible.

7. If I/O is a challenge in the SAS environment, it may be best to use data steps to improve I/O. However, if I/O is not a challenge, then data step code that accomplishes the task could be benchmarked against PROC SQL code that accomplishes the same task, and whichever ran faster would then be used.

Chapter 2

1. The XPT format is a SAS format used for reducing the size of `*.SAS7bdat` datasets so they can be transported and extracted into another SAS system.

2. The `GUESSINGROWS` option in `PROC IMPORT` allows the user to set how many rows SAS reads in order to guess the `informat`, `format`, and `input` code `PROC IMPORT` automatically generates for the `infile` statement.

3. Using `PROC IMPORT` with a dataset induces SAS to guess at building `infile` code. SAS automatically builds this code, then outputs to the log file. Even if this code has errors, much of the generated code is helpful to the programmer because it already has `format`, `informat`, and `input` lines for each variable in the source dataset. The programmer can copy this generated code from the log file into a code file and edit out the errors.

4. Because fixed-width files do not have delimiters, and the analyst must include in SAS `infile` code the positions for each character of each variable in each row (or record). Without documentation as to this information, the SAS analyst needs to use strategies to research the data and build `infile` code that reads in characters from the proper locations in the source file into the SAS dataset.

5. First, you would use a `LIBNAME` statement to direct SAS as to the physical location of the source `*.csv` so SAS will know where to look when reading it into SAS. Next, you would use a `LIBNAME` statement to direct SAS where to output the `*.SAS7bdat` file onto static media.

6. The datasets should be stored in a SAS format, since they will be analyzed in SAS and no other program. Given that old datasets are rarely used, they can be stored in XPT format to take up the least space.

7. If you are given a choice as to the data format, and you are working at a SAS data warehouse, the best format to request is `*.SAS7bdat`, because the data will be read in most easily by SAS. However, there can be challenges when transferring data in this format due to the large size of the files.

8. If a `*.SAS7bdat` dataset is too large to fit on static media, and the file is being transferred to a SAS data warehouse, the next choice would be XPT format. This would hopefully reduce the size of the file while retaining the SAS metadata contained in the file. If this does not reduce the size enough, converting it to a `*.csv` or `*.txt` file will likely reduce it more, and may even make it small enough to transfer. However, it may introduce extra work on the receiving end in terms of reading the data into SAS.

Chapter 3

1. The PROC CONTENTS option for this is VARNUM.

2. These are criteria for when using SAS operators, meaning *greater than or equal to 10*, and *less than or equal to 20*.

3. A programmer may want to attach a user-defined format to a categorical variable with five levels coded 1, 2, 3, 4, and 5 in order to decode each level. That way, if the variable is used in processing, its levels do not display as a code but instead display as a text string, describing what the code means.

4. The main reason could be that the data is being displayed in an application other than SAS, and therefore, that application could not use SAS labels and formats. However, even in a SAS warehouse, another reason could be that the overhead in maintaining labels and formats is higher than maintaining such metadata another way, so the leaders have chosen to use alternatives.

5. The programmer can choose whichever way they want to view data. If they have SAS experience, they may choose PROC PRINT, but if they have SQL experience, they may choose PROC SQL.

6. The main problem with adding a format to make a continuous variable appear categorical is that the actual underlying data is still continuous. The category imposed by the format only impacts the appearance of the data, not the underlying values. Therefore, if a programmer is doing an operation that requires a numerical code for BMI category, this variable will need to be developed and saved as data.

7. The observation number in SAS datasets can act like an index, and can be used to look up specific rows.

Chapter 4

1. Data transfers direct from the data provider must be stored in a highly secure environment because they often contain sensitive data that is necessary to have during transformation, but will be removed before the data is loaded into the warehouse, mart, or lake.

2. Maintaining modular, systematically named code files, rather than one long code file, allows the opportunity for manual rollback. This makes troubleshooting to repair variables easier if a problem is identified after processing.

3. The main advantage of naming these variables COST1, COST2, and so on up to COST10 is that they could be easily declared in an array by using the range COST1-COST10. The main disadvantage of renaming these variables with names such as COST1 and COST2 is that the variable names would become unintuitive and documentation would be needed in order to understand what they mean.

4. Formats can be created in SAS using PROC FORMAT and saved outside of the data step code. In contrast, arrays can only be called up from inside data step code and cannot be run and saved independently as formats can.

5. Programmers in the SAS data warehouse who have access to raw data from data providers and participate in ETL typically have a deeper understanding of the dataset and are good at using the documentation associated with the data. Therefore, they are most likely to know the appropriate text to place in SAS labels and formats for this data.

6. For both developers and end users, standardizing the names of variables, such as adding _GRP to the suffix of all grouping variables, can help with an intuitive understanding of the variables. For developers, having a naming convention policy for variables provides guidance when developing names for novel variables.

7. Senior programmers who serve as SMEs for particular datasets can be very helpful because they typically have advanced knowledge of the subject of the data as well as the underlying data system from which the data was transferred. They may also have an advanced understanding of data documentation and, if so, are perfectly situated to maintain SAS labels and formats and other metadata for the dataset. They are also perfectly situated to develop or provide consultation on the development of ETL code for the dataset.

Chapter 5

1. PROC TABULATE is for developing well-formatted tabular reports.

2. The ODS allows the programmer to tell SAS which internal tables generated during the PROC should be saved outside the PROC, and where to save them.

3. No special code is needed. For the PROCs that use GTL – PROC SGPLOT, PROC SGPANEL, PROC SGSCATTER, and PROC TEMPLATE used with PROC SGRENDER – the ODS is automatically deployed when running the PROC.

4. Since the department leader wants to write paragraphs next to the tables, the best format in which to produce the SAS report would be `*.rtf`, which opens in Microsoft Word. `PROC TABULATE` could be used to format the tables as per the department leader's specifications.

5. To output a list of variable names using the ODS, first, the programmer would need to recall what `PROC` would produce in its output. `PROC CONTENTS` is the commonly used `PROC` for outputting a list of variables in a dataset. Once the programmer identifies the appropriate `PROC`, the programmer would need to identify the internal table in the `PROC` to output using the ODS (through using a `TRACE` command, or referring to support documentation). Once the title of the correct internal table is identified, the programmer would need to use ODS programming with the `PROC` to output the internal dataset to a specified location.

6. When making a histogram in `PROC UNIVARIATE`, the histogram is requested through an option, and the output is not very customizable. When making a histogram using `PROC SGPLOT`, since the `PROC` itself utilizes the ODS and was designed specifically for plotting, more customizations can be applied to the histogram. Finally, making a histogram using `PROC TEMPLATE` and `PROC SGRENDER` would require the programmer to first build a customized template in `PROC TEMPLATE` for the histogram, and then to connect the template to the data and render the histogram using `PROC SGRENDER`.

7. Because the team is interdisciplinary, they may not all be familiar with histograms. Therefore, it will be very important to make the histogram clear. The programmer will need to be able to make clear labels and choose the size of bins. Therefore, the histogram from `PROC UNIVARIATE` would not be a good choice. A histogram generated through either `PROC SGPLOT` or through `PROC TEMPLATE` with `PROC SGRENDER` would be superior in this case.

Chapter 6

1. It is good to include `_list` at the end of an array name so that it is clear by its name that it is an array, and not a variable.

2. Adding a condition to an array allows it to process certain parts of the input array one way, and other parts of the input array another way, depending on the criteria that's been set for the input array.

3. The reason why it is often necessary to rename variables for arrays is that array processing goes faster if the input variables are named according to a naming convention, where each one has the same prefix and is followed by an incrementing number, such as CM1 through CM11 for the co-morbidity variables, as demonstrated in this chapter.

4. SAS data warehouse managers need to make the following considerations about serving up index variables to users: whether the index variable was passed on from the native data provider or calculated at the SAS warehouse, how the index variable is documented, what index variables should be made available, and what inputs to index variables should be made available.

5. MCHP and other organizations who publish their SAS arrays and algorithms are generally trying to assist users of their data in developing optimal analyses. This includes providing support and documentation for index variables created by SAS arrays and algorithms. A side effect of publishing this information in the public domain is that it can help other organizations using similar arrays and algorithms by providing standards and guidance for them.

6. The first reason why this problem would be a good candidate for an array is that the dataset is being transferred often, meaning that there will be a return on investment for the effort of programming the array. Another reason is that there are 20 survey questions all coded the same way, so the array will be easy to program without any conditionals. Finally, the report is produced weekly, which requires a quick ETL protocol to be executed every week.

7. Technically, the answer is "no" – it is not necessary to include array processing in data steps in ETL code for a SAS data warehouse – as long as the native data being processed is sufficiently small, where using array programming would not be able to improve the efficiency of code execution. This situation is actually surprisingly common in SAS data warehouses, where some datasets that have been transferred represent summary data and are actually relatively small. An example of such a dataset could be a financial report that provides a few financial metrics for all 50 states in the United States every month. If array processing is not necessary to gain efficiency, this should be avoided due to its unnecessary complexity and the complexity it introduces in supporting and maintaining SAS data warehouses.

Chapter 7

1. An analyst gains knowledge of what each variable means and what each level of each categorical variable means in a dataset from reviewing the **data dictionary** for that dataset. If the dataset is derived from survey responses, a **codebook** should also be included in the warehouse **data curation**.

2. When preparing continuous data for processing into a data warehouse, it is necessary to know all possible values in the continuous variables, and that information is not available in PROC UNIVARIATE output. Using PROC FREQ on a continuous variable allows the analyst the ability to view the existence and distribution of every single value in the continuous variable. Although the output is very long, viewing the top and bottom of the output (extreme values) can provide instructive information about how to optimally process the variable into the warehouse.

3. It is helpful to plan transformed variables in a data dictionary before creating ETL code for them for a few reasons. First, it allows the analyst to do research on the native variables involved, and come up with a data-driven rationale for transforming the new variables. Next, it provides the opportunity for other programmers and leaders to weigh in on plans for coding the new variables. Third, it creates the beginning of documentation that can continue to be updated and shared with data warehouse managers as well as users. Finally, it can provide analysts developing ETL code with guidance so that they have a roadmap to help them plan their SAS ETL code as efficiently as possible.

4. PROC UNIVARIATE only operates on numeric variables and will include all values of the numeric variable in summary statistics. This can be a problem if codes (such as 77 or 99) are included in the data. It can also be a problem if missing data is coded as zero because the zero will be seen as a valid value by PROC UNIVARIATE, while the PROC will skip over all missing values. That is why if a *cleaned-up* version of a variable is created (the way we created SLEPTIM2 from SLEPTIM1), this variable can suppress unwanted numeric values by coding them as missing, and then they will not be considered in PROC UNIVARIATE calculations.

5. Stock market data containing the value at closing every day would consist of a continuous variable. One classification variable often paired with stock data values is the direction of change from the previous day's closing value (up, down, or the same). Another classification variable could compare it to last year on the same day in terms of its closing value (up, down, or the same). There are many different classification variables that could be made from continuous variables, so it is important to study warehouse user needs in order to serve up the most useful variables in the data warehouse.

6. The problem with serving up the TEMP variable in the data warehouse is that it represents some accurate readings of temperature mixed with some erroneous ones. If the erroneous ones could be identified by the scientists empirically, another variable (perhaps TEMP2) could be created that suppressed the erroneous values. If the erroneous values could not be identified by the scientists, the warehouse could make business rules, and simply suppress the values outside of the business rules. A third approach would be to serve up all the data but include a two-state flag indicating which data is likely to be valid (as we did with slp_valid). If any version of TEMP is served to users, it should be well documented, so users can use it correctly.

7. First of all, because the name agrp ends in grp, you would immediately assume it was a grouping variable created by the data warehouse (and not a native variable from a data provider). That would mean the warehouse would need to troubleshoot its own programming. Next, if you saw the order in which agrp was created, you should easily be able to locate the modular code that created agrp and rewrite it. Or, you could simply add code to the end of the ETL process, and transform brfss_d into a new dataset, brfss_e, with agrp correctly coded in it, and then export out vet_analytic. Naming conventions of code, datasets, and variables would help in this troubleshooting.

Chapter 8

1. In SAS, a macro variable is a blank variable that can be established and set to a particular value. By contrast, a macro is a snippet of code that can be automatically launched to run. SAS macros commonly use macro variables, but macro variables can be used outside of macros.

2. The NONOTES, NOSTIMER, and NOSOURCE options suppress information from the log file. NOSYNTAXCHECK prevents the automatic check of code syntax SAS does when code is run.

3. The reason why SAS macro code is developed in a careful, step-by-step process is that it is very hard to troubleshoot errors that occur. By creating code that runs without any macro programming, and using a step-by-step method to convert it to a macro, you can cut down on time spent troubleshooting.

4. Conditions are added to macros to address anomalies in underlying coding. In our case, all of our variables except one were coded according to a system, and one was coded according to another system. The conditional programming allowed us to handle these differences without having to make separate macros, allowing fuller automation.

5. When macros are stored in separate files to the SAS code that calls them, it allows flexibility in splitting up the work at a data warehouse. One team can be focused on maintaining up-to-date macros and documentation, and another team can actually launch the macros to conduct ETL. That way, the integrity and standardization of the ETL protocol are maintained, while data can be loaded at any time.

6. If you are an analyst finding out you will need to develop an ETL protocol for historical data files, first, it is important to ask how many there are. If there is only one long, huge file, then loading it will present a different challenge than loading hundreds of smaller files. If you find that there are a lot of files (for example, 20 or more), you might consider developing a loading macro. If the files were mostly in the same structure and were named according to a predictable convention, the situation would lend itself even more to macro development. However, if the files are named unpredictably, and the data within them is unpredictable, it might be better to develop data step code manually and improve it before considering turning it into a macro to load data files automatically.

7. In this scenario, the answer is definitely "no" for a number of reasons. First, it appears you will only be loading two sets of data in the near future. This means that building macros will probably take more time than actually conducting the rest of the ETL functions on the data manually. Secondly, as was shown in the chapter, macro code is built from code that runs without macros. Therefore, the first step in making the macros would be coding the ETL by hand anyway. If macros are needed later, that code can be converted. Third, only some of the variables are similar from survey to survey. Therefore, little advantage could be gained by adding automated processing. Finally, while SAS macros can be very powerful, if they need to be replaced, it can be a very complex process. As an example, functions are to R what macros are to SAS, and it can take many hours to deconstruct a SAS macro and rebuild it as an R function.

Chapter 9

1. Even though code runs the same way no matter how it is formatted, the answer to this question could be *false* if you are working in a team producing ETL code. In that case, making well-formed and well-formatted code could matter a lot, in that it could save hours of troubleshooting and debugging. However, if you are working on a one-person, one-time project (such as a homework assignment), then the answer to this question could be *true*.

2. A good place to start troubleshooting an error in a transformation step is to start by going to the first error registered in the log file and ensuring that semi-colons are in the right place. A next step could be to break the data step apart and see if you can get a smaller piece of it to run.

3. The programmer uses a PUT statement to ask SAS to print certain values to the log file. A programmer can use the PUT statement during a data step to print certain values to the log file to help with data step troubleshooting.

4. If you need to do 10 transformation steps, it is best if you work out the code for each transformation step individually. Depending on the step, you may need to do substudies on parts of the dataset to work out all the steps (including arrays and conditions). But once each step is worked out separately, in order to maintain efficiency, all steps can be put together is one big data step with one RUN command.

5. Setting BREAK on a loop in the data step debugger allows the user to step through the data step and see the resulting values in each iteration of the loop.

6. If you are in a hurry, and you think you can program a macro from scratch and get it to run, you might be correct. However, people are not usually successful at this, so it might end up taking more time that way, rather than using the step-by-step method. Also, when you program a macro from scratch and it runs the first time, you may not be convinced that it is executing correctly, because you did not carefully test each piece. It would be necessary for you to go back and verify your work at some point.

7. The %PUT _local_ statement in a macro tells SAS to print the current value of all the macro variables involved in the macro to the log for each time the statement is processed. Using this statement can help you monitor whether the macro variables are evaluating properly as you step through the lines and the loops in the macro, as they will print out sequentially in the log file.

Chapter 10

1. One reason why it is helpful to consider both analysts and developers as user classes in a data warehouse or data lake is because they both have unique sets of needs that should be served. Serving their needs will increase their satisfaction, which is important for different reasons. Increasing the satisfaction of analyst users will increase their ability to provide analytic output, and that will increase the value of the data system. Increasing the satisfaction of developer users will improve their ability to do their job efficiently, which will also improve the data system. Another reason to consider them user classes is that, as a practical matter, permissions should be granted on the basis of subclasses of user functions. Since these subclasses apply to both analysts and developers, it is helpful to have user classes of both analysts and developers.

2. There is a similarity between analyst users of data lakes and developer users of data warehouses, and that is both of these user classes work with raw or mostly raw source data. Because the data is raw or mostly raw, the only documentation is that which comes from the source. Therefore, for the data to be usable for both user classes, extensive source documentation is necessary.

3. If analysts access a server in a data lake directly, it can reduce the need for support from developers in the data system, because the analysts can assemble their own datasets. They can also do their own source documentation research. However, a disadvantage is that they will have access to a lot of data, and therefore there are more privacy risks, and more of an honor system. If analysts do not access the data lake server directly, the main advantage is the data lake can ensure the privacy of data, and also the appropriate use of the data. This is through the developer controlling the extraction and use of the data by the analyst. The disadvantage of this model is that it requires a lot of extra effort to provide support to analysts. Highly trained personnel who have customer service, management, and programming skills are required for such roles.

4. Having more foreign keys in the data warehouse means that there is more flexibility with respect to adding data to existing data in the warehouse. In the chapter, we showed how to use an index such as a state FIPS code to hook on state data from other datasets. The more foreign keys we have for states, the more datasets about different states we will be able to hook into our warehouse.

5. If we made a categorical crosswalk variable for ADJPRICE and ADJPRICE2, the crosswalk variable would need to include a level to indicate that the variable did not exist at the time of the dataset. That level would be used to code the datasets before the first dataset with ADJPRICE in it. Next, the crosswalk variable would need to be able to handle all the levels of the coding in both ADJPRICE and ADJPRICE2. That way, both variables could be crosswalked into the crosswalk variable system as the datasets are processed during ETL.

6. As described in *Chapter 7, Designing and Developing ETL Code in SAS*, SAS warehouse developers often do studies of historic data they receive from source data systems that inform how they develop ETL protocols. These studies should be written up and maintained as part of the curation around the ETL protocols for these datasets. This provides just one example of curation that would only be applicable to warehouse developers and not analyst users. On the other hand, the analyst user would need to understand how transformed variables in the SAS data warehouse were derived from warehouse ETL protocols, so curation files about the resulting ETL protocol would need to be available to both developer and analyst users.

7. Different factors can be considered in the decision to devote time and energy to making particular curation files. Logically, curation files that seem to be needed given the tasks immediately at hand that are relatively quick and easy to make should definitely be developed, because they can be useful right away. But there are other reasons to devote effort to developing curation files. Data dictionaries that can help serve as design documents as well as management tools as described in *Chapter 7, Designing and Developing ETL Code in SAS*, are time-consuming and challenging to develop, but are necessary files because they facilitate management processes. In general, if the use of a curation file by developers or analysts would greatly reduce their expended effort on a task, it is worth making the curation file. This is because analyst and developer time tends to be very costly, and if spending a few hours developing a creation file can save many hours of future analyst and developer time, it is worth developing the file.

Chapter 11

1. In de-identification, identifiers need to be placed on a server. If they are placed on a server not connected to the internet, there is no chance for a hacker to steal the identifiers by connecting to the server through the internet. Therefore, if real identifiers are placed on a server not connected to the internet, and while on that server, they are replaced with identifiers that are consistent but cannot be decoded to actual people, then the resulting data is safe. Even if they are placed on an internet server and stolen, the identifiers cannot be decoded by whoever stole the data. If no other identifying data is in the stolen data, then the privacy and confidentiality of the people represented in the database records are preserved.

2. As described throughout this book, data warehouse and data lake projects are enormously expensive and effort-intensive, so it is necessary to make them valuable so that analyst users use them. When data systems are set up such that analysts and developers work remotely, it greatly expands the pool of talent that can be deployed to work on the database. It also greatly expands the pool of potential analyst users, or *customers*, of the data system. In order to achieve a **return on investment (ROI)** for setting up the data system, it should be easily accessible to as many customers as possible.

3. In data warehousing, a star schema represents relatively flattened data. That means throughout the data structure, identical data is often repeated in different tables, and metrics that would usually be derived from a formula, such as a maximum temperature for a day, are already hardcoded in tables. While this configuration speeds up the retrieval of data in a query, the storage of repeated data and the storage of the additional hardcoded data requires expanded space.

4. An ODBC connection using SAS/ACCESS allows the SAS user to connect to other databases through the internet. These databases can be in other formats, including SQL. Creating an ODBC connection is particularly useful in a SAS data warehouse to facilitate importing and exporting data in and out of the warehouse over the internet.

5. The first advantage of using PROC SQL views in a SAS data warehouse is that saving the code associated with the view takes up very little space compared to saving an actual data table. Another advantage is that running a view allows you to apply dynamic criteria to the remote data and actually view the data you are planning to copy before copying it. It allows you to carefully limit the data in an informed way before actually copying it from the remote system to the local SAS data warehouse.

6. While it is excellent that SAS offers the choice of using the ODBC connection to transfer data from non-SAS data stores, there can be disadvantages with this method that improve the desirability of a more manual transfer method. First, an ODBC connection is made over the internet, so if sensitive data is being transferred, it might be safer to do so in a more manual way. Next, the remote SQL database may be filled with complex tables, and it may be difficult in a practical sense for the SAS programmer to find their way around and set up the appropriate query so that they extract the correct data. Worse, they could cause an error in the SQL database by accidentally erasing something if permissions are not set properly. This is why regular data transfers, whether manual or by ODBC, should be governed by written policies and procedures that provide clear guidance.

7. In order for PROC SQL to be used to export a dataset from a SAS environment into a SQL environment, first, an ODBC connection would have to be set up between the SAS and SQL environments. A libname statement referring to the report SQL database would have to be designated. Then, using that libname, a CREATE TABLE command could be used to copy the table from the SAS data warehouse to the remote SQL location.

Chapter 12

1. It is important to build SAS reports into the ETL process that report metrics about the ETL and can help us troubleshoot if anything goes wrong during the ETL protocol. In one of the examples in this chapter, we said if we knew that the state of Alaska only has 10 hospitals, and California has over 300, then we would never expect to see Alaska as the state with the most staffed hospital beds throughout the whole of the US. Therefore, if we had a SAS report among our ETL reports that selected the state with the most staffed beds after one data load, and that happened to be Alaska, it would alert us to look more closely at what happened during ETL, because Alaska would be an illogical result.

2. The source of the limitations we saw in the SAS implementation that provides a query tool for the BRFSS is that SAS has added features over the years to allow data to be displayed over the web. In other words, these features are add-ons to SAS. Because they are add-ons, these features are not at the level of quality that we would see from applications developed specifically for visualizing data on the web, such as Tableau, as well as packages built for this purpose in R.

3. If you were running a data warehouse and had to regularly use SAS/ACCESS to connect to multiple SQL databases for importing data, the SAS Enterprise Guide could help you maintain a dashboard of connections to these data sources. You could navigate and modify these connections using tools in the SAS Enterprise Guide.

4. Since until now, the SAS data warehouse has not implemented the SAS Enterprise Guide, the warehouse should take a closer look at implementing SAS Viya instead of the SAS Enterprise Guide. That is because the SAS Enterprise Guide is an older application that is still being supported. New implementations of the functions in the SAS Enterprise Guide are better done using SAS Viya. Since SAS Viya is web-enabled, it may be strategic to incorporate cloud storage using Snowflake for SAS data that is visualized and managed through Viya.

5. Even if you have access to the most up-to-date visualization software from SAS, since SAS has historically been optimized for data management and analysis, even the best visualization tools from SAS cannot compete with tools optimized for data visualization today, such as Tableau. Therefore, in the scenario described, it is best to first specify the type of visualization needed, and then consider the different products available. If the visualizations and reporting resources in SAS are not adequate for the function, it is possible to turn to Tableau, R, or another software optimized for reporting for a potential solution.

6. In order to give a data warehouse good advice on whether or not they should rebuild a SAS report in R or Tableau, a few items need to be considered. First, if the SAS report is relatively simple, and does not include arrays, macro variables, or macros, it might be very easy to deconstruct and rebuild in another program. Second, if the SAS coding is complex, but the data warehouse is facing significant problems with keeping the SAS report code updated, then the SAS data warehouse might have to consider the situation of a forced migration. If the issue is with I/O, Snowflake can be incorporated as a temporary fix, but a bigger picture solution will need to be designed. If the appearance of the reports in a printed format is the issue, or there are I/O or display problems with reporting to the web, other resources might need to be brought in, such as R or Tableau. In any case, even if the answer to whether or not a SAS report needs to be rebuilt in another program is "no" today, it might easily change to "yes" tomorrow. This is why it is important to proactively consider legacy reporting functions coming out of SAS in a data warehouse, and strategize what sort of improvements can be done before serious issues arise.

7.   Starting a data warehouse today is much different from maintaining a legacy data warehouse. Today, SAS's analytics components, which are available through SAS Viya, would be a definite win to include in a modern data warehouse that supports analytic functions. These components by SAS are unmatched by other software, and SAS Viya makes it easy to take SAS's analytics tools and connect them to a cloud data store in Snowflake. But if the data warehouse does not have an analytic function, then there is a question as to what tools would be sensible to implement anew from SAS. There are likely situations today where big data processing might still be best accomplished by a protocol involving Snowflake and SAS Viya, rather than any other approach. However, the new use of SAS components should be done on a case-by-case basis, responding to the specific needs of data warehouse users. The historic approach of implementing a complete suite of SAS components on a SAS server setup as a way to build SAS data warehouse from the ground up is now obsolete.

Other Books You May Enjoy

If you enjoyed this book, you may be interested in these other books by Packt:

R Statistics Cookbook

Francisco Juretig

ISBN: 978-1-78980-256-6

- Become well versed with recipes that will help you interpret plots with R
- Formulate advanced statistical models in R to understand its concepts
- Perform Bayesian regression to predict models and input missing data
- Use time series analysis for modelling and forecasting temporal data
- Implement a range of regression techniques for efficient data modelling
- Get to grips with robust statistics and hidden Markov models
- Explore ANOVA (Analysis of Variance) and perform hypothesis testing

SQL for Data Analytics

Upom Malik, Matt Goldwasser, Benjamin Johnston

ISBN: 978-1-78980-735-6

- Perform advanced statistical calculations using the WINDOW function
- Use SQL queries and subqueries to prepare data for analysis
- Import and export data using a text file and psql
- Apply special SQL clauses and functions to generate descriptive statistics
- Analyze special data types in SQL, including geospatial data and time data
- Optimize queries to improve their performance for faster results
- Debug queries that won't run
- Use SQL to summarize and identify patterns in data

Leave a review - let other readers know what you think

Please share your thoughts on this book with others by leaving a review on the site that you bought it from. If you purchased the book from Amazon, please leave us an honest review on this book's Amazon page. This is vital so that other potential readers can see and use your unbiased opinion to make purchasing decisions, we can understand what our customers think about our products, and our authors can see your feedback on the title that they have worked with Packt to create. It will only take a few minutes of your time, but is valuable to other potential customers, our authors, and Packt. Thank you!

Index

W

X

www.ingramcontent.com/pod-product-compliance
Lightning Source LLC
Chambersburg PA
CBHW081454050326
40690CB00015B/2799